Sustainable Rural Development Perspective and Global Challenges

Edited by Orhan Özçatalbaş

Published in London, United Kingdom

Sustainable Rural Development Perspective and Global Challenges
http://dx.doi.org/10.5772/intechopen.98050
Edited by Orhan Özçatalbaş

Contributors
Orhan Özçatalbaş, Innocent Chirisa, Jeofrey Matai, Tafadzwa Mutambisi, Precious Tirivanhu, Nenavath Sreenu, Kondru Sunda Sekhar Rao, Muhammad Imran, Shamsheer Ul Haq, Orhan Ozcatalbas, Marisol Medina-Sierra, Mario Cerón-Muñoz, Luis Galeano-Vasco, Ifije ohiomah, Clinton Aigbavboa, Nita Sukdeo, Pham Van Kien, Nguyen Hay, Le Quang Huy, Simon M. Munthali, Jeremiah Machavi, Jonas Mongoè, Adedeji A. Adebukola Adelodun, Ochuko Mary Ojo, Temitope M. Olajire, Paul L. Woomer, Dries Roobroeck, Didier Yelognisse Alia, Luz María Castro, Diana Encalada, Luis Rodrigo Saa, Meike Henseleit, Sandra Venghaus, Wilhelm Kuckshinrichs, Eriola Marius Charlot Adenidji, Emmanuel A. Onsay, Kevin C. Baltar, Eleanor R. Galicia, Ivan Ruzzel C. Pesino

First published in London, United Kingdom, 2023 by IntechOpen
IntechOpen is the global imprint of INTECHOPEN LIMITED, registered in England and Wales, registration number: 11086078, 5 Princes Gate Court, London, SW7 2QJ, United Kingdom

British Library Cataloguing-in-Publication Data
A catalogue record for this book is available from the British Library

Additional hard and PDF copies can be obtained from orders@intechopen.com

Sustainable Rural Development Perspective and Global Challenges
Edited by Orhan Özçatalbaş
p. cm.
Print ISBN 978-1-80355-420-4
Online ISBN 978-1-80355-421-1
eBook (PDF) ISBN 978-1-80355-422-8

For EU product safety concerns:
IN TECH d.o.o., Prolaz Marije Krucifikse Kozulić 3, 51000 Rijeka, Croatia,
info@intechopen.com or visit our website at intechopen.com.

Meet the editor

Dr. Özçatalbaş graduated with a degree in Agricultural Economics from Çukurova University. He became an assistant professor at Akdeniz University, Turkiye, in 1998, and a professor in 2011. He is the director of the Institute of Natural and Applied Sciences at Akdeniz University. His research is in rural development and extension, energy policy and ICT in agriculture. In 1999–2000, he was a visiting scientist at the Institute of Horticultural Economics, Hannover Leibniz University, Germany, as well as for vocational courses in Italy, the Erasmus program in Poland, and participation in congresses in other countries. Dr. Özçatalbaş has professional experience in rural development/policy/extension as an expert and coordinator in national and international projects.

Contents

Preface

In the last two centuries, there has been rapid industrialization with the use of electrical energy in all fields, including agriculture, following the discovery of new inventions like steam-powered machines that convert potential energy into kinetic energy. This process has led to great changes in the residential areas of the world's population, especially in developed industrial societies, reflected by a change in the distribution of the population from rural to urban areas.

The rapid increase in the world's population (estimated to reach 10 billion in 2050) necessitates increasing food production on a global scale. According to the United Nations Food and Agriculture Organization (FAO), world food demand will increase by 70% by 2050. In addition to food insecurity, the world is experiencing climate change and global warming. The predicted effects of global warming will affect settlements, lifestyles, and production systems. As such, new production systems that are compatible with increasing temperatures brought about by global climate change should be developed.

Agriculture is vitally important to humanity. Climate change, environmental pollution, global warming, and the COVID-19 pandemic have highlighted the importance of food safety and food security. Although awareness of these challenges is greater today than it has ever been, there is still much work to be done to increase sustainable food production. Today's existing agricultural production systems are not sufficient to ensure food security and overcome the negative effects of supply chain issues. Thus, new production approaches and technologies in agriculture, such as Industry 4.0 and Society 5.0, are imperative.

Sustainability is a concept that is spreading widely to various fields and industries. It encompasses notions of zero waste, renewable energy, and green, circular, and blue economies. This book discusses sustainability in terms of rural activities.

Orhan Özçatalbaş
Department of Agriculture and Economics,
Institute of Natural and Applied Sciences,
Akdeniz University,
Antalya, Türkiye

Section 1

Perspectives

Chapter 1

An Evaluation of the Transition from Linear Economy to Circular Economy

Orhan Özçatalbaş

Abstract

The agriculture sector and food industry have always been essential sectors of the economy and countries. Factors such as limited resources in the face of a growing population as well as climate change and even the COVID-19 pandemic, and between Russia with Ukraine conflict have maximized the strategic importance of the agro-food industry on a global scale. Since it is a global issue affecting almost every country adversely, especially the fight against climate change requires institutional establishment of international cooperation. The aim of this study is to evaluate the current dynamics of existing economic theories and economic models, highlighting the transition from linear economy to circular economies. To do this, the concept of the circular economy and the economy is discussed in relation to the SDGs to better understand the problem. To evaluate the transition from a linear economy to a circular economy, again, the sustainability publications were searched using bibliometric methods. Mostly four concepts related to sustainability, sustainability and rural development, circular economy, and green economy were analyzed by using VOSviewer software. These four concepts are closely associated with sustainability.

Keywords: sustainability, linear economy, green economy, green deal, circular economy

1. Introduction

With its vital functions for humanity, the agro-food industry has always been an essential sector of the economy. Factors such as limited resources in the face of a growing population as well as climate change and even the COVID-19 pandemic have maximized strategic importance to the agricultural industry on a global scale, thus the sustainability of development and rural life.

Sustainability stands out as a concept whose frequency of use is constantly increasing in academic literature and socioeconomic life and which remains to be one of the major agenda items because it represents an important approach to the future of the world, humanity, and the ecosystem. With its respective methodology in almost all fields, sustainability has been considered a go-to approach for the earth.

International agreements based on sustainability have been concluded in order to leave a habitable planet to future generations, given the adverse impact of the global climate change. In this context, in September 2000, the United Nations announced the Millennium Development Goals (MDGs) consisting of eight items, highlighting "eradication of extreme poverty and hunger" as the first goal. The remaining seven goals are as follows: promote gender equality and empower women; reduce child mortality; improve maternal health; combat HIV/AIDS, malaria and other diseases; ensure environmental sustainability; and develop a global partnership for development [1].

Later, the Millennium Development Goals were restructured in 2015, increasing the number of goals to 17. The first two goals among the Sustainable Development Goals (SDGs) announced by the UN in 2015 are no poverty and zero hunger. The SDGs went into effect in January 2016 and will guide UNDP's policies and financing until 2030. UNDP, the leading development agency of the United Nations, represents a crucial potential for the implementation of these goals as it carries out its operations in more than 170 countries and regions. Although sustainability has not dominated all sectors in all of the 17 goals of the UN, efforts are underway to be successful at the international level. In particular, some important studies and applications need to be carried out in the field of combating inequality and injustice, especially to "eradicate extreme poverty (no poverty and zero hunger)" on a global scale. Note that the sustainability of rural life and production in rural areas, which is still of vital importance for food production, and thus the increase in welfare in rural areas remain to be major political agenda items [2]. To achieve the Sustainable Development Goals, the UN Development Programme (UNDP) needs the involvement of governments, the private sector, civil society, and citizens. However, unfortunately, it seems that it will not be possible to achieve these goals until 2030 due to the impacts of the COVID-19 pandemic.

It is clear that it concerns all countries and industries. Therefore, it is no doubt that it would be appropriate to relate the SDGs to the current realities of sustainable development.

2. Development, climate change, and agricultural policies

The position of communities in the development process can be defined to a significant extent by sociocultural, environmental, and economic indicators. Undoubtedly, the initial phase of the social and economic development of humanity was dominated by rural and agricultural production. The transition from this initial phase, namely, agrarian society, to industrial society was not easy, whereas the transition from industrial society to information society has been far more complicated. The transition from agrarian society to industrial society began in the second half of the 18th century when the steam engine was invented and started to be used as a source of energy. Therefore, in the last two centuries, there has been a rapid industrialization process with the use of electrical energy in all fields, including agriculture, following the advent of new inventions, especially steam-powered machines that convert potential energy into kinetic energy [3].

This process has led to great changes in the settlement areas of the communities around the world, particularly in developed industrial societies. Skyrocketing population growth has brought with it urbanization and new economic activities, resulting in alteration and diversification of consumption patterns. This, in turn, has increased the pressure on the environment and natural resources and the rapid development process has led to new social and environmental challenges.

Even the projections for the world population keeping on rising and reaching 10 billion by 2050 entail expansion of food production on a global scale. According to the estimations of the Food and Agriculture Organization (FAO) of the United Nations, the global demand for food needs to be increased by 70% by 2050. Hence, would require raising overall food production by some 70% between 2005/2007 and 2050 [4] related another study; the total global food demand is expected to increase by 35 to 56% between 2010 and 2050, while the population at risk of hunger is expected to change by −91 to +8% over the same period. Moreover, if climate change is taken into account, the ranges change slightly (+30 to +62% for total food demand and − 91 to +30% for the population at risk of hunger) but with no statistical differences overall [5]. As shown, the anticipated changes will be so great that they will affect the settlements, lifestyles, and production and consumption systems.

The fact that all these challenges threaten human life and our planet at a global scale and the threats of climate change and global warming have made the agricultural industry central to life and production sectors. Considering it from the perspective of climate change, it is expected that global warming will lead to global temperature increases every year, resulting in extinction of one quarter of (or more than 1 million) plant and animal species [6] by 2050. According to the United Nations Intergovernmental Panel on Climate Change (IPCC) report, each country will be affected by global climate change at varying levels [7]. This is why the issue of global warming has been addressed not only conceptually, but also concretely by governments, economic sectors, and communities. In this context, the UN Framework Convention on Climate Change (UNFCCC) was adopted in 1990, through which the signatory countries agreed to reduce the emissions of carbon dioxide and other greenhouse gases [8].

Considering the climate change projections, the risks posed by global climate change entail measures for increasing production and productivity in the face of population growth as well as the development of novel technologies and production systems based on adaptation to increases in temperature [3]. In this framework, R&D and technology policies and research and extension (R&E) policies need to be highlighted.

As is known, worldwide experience shows that new technologies have been the driver of social and economic development. As shown in **Figure 1**, “Cumulative Adoption of Technology” refers to the sum of all countries. Generally speaking, countries compete to take, use and utilize the technology by their own means (locally) or under the influence of international scientific research. Naturally, countries’ ability to produce technology and adoption of innovations lead to faster utilization of the positive effects of such technologies, but not all countries have the capacity to develop and transfer from outside and uphold the technologies they need [9]. Therefore, it is necessary to define a new approach in which the technologies produced for climate change are considered, in a sense, the “common property of humanity” for countries that are unable to produce or transfer technology and have no means to compete with others. It is a fact that the creation of such a culture of sharing will serve all of the 17 Sustainable Development Goals set forth by the UN. So, in order to reduce the effects of climate change, which is a global threat to the earth, it will be important that the UN develops a mechanism that will ensure the exchange of existing and new technologies to be developed among countries, regardless of their ability to produce technology.

It is clear that the global climate change will disrupt the current balances, resulting in adverse impacts on the sustainability of agro-food industry and rural development. Hence, the UN specifically addresses this issue in its SDG 13 (Take urgent action to combat climate change and its impacts) and highlights SDG 17 (Partnerships for the

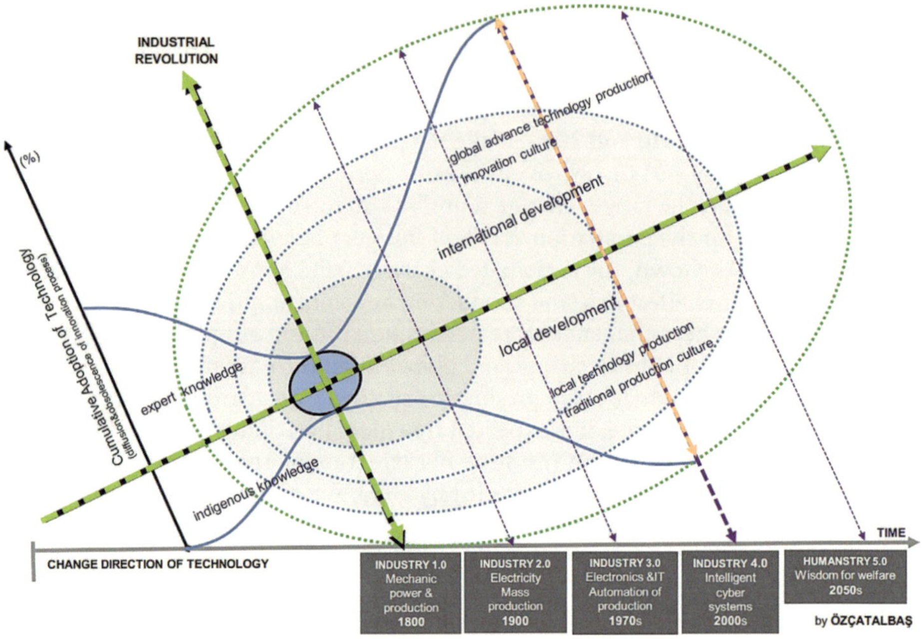

Figure 1.
Processes and factors determining the direction of change in development and technology.

goals: Strengthen the means of implementation and revitalize the global partnership for sustainable development) as a solution [2]. Since it is a global issue affecting almost every country adversely, the fight against climate change requires institutional establishment of international cooperation.

3. Conceptual history of sustainability

In academic circles, there is agreement that sustainable development requires simultaneous improvement of environmental, social, and economic outcomes [10]. Overall, sustainable development takes a global and long-term perspective for a prosperous, just, and secure future [11]. Sustainability science has been defined as "an emerging field of research dealing with the interactions between natural and social systems, and with how those interactions affect the challenge of sustainability. Sustainability is one of the most used concepts in almost every field today. This situation is related to today's global and local conditions. Here, we look at the historical background of the concept and discuss the current situation. The derivation of technical and academic terms and concepts relates to a use, application or invention. When concepts express approaches, they reflect deeper content, so is the concept of sustainability [10].

Looking at the past, it can be predicted that there were practices within the scope of sustainability in different civilizations in the past centuries, although they do not exactly overlap with today's terminology and content. This area is a subject of a separate study, and it is possible to find various examples that can inspire today when researched. However, when the available sources are examined, the first use of the term and concept of sustainability dates back to the 18th century. Historically, depletion of natural resources is not new and hits renewable resources first. The scarcity of wood was the concern of feudal Europe and led to the introduction of sustainability principles in forest management. It was in order to preserve wood supplies for the Kingdom of Saxony where in 1713 it was for the first time clearly formulated that forestry had to be "sustainable", which meant that logging and reforestation had to be in balance. Similar concepts arose in France and Japan [12].

In the same sense as today's literature, the term "sustainability" itself was used for the very first time in 1953 by JL. Fisher [13, 14]. However, as an original idea "limits to growth", which underlies the whole concept of sustainability and sustainable development, and which would later take the form of the contemporarily well-known multidimensional sustainability idea, was expressed much earlier by Malthus's population growth theory, in the end of the 18th century. The United Nations Conference on the Human Environment was an international conference convened in 1972, and brought the industrialized and developing nations. It was the UN's first major conference on international environmental issues and marked a turning point in the development of international environmental politics [15]. The present and modern understanding of this concept have been though fully and explicitly articulated for the first time in 1987, within the frame of the so-called Brundtland Report on Sustainable Development, which is sometimes termed shortly "Our Common Future". As the term sustainability is considered not only a scientific idea, but a moral value, a normative goal based on this value, and a pathway for the international policy or social movements as well [14]. Generally in its most widespread perception, sustainability is defined as "a development that meets the needs of the present without compromising the ability of future generations to meet their own needs" [11]. In this context, sustainability stands out as a concept whose frequency of use is constantly increasing in academic literature and socioeconomic life and which remains to be one of the major agenda items because it represents an important approach to the future of the world, humanity, and the ecosystem.

So, the concept of sustainability has been in use for more than three centuries. Therefore, after emerging out of necessity, it has made its way to today's lexicon, getting enriched in content and being used in conjunction with different concepts.

4. Towards sustainability and circular economy

During the industrial revolution, the increasing demand for inputs and the production factors put into commercial use paved the way for the development of economics and new economic models. The journey that started with the classical economics still continues. However, in addition to factors such as the scarcity of resources, challenges related to resource utilization efficiency, misuse of natural resources, limitedness of production factors and unjust distribution of resources, threats such as climate change have also paved the way for discussions around new models as an alternative to linear economy. This has, in a sense, rendered the SDGs central to sustainability discussions, and the developments that support the SDGs, such as the green economy, circular economy, and bioeconomy have become popular narratives in macro-level sustainability discussions in policy, scientific research, and business. The 17 Sustainable Development Goals set forth by the United Nations in 2015 have renewed a global vision to address sustainability challenges and emphasized the urgency for concerted efforts by multiple societal actors. Over the past decades, "[s]ustainability science" has attracted tens of thousands of researchers, practitioners, knowledge users, teachers, and students from diverse institutions and disciplines from across the world [16]. However, circular economy (CE) is currently a popular concept promoted by the EU, by several national governments and by many businesses around the World [17]. Therefore, the process of change that we have experienced demonstrates the UN will play a key role in achieving the Sustainable Development Goals. According to Velenturf and Purnell [10], economic growth is still part of the goals, e.g., SDGs 8 and 9, which relate economic growth to resource efficiency, and the main mechanism proposed to achieve a balance is through decoupling. SDG 12 does not mention circular economy literally, but it does cover the sustainable management and efficient use of natural resources, reducing and preventing wastes, uptake of reuse etc., and changes in lifestyles, procurement policy and business reporting, all of which are closely aligned with circular economy principles. Targets under sixteen out of the seventeen SDGs are related to a circular economy (**Figure 2**).

However, the circular economy can contribute positively to most of the sustainable development goals, but sustainable development and circular economy are on diverging pathways. While the sustainable development agenda puts people front and center with economic prosperity recognized as a means for living fulfilling lives in harmony with nature, circular economy remains fixated on technological solutions, the implementation of which is driven by a promise of traditional economic growth. Circular economy also must be fully integrated with sustainable development. Circular economy should be understood as an emerging practical ideology that lacks an evidence-based theoretical framework to guide implementation. It lacks an economic theory that can pragmatically guide the transition from the prevailing neoclassical model toward one that would drive the transition toward a sustainable circular economy and be palatable for governments. The critique on circular economy should be understood as much as a critique on sustainable development itself, and both require research and constant learning to ensure progress towards sustainability [10]. It is an inevitable necessity to manage the process taking into account the contribution of local and international cooperation efforts to sustainability.

As is known, circular economy is an economy constructed from societal production-consumption systems that maximize the services produced from the linear nature-society-nature material and energy throughput flow. This is done by using cyclical materials flows, renewable energy sources and cascading[1]-type energy flows. Successful circular economy contributes to all three dimensions of sustainable development. Circular economy limits the throughput flow to a level that nature tolerates and utilizes ecosystem cycles in economic cycles by respecting their natural reproduction rates [17].

Previous research shows that Circular Economy (CE) is a concept that stems from the need to address environmental degradation, social unrest and inequalities, institutional instability, resource scarcity, and economic challenges caused by the linear nature-society-nature systems that a large portion of society operates on [18]. In this framework, the acceptance of circular economy on a global scale gives rise to a need for novel technologies such as renewable energy generation technologies, which will be among the major topics of discussion in the medium term. It would be appropriate to explore academic research to clarify this issue. The following sections deal with the extent to which topics such as sustainability, green economy, and circular economy are addressed and what other topics are highlighted in scientific research and publications. It is of particular importance in that it shows the direction of change in scientific studies with respect to the concepts in the specified fields. Therefore, a bibliographic assessment of sustainability is provided below.

Source: Velenturf and Purnell 2021

Figure 2.
Circular economy can enable a significant number of targets under the UN sustainable development goals (legend: Fraction of targets under each goal that would be strongly (red) and partially (orange) enabled by the implementation of circular economy measures).

5. An assessment of sustainability approach

Bibliometric analyses provide a big and profound picture of the current knowledge base. Also known as science mapping, bibliometric analyses have been applied in a variety of fields and disciplines. However, studies on the concepts of circular economy and sustainability are also being conducted. Therefore, this section analyses the global knowledge base on sustainability and new concepts associated with it.

The concept of sustainability has been used continuously in academic literature and socio-economic life on a global and local scale since the day it was introduced. Pollution and the threat of global climate change, especially as a result of industrialization, led to controversies surrounding classical economics and the search for alternative methods. In this context, they paved the way for the emergence of circularity-based concepts and approaches as an alternative, such as green and blue economies. To provide a better understanding of this development process, the concepts of sustainability, green economy, circular economy, and green deal are addressed.

6. Material and methodology

In the bibliometric analysis, VOSviewer was used to retrieve the publications on sustainability. VOSviewer is a software tool for constructing and visualizing bibliometric networks. At the same time, it can be used to construct maps of authors or journals based on co-citation data or to construct maps of keywords based on co-occurrence data [19, 20]. To identify the most commonly used themes associated with the theme of the research, an analysis of the networks was conducted based on the keywords. There are several network techniques used for this purpose. Among them, researchers usually employ methodologies based on distance, graphs, and timelines [21]. Keywords refer to the basic concepts used in a study. They are the core and essence of a document, which is a high-level summary of the article content [22]. In this paper, the bibliometric data come from Scopus, one of the world's leading academic databases. What is conducted here is network of fractional order research on the scientific literature concerning sustainability, green economy, and circular economy. The figure shows the distribution of hot keywords in fractional order research, and different colors in the figure represent different clusters of research hotspots. Figures are mainly composed of nodes and connecting lines. The larger the keyword node, the higher the frequency of occurrence, and the thicker the line between nodes, the higher the co-occurrence frequency. Moreover, the distribution of the distance between nodes is positively correlated with the relevance of keywords [23]. Accordingly, the analysis of the keywords provides strong data about the content of the subject discussed in the research articles.

7. Bibliometric analysis of keywords

7.1 The concept of sustainability with related main keywords

A keyword search carried out on the subject of "TITLE-ABS-KEY (sustainability)" generated almost 286,025 keywords in the Scopus database. From this database, by specifying exclusively the articles, a collection of 188,662 documents was listed using the command TITLE-ABS-KEY (sustainability) AND (LIMIT-TO (DOCTYPE, "ar"). Finally, from this list of articles, the first 2000 articles were considered for keyword analysis. VOSviewer was used to generate the keyword maps. **Figure 1** below shows all the keywords according to the topics covered in this presentation.

As shown in **Figure 3**, the major concepts found to be associated with the concept of sustainability in the articles published between 2015 and 2017 include keywords such as corporate sustainability, sustainability assessment, sustainability indicators, sustainability reporting, urban sustainability, higher education, sustainability education, life cycle sustainability, energy, decision making, and sustainability criteria. In addition, new concepts such as COVID-19, firm performance, organizational sustainability, theory building, waste, and bio-based products have been commonly used in 2020s. It suggests that the research on sustainability keeps on developing by deriving new concepts.

7.2 The concepts of sustainability and rural development with related main keywords

For "Sustainability and Rural Development", the same collection procedure was used. A document search was carried out in the Scopus database. The finding was 7047 documents from which almost 5179 documents were extracted by the following command: TITLE-ABS-KEY (sustainability AND rural AND development) AND (LIMIT-TO (DOCTYPE, "ar")). Among these articles, the first 2000 were selected for analysis. A total of 5169 keywords were counted, with a minimum occurrence of one. The graph above presents the occurrence of keywords by year. Thus, words that have made up the lexical field of the topic before 2010 (environment, climate, biodiversity, etc.) are colored darker. The recent keywords are colored yellow (agritourism, circular economy, etc.). The keywords' importance is measured by the size of the diameter of their circle. The larger the circle is, the more the keyword is used. Sustainability appears to be the most used keyword with a larger diameter. **Table 1** shows the top ten keywords for this category.

The analysis shows what concepts are most widely used by publications in which years (**Figure 4**). Accordingly, the major concepts used in association with the concept of "sustainability and rural development" in the articles published between 2015 and 2017 include keywords such as sustainability, rural development, sustainable development, rural tourism, rural areas, China, rural, agriculture, development,

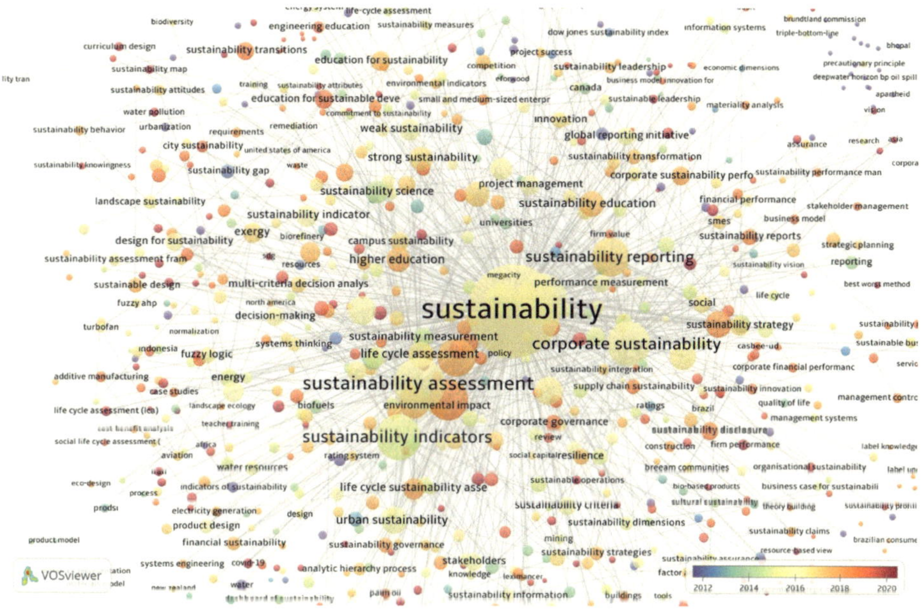

Figure 3.
Sustainability main keywords based on Scopus data.

Topics	Sustainability	Sustainability and rural development	Circular economy	Green deal
Keywords	Sustainability Sustainability assessment Sustainable development Sustainability indicators Corporate sustainability Social sustainability Environmental sustainability Sustainability reporting Economic sustainability indicators	Sustainability Rural development Sustainable development Rural tourism Rural areas China Rural Agriculture Development Rural electrification	Circular economy Sustainability Sustainable development Recycling Waste management Industry 4.0 Resource efficiency Waste Business model	Green economy Sustainable development Circular economy Sustainability Green growth Climate change Renewable energy Environment Green jobs China

Table 1.
Top keywords according to the topics.

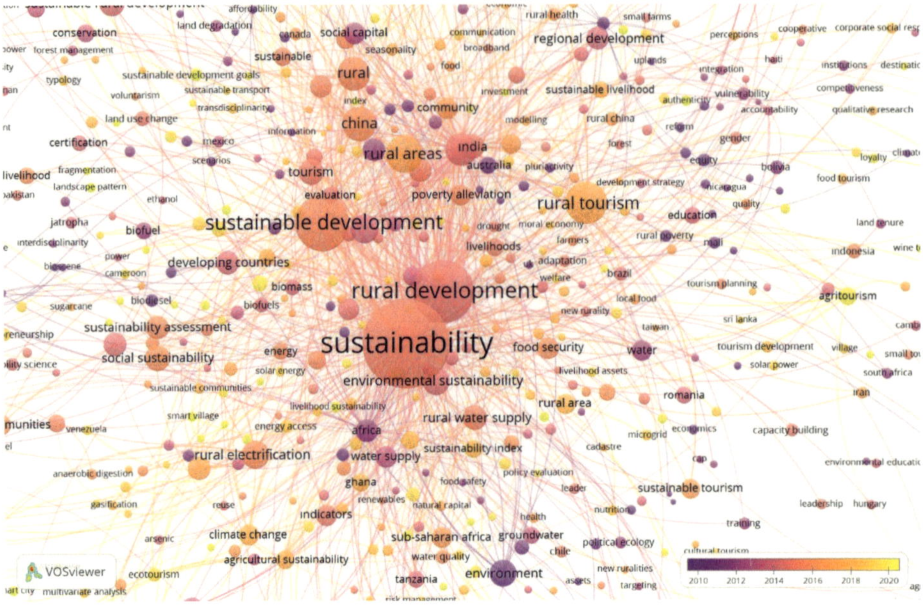

Figure 4.
Sustainability and rural development.

rural electrification, regional development, India, social sustainability, sustainability assessment climate change, agricultural sustainability, sustainable tourism, and food security. Furthermore, concepts such as SDGs, agritourism, ecotourism, sustainable development goals, smart village, loyalty, and landscape pattern have been introduced in the 2020s. It suggests that the research on sustainability and rural development keeps on developing by deriving new concepts.

7.3 The concepts of circular economy with related main keywords

A keyword search on the circular economy was conducted through 15,098 documents retrieved by the string TITLE-ABS-KEY (circular AND economy). The Scopus database was used to collect the large number of keywords. A random selection of 2000 words was made from nearly 9606 documents generated by the search TITLE-ABS-KEY (circular AND economy) AND (LIMIT-TO (DOCTYPE, "ar")). On this basis, and with the assistance of VOSviewer, the keyword map depicted above was generated. The top keywords include circular economy sustainability, sustainable development, recycling, waste management, industry 4.0, resource efficiency, waste, and business model (**Figure 5** and **Table 1**).

The analysis shows what concepts are most widely used by publications in which years (**Figure 5**). The major concepts used in association with the concept of circular economy in 2019 include circular economy, sustainability, sustainable development, recycling, waste management, industry 4.0, resource efficiency, waste, business model, linear economy, and environment. Moreover, concepts such as industry 4.0, Covid-19, digital twin, posthumanism, food supply chain, circular cities, and goals were used in association with the concept of circular economy. It is important in that it shows the new concepts that are increasingly used in association with circular economy are the topics of scientific research.

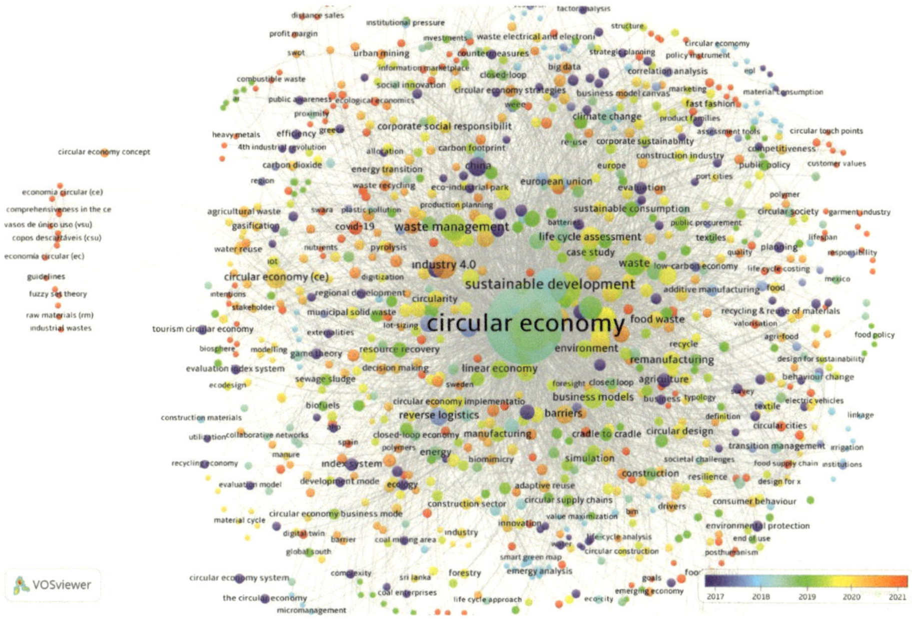

Figure 5.
Mapping the circular economy with main keywords based on Scopus data.

Figure 6 shows the frequency of use of the keywords when circular economy is selected as the main concept. Accordingly, the top six concepts that are most used with circular economy are sustainability, sustainable development, recycling, waste management, industry 4.0, and resource efficiency In addition, other related keywords that are used with it, including in the studies discussing linear economy, are also given. The period in which other keywords are used with circular economy is again 2015–2021. What we see here is that the top five keywords that are most used with circular economy are in the period 2018–2021. The keywords that are often used with circular economy as of 2020 include resource efficiency, business models, reverse logistics, industrial symbiosis, and zero waste. It is important in that it provides information about the topics of new research.

7.4 The concept of green economy with related main keywords

A total of 8191 documents were listed by the search TITLE-ABS-KEY (green AND deal) on Scopus database. By searching only articles, 5036 articles were listed. VOSview helps to generate the network of keywords. The most used keywords are in **Figure 5** below. The period from 2016 to 2018 seems to be the one where the keyword "green deal" was most used. To date, it is the most used keyword given its size through the node. The network represents this period in green color for nodes (**Figure 7**).

With regard to what concepts are most used in which years (**Figure 5**), the analysis shows that the keywords that are most used with the concept of green economy in 2014–2015 include low-carbon economy, green chemistry, green supply chain, green building, green economics, atom economy, well-being, and Rio + 20. Among the other major concepts used in conjunction with the concept of green deal in the period 2016–2018 are keywords such as green economy, sustainable development, circular economy, sustainability, green growth, climate change, renewable energy,

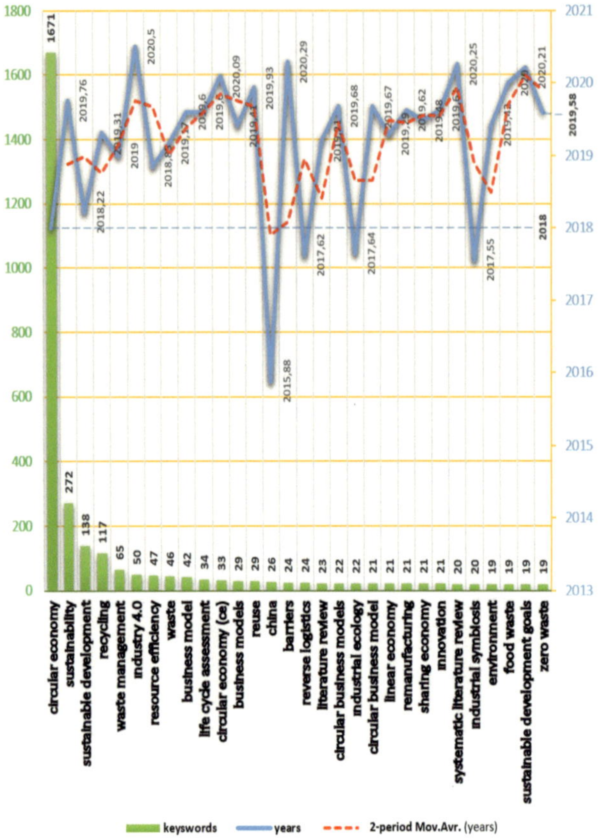

Figure 6.
Annual occurrence of relevant keywords.

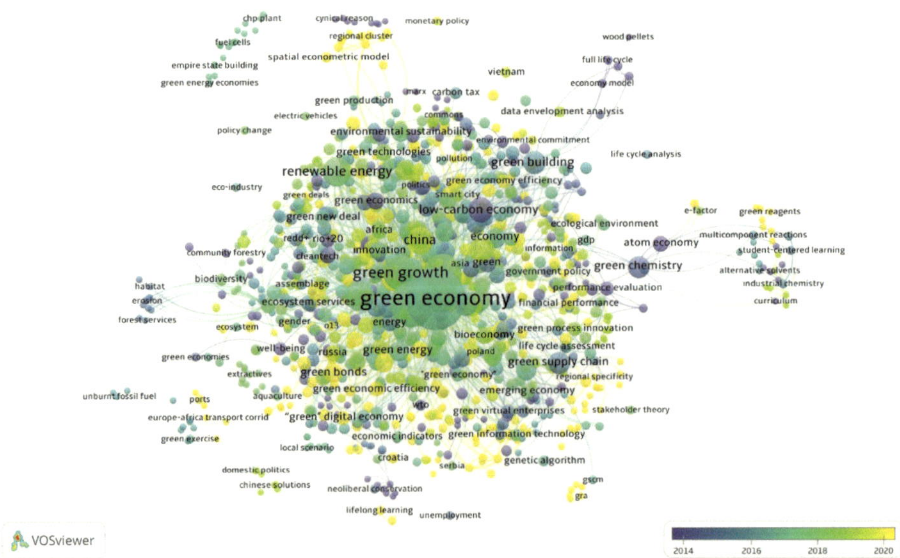

Figure 7.
Green deal-related keywords' network.

environment, green jobs, and China. Moreover, concepts such as green economic efficiency, pollution, Russia, Croatia, ports, lifelong learning, spatial econometric model, regional cluster, and green reagents have been widely used in the 2020s. It is important in that it shows the number of studies involving new concepts related to the concept of green deal and their frequency of use has increased.

8. From linear economy to circular approaches

Rising demand for food on a global scale remains to be a challenge due to the increase in world population. However, factors such as global climate change and the COVID-19 pandemic as an extreme threat have led to a strong perception of the importance of agricultural production by all societies. It is understood that the demand for agricultural products that are healthy, high in nutritional quality and produced in an environment-friendly manner will increase gradually.

Today, development policies at the international level are focused on sustainability. They are usually aimed at establishing the conditions that will ensure the sustainable use of resources and the sustainability of human life in the future. Efforts are underway to end poverty and improve and expand fundamental rights and freedoms by reducing welfare differences between countries and between segments of societies, especially through the adoption of a sustainability approach on a global scale [24].

9. Evaluation of the top keywords according to the topics

The results of the conceptual analysis are given in **Table 1**. Accordingly, the four concepts that are addressed here are closely associated with sustainability. All of the keywords related to sustainability are accompanied by the concept of sustainability. The analyses carried out concerning the concepts of sustainability and rural development, circular economy, and green deal involve the concept of sustainability twice for each of them. It is important in that it shows the concept of sustainability continues to be used in conjunction with emerging concepts. It suggests that although new concepts are derived and new approaches are put forth, the concept of sustainability maintains its place in the literature.

While the 'take-make-dispose' model suggested by the classical economics (i.e., linear economy) has dominated all economies for the last two centuries, the sustainability approach has always existed despite the fact that it does not fit well with the assumptions of classical economics and its level of effectiveness is low.

Accordingly, the linear "take-make-dispose" model relies on large quantities of easily accessible resources and energy, and as such is increasingly unfit for the reality in which it operates. Working towards efficiency alone—a reduction of resources and fossil energy consumed per unit of manufacturing output—will not alter the finite nature of their stocks but can only delay the inevitable [25]. As is known, circular economy, as opposed to the current linear economy, can be seen as a sustainable economic system where economic growth is decoupled from resources use, through the reduction and recirculation of natural resources. The circular economy concept attracts increasing attention of governments, scholars, companies, and citizens as a necessary step to achieve sustainable development [26]. Again according to The Ellen MacArthur Foundation [25], the circular economy draws a sharp distinction between the consumption and use of materials: circular economy advocates the need for a "functional service" model in which manufacturers or retailers increasingly retain the ownership of their products and, where possible, act as service providers—selling the use of products, not their one-way consumption. This shift has direct implications for the development of efficient and effective take-back systems and the proliferation of product and business model design practices that generate more durable products, facilitate disassembly and refurbishment, and where appropriate, consider product/service shifts. As circular economy thinker Walter Stahel explains, "The linear model turned services into products that can be sold, but this throughput approach is a wasteful one. In the past, reuse and service-life extension were often strategies in situations of scarcity or poverty and led to products of inferior quality. Today, they are signs of good resource husbandry and smart management." As a result, circular economy is also an industrial system that is restorative or regenerative by intention and design. It replaces the "end-of-life" concept with restoration, shifts towards the use of renewable energy, eliminates the use of toxic chemicals, which impair reuse, and aims for the elimination of waste through the superior design of materials, products, systems and, to develop within this, business models.

Again, considering the European Green Deal (EGD), it is clear that at the heart of the main components of the Deal is the aim of transforming the EU economy for a sustainable future. In line with this, the EGD is intended to reduce greenhouse gas emissions within a certain program in order to make the EU climate-neutral by 2050. In order to reduce carbon leakage from the EU, efforts are underway to develop a new system built with new taxes and non-tariff barriers in trade through the border carbon adjustment (BCA) mechanism. Again, by 2030, the EU aims to transform 25% of the agricultural lands in its territories into organic farming land. Organic production in the fresh fruit and vegetable industry and dried and frozen products industry is expected to increase in the coming years. In addition, organic products will be in demand in other agricultural product markets such as medicinal and aromatic plants and mushrooms, so plans are being made to ensure product safety. Another important aspect of the European Green Deal is that it gives priority to energy saving and renewable energy sources. In general, renewable energy sources such as solar energy are supported. The Green Deal also aims to increase the amount of fruit and vegetable consumption in human nutrition by regulating the consumption habits of consumers and combating obesity and diseases such as cancer, thereby eliminating the serious economic burden on the healthcare system [27].

Reaching and indeed exceeding national or EU GHG reduction targets, such as those proposed by the EU Green Deal, is not an easy task. However, many municipalities across Europe are aware of their crucial role and are developing strategies toward carbon neutrality by 2050. The experience of those cities that have already joined the CoM 2030 initiative with ambitious GHG reduction targets could serve as an example for others [28].

The EU countries have taken the lead in the transition to a sustainable food system, and this process will affect many countries. The technical and financial support of the EU instruments such as harmonization funds and the European Agricultural Fund for Rural Development will facilitate this transition.

10. Conclusions

While the journey that started with linear economy still continues, challenges related to resource utilization efficiency, misuse of natural resources, limitedness of production factors as well as threats such as climate change have given rise to discussions around new models as an alternative to linear economy. In particular, circular economy is addressed frequently. Circular economy is a concept that stems from the need to address environmental degradation, social unrest and inequalities, institutional instability, resource scarcity, and economic challenges caused by the linear nature-society-nature systems that the large portion of society operates on [18]. The acceptance of circular economy on a global scale will probably be among the major topics of discussion in the medium term, as there is need for novel technologies such as renewable energy generation technologies.

In this framework, this paper deals with the conceptual developments in academic studies in a bibliometric fashion. It shows that the concepts have both varied and increased. The use of the concepts of sustainability, green economy and circular economy in academic studies has been more widespread. Therefore, it is seen that the process of transition from linear economy to circular economy is supported by academic studies, which derive new concepts that are addressed as further research topics by new studies. In conclusion, it can be said that the process of change in this field will contribute to the achievement of the Sustainable Development Goals of the UN. With the global cooperation efforts that are being carried out and expansion of practices focusing on sustainability, academic studies will play a key role in, and accelerate, the transformation in question.

Author details

Orhan Özçatalbaş
Department of Agriculture and Economics, Faculty of Agriculture, Akdeniz University Institute of Natural and Applied Sciences, Antalya, Turkey

*Address all correspondence to: ozcatalbas@akdeniz.edu.tr

References

[1] MDG. The Millennium Development Goals Report 2015. United Nations; 2015. ISBN: 978-92-1101320-7. Available from: https://www.un.org/millenniumgoals/2015_MDG_Report/pdf/MDG%202015%20rev%20(July%201).pdf

[2] UN. The 17 Goals: Sustainable Development. UN Department of Economic and Social Affairs. New York: United Nations Statistics Division Development Data and Outreach Branch; 2022. Available from: https://sdgs.un.org/goals

[3] Özçatalbaş O. Possible Effects of Global Climate Change on Agricultural Extension and Politics (Küresel İklim Değişikliğinin Tarım Yayımı ve Politikaları Üzerine Olası Etkileri) (in Turkish). Ulusal Aile Çiftçiliği Sempozyumu Bildiri Kitabı, 30-31 Ekim 2014, Ankara; 2014

[4] FAO. How to Feed the World in 2050 (High-Level Expert Forum, 2009). 2009. Available from: https://www.fao.org/fileadmin/templates/wsfs/docs/Issues_papers/HLEF2050_Global_Agriculture.pdf

[5] van Dijk M, Morley T, Rau ML, Saghai Y. A meta-analysis of projected global food demand and population at risk of hunger for the period 2010-2050. Nature Food. 2021;**2**:494-501. DOI: 10.1038/s43016-021-00322-9

[6] Thomas CD, Cameron A, Green RE, Bakkenes M, Beaumont LJ, Collingham YC, et al. Extinction risk from climate change. Nature. 2004;**427**:145-148. DOI: 10.1038/nature02121

[7] ÇŞB. Türkiye İklim Değişikliği Stratejisi 2010-2020 (Turkiye's Climate Change Strategy 2010-2020, T. R. Environment and Urban Ministry). Ankara: T. C. Çevre ve Şehircilik Bakanlığı; 2012

[8] UNFCCC. The United Nations Framework Convention on Climate Change (UNFCCC), The Secretariat of the United Nations Framework Convention on Climate Change is Located in Bonn, Germany. 2014. Available from: http://unfccc.int/2860.php

[9] Özçatalbaş O. Rural development policies and strategies in Turkiye (Türkiye'de Kırsal Kalkınma Politika ve Stratejileri) (in Turkish). Journal of the Yeni Türkiye Agricultural Policies Special Issue I. 2020;**26**(113):392-408

[10] Velenturf APM, Purnell P. Principles for a sustainable circular economy. Sustainable Production and Consumption. 2021;**27**:1437-1457. DOI: 10.1016/j.spc.2021.02.018

[11] WCED. Report of the World Commission on Environment and Development: Our Common Future [Technical Report]. New York, NY, USA: World Commission on Environment and Development; 1987. Available from: https://sustainabledevelopment.un.org/content/documents/5987our-common-future.pdf

[12] Höfer R. Sustainable Solutions for Modern Economies. RSC Green Chemistry Series: 4. Chambridge, UK: RSC Publishing; 2009. ISBN: 978-1-84755-905-0

[13] Fisher JL. Natural resources and technological change. Land Economics. 1953;**29**(1):57-71. DOI: 10.2307/3144285

[14] Kamińska AM, Opaliński Ł, Wyciślik Ł. The landscapes of

sustainability in the library and information science: Systematic literature review. Sustainability. 2022;**14**:441. DOI: 10.3390/su14010441

[15] UN. Report of the United Nations Conference on the Human Environment. 1972. Available from: https://sustainabledevelopment.un.org/milestones/humanenvironment and Printed in Switzerland. https://www.un.org/ga/search/view_doc.asp?symbol=A/CONF.48/14/REV.1

[16] D'Amato D, Korhonen J. Integrating the green economy, circular economy and bioeconomy in a strategic sustainability framework. Ecological Economics. 2021;**188**:107143. DOI: 10.1016/j.ecolecon.2021.107143

[17] Korhonen J, Honkasalo A, Seppälä J. Circular economy: The concept and its limitations. Ecological Economics. 2018;**143**:37-46. DOI: 10.1016/j.ecolecon.2017.06.041

[18] Muchangos LSD. Mapping the circular economy concept and the global south. Circular Economy and Sustainability. 2022;**2**:71-90. DOI: 10.1007/s43615-021-00095-0

[19] van Eck NJ, Waltman L. Citation-based clustering of publications using CitNetExplorer and VOSviewer. Scientometrics. 2017;**111**:1053-1070. DOI: 10.1007/s11192-017-2300-7

[20] van Eck NJ, Waltman L. Software survey: VOSviewer, a computer program for bibliometric mapping. Scientometrics. 2010;**84**:523-538. DOI: 10.1007/s11192-009-0146-3

[21] van Eck NJ, Waltman L. Visualizing bibliometric networks. In: Ding Y, Rousseau R, Wolfram D, editors. Measuring Scholarly Impact: Methods and Practice. Springer; 2014. pp. 285-320

[22] Wang W, Lu C. Visualization analysis of big data research based on Citespace. Soft Computing. 2020;**24**:8173-8186. DOI: 10.1007/s00500-019-04384-7

[23] Yang Y, Lv K, Xue J, Huang X. A bibliometric analysis and visualization of fractional order research in China over two decades (2001-2020). Journal of Mathematics. 2021;**2021**:7996776. DOI: 10.1155/2021/7996776

[24] Özçatalbaş O. Technology transfer and change management. In: Leal Filho W, Azul A, Brandli L, Özuyar P, Wall T, editors. Zero Hunger. Encyclopedia of the UN Sustainable Development Goals. Cham: Springer; 2020. DOI: 10.1007/978-3-319-69626-3_53-1

[25] EMAF. Towards The Circular Economy: Economic and Business Rationale for an Accelerated Transition. United Kingdom: The Ellen MacArthur Foundation; 2020

[26] Corona B, Shen L, Reike D, Carreón JR, Worrell E. Towards sustainable development through the circular economy—A review and critical assessment on current circularity metrics. Resources, Conservation and Recycling. 2019;**151**:104498. DOI: 10.1016/j.resconrec.2019.104498

[27] EU. Policy Document: The European Green Deal. Brussels: European Environment Information and Observation Network (Eionet); 2019. Available from: https://www.eea.europa.eu/policy-documents/com-2019-640-final

[28] Rivas S, Urraca R, Bertoldi P, Thiel C. Towards the EU Green Deal: Local key factors to achieve ambitious 2030 climate targets. Journal of Cleaner Production. 2021;**320**:128878. DOI: 10.1016/j.jclepro.2021.128878

Section 2

Theory and Approaches

Chapter 2

Blending Climate Action and Rural Development in Africa's Sahel

Paul L. Woomer, Dries Roobroeck and Didier Yelognisse Alia

Abstract

This paper describes the opportunity for combining climate action and improved food and nutritional security as mutual elements of rural development projects, with particular reference to the situation in the African Sahel. This progress is achieved by identifying climate-smart agricultural production technologies and bundling them into solutions for inclusion within larger projects and programs. Seventeen (17) such technologies are offered in this chapter that represent genetic innovations, improved soil and water management, and directed improvement across landscapes. Examples of the efficacy of these technologies are presented based on results from the African Agricultural Transformation Program (TAAT) with specific reference to improved cereal production. An example of the deployment of TAAT technologies for millet and sorghum involving 83,620 households managing 123,863 ha led to nearly 200,000 MT of increased food production worth about $42 million. This effort led to an estimated annual increase of 177,279 MT CO_2e in biomass and soil worth $3.9 million, assuming buyers could be found. The relationship between three principal drivers of agricultural transformation, the public, private, and farming sectors, is considered in terms of how these different technologies are mobilized and deployed. The potential for increasing food supply and carbon gains under current agricultural investment levels across the Sahel by International Financial Institutions, about $683 million per year, is described. This chapter then offers recommendations in how improved rural development projects combining climate action and food security in the Sahel may be designed in the future.

Keywords: African drylands, cereal crops, climate change, IITA, soil and water management, TAAT Program, technology deployment, transferable assets

1. Introduction

Adaptation to climate change by small-scale farmers is considered an important part for the climate solution agenda [1, 2]. This is specially the case in the Sahel where food security is tenuous and becoming more so due to rising temperatures and more episodic precipitation [3, 4]. Awareness of this situation is not new, and several farming technologies were identified and modified that allow rural households to cope with increased risks through reliance upon improved crop varieties, more efficient water harvesting, protection of soil quality and participation in well planned, systems-level improvements to their agro-ecosystems [5]. Indeed, isolated cases of successes are documented and used as the basis of designing larger, subregional projects [6] intended for the joint purpose of increasing food and nutrition security in ways that constitute climate action by legions of small-scale farming households [7, 8].

All rural development projects require inclusive and active participation by the public and private sectors, and the client farmers themselves, because local organizations acting through public works and as customers of proven production inputs represent a complete package toward change. Rural development projects are often financed by sovereign loans from International Financial Institutions (IFIs). It is the design and implementation of these projects that prove difficult. In some cases, countries receiving sovereign country loans rely upon suboptimal, existing technologies and are reluctant to involve what they perceive as overly expensive international partners. In other cases, it is not the technologies that are flawed, but rather the manner that they are bundled as solutions, because effective interventions seldom require only one new technology but rather balanced sets of accompanying production inputs and innovative practices [9]. In yet other cases, it is not the solutions that are inadequate, but rather their manner of deployment, often in expectation of too rapid adoption [10]. Complicating this arena is the growing recognition that small-scale farming households are both victims of climate change yet offer the means to effect corrective actions when offered the opportunity and incentive to do so [1].

2. The Sahelian situation

Dryland farming is the dominant mode of livelihood across the Sudano-Sahelian zone of Africa, a transition zone about 400–600 km wide that stretches from the Atlantic Ocean in Senegal to the Red Sea in Djibouti and Indian Ocean in Somalia [11]. Climate-smart solutions and modernization of technologies are critical to improving agriculture in the zone. The Sahel is home to a population of about 110 million persons, the majority of whom rely upon agriculture through the cultivation of about 30 million ha. Landscapes are flat to gently undulating and rainfall at theses latitudes is concentrated in a single growing season between June and September, with a total annual precipitation of only 150–600 mm that is often deposited by only a few heavy storms. Daytime temperatures often exceed 40°C. The natural vegetation ranges from semi-desert in the north to woody grassland in the south. Millet is widely grown in the Sahel and Sudanese zones, but so too is sorghum and maize. New varieties of wheat can be grown too, particularly during the cooler months [12]. Semi-nomadic pastoralism is widely practiced and overgrazing has led to extensive land degradation and desertification. Rice cultivation is possible in some areas, most notably the valleys of major rivers, and represents an important crop in household diets and livelihoods. The adjoined Sudanese Zone receives greater rainfall (600–1200 mm per year) but is confined to a 2–3 month window and its farmers are faced with similar challenges to crop production as their neighbors in the Sahel [5].

Agricultural production in the Sahel is perilous because of severe and cyclical droughts [13]. Other soil limitations exist due to low water-holding and nutrient retention capacities and soils are often sandy and acidic [14]. Because of their unfavorable soil physical properties and low nutrient reserves, soils of the African drylands present a challenge to farmers [15]. Clearly, farmers in the Sahel are acutely aware of drought as a chronic risk and are prepared to adjust their cropping strategies accordingly. Population densities in the agricultural areas remain relatively low, with

0.5–1.5 ha available per capita. Land availability alone does not assure rural prosperity in the Sahel owing to the poor crop productivity resulting from low rainfall and chronic risk of drought. Despite the severe conditions experienced by farmers in the Sahel, large opportunities are available for employing improved soil and water management technologies, including those important to climate actions [5].

The Technologies for African Agricultural Transformation Program (TAAT) deploys proven technologies to African farmers, including those in the Sahel. TAAT arose as a joint effort of the International Institute of Tropical Agriculture (IITA) and the African Development Bank (AfDB) and is a crucial component of the latter's Feed Africa Strategy [10]. It is organized around 15 "Compacts" that represent priorities and partnerships to achieve food security in Africa and advance its role in global agricultural trade [16]. TAAT operates a Regional Technology Delivery Infrastructure that offers a menu of tested and proven food production technologies for nine priority commodities to program partners and stakeholders. These technologies are bundled into "technology toolkits" [17] that are included within country projects and deployed through extension campaigns. These technologies include improved crop varieties, seed systems innovations, accompanying soil fertility and pest managements, harvest and postharvest handling, digital applications, and value addition processes [18], providing Regional Public Goods that attract broad public interest and recognizable benefits. TAAT offers a unique collaborative platform where government, international donors, private actors, and nonstate actors committed to advancing transformative agricultural technologies connect with those who need them most, particularly within programs addressing agricultural production and rural development. It offers a mechanism for the development community to buy into proven technical advances [19]. This paper describes how TAAT's technologies are of benefit to the Sahel and how they may be better integrated within climate action efforts.

3. Appropriate solutions

Solutions are available that assist farmers in the Sahel to increase productivity and achieve food security while also being able to tackle environmental challenges posed by drought, land degradation, and climate change. The solutions are based on greater access to proven technologies that remain under-recognized, inadequately delivered or too difficult to access. Once mobilized, however, key technologies may be bundled into toolkits offering solutions to those seeking to modernize and transform dryland agriculture by combining improved crop varieties, more effective water conservation practices and proven approaches for soil fertility management [9, 17]. Cereal improvement in the Sahel focuses upon millet, sorghum, maize, and wheat that are both drought- and heat-tolerant [20]. Better water management achieves water storage from contour bunds, water harvesting within zaï pits, diversion of seasonal floods, and small-scale irrigation schemes [21, 22]. Practices for integrated soil fertility management involve rotation with legumes, fertilizer micro-dosing, strategic timing of nitrogen application and effective use of organic resources [14]. Larger-scale impacts are achieved through transition from open fields to agroforestry parklands, improved rangeland management and other climate actions specifically targeted to semiarid agro-ecologies. It is essential that these technologies become incorporated into larger rural development projects, but first they must be readily understood by development planners, extension supervisors, and business persons seeking to enhance the lives and livelihoods of farmers. The Sahel is one of the areas of the world that is unfairly penalized by industrial polluters in developed countries, and the impacts of climate change it suffers are not of its own making. Inclusion of these technologies into rural development projects, including those financed with sovereign loans from International Financial Institutions, and embedding them into country-level climate actions serve to correct this disparity.

TAAT offers 17 technologies useful to both rural development and climate action (see **Table 1**). These technologies are grouped according to their relationship to improved field crop varieties (four crops), better management of water resources (four technologies), relationship to integrated soil fertility management (four technologies),

Technology objective	TAAT holder[1]	Approach	Mechanism
New varieties			
Improved pearl millet	ICRISAT	Conventional breeding	Community-based seed production
Improved Sorghum	ICRISAT	Conventional breeding	
Drought-tolerant maize	AATF/IITA	Conventional breeding	Commercial hybridization
Heat-tolerant wheat	ICARDA	Conventional breeding	Public-private partnership
Water management			
Bund walls	IFDC	Soil & water conservation	Community-based action
Zaï pits	IFDC	Soil & water conservation	Farmer action
Spate irrigation	IWMI	Seasonal water harvesting	Community-based action
Small-scale irrigation	IWMI	Year-round cultivation	Commercial suppliers
Soil quality			
Fertilizer micro-dosing	IFDC	Better fertilizer placement	Extension information
Strategic timing of N	IFDC	Improved fertilizer timing	Extension information
Inoculation and BNF	IITA	Symbiotic N fixation	Commercial suppliers
Organic resource management	IITA	Farmer-available resource	Farmer action
Systems transformation			
Control insect invasions	IITA	Combat episodic pests	Public project leadership
Overcoming *striga*	AATF/IITA	Eliminate soil infestation	Public project leadership
Transition to parklands	IITA	Agroforestry intervention	Public project leadership
Improved range management	ILRI	Combat land degradation	Public project leadership
Local biogas	Clearinghouse	Alternative rural energy	Commercial suppliers

[1]*TAAT lead partner organization (see TAAT website https://taat-africa.org).*

Table 1.
A summary of TAAT's 17 climate-smart dryland technologies.

and possibilities for system-level improvement (five technologies). Not considered among these technologies is rice (*Oryza sativa*), an important irrigated crop of Sahelian river basins, and animal enterprises that are extremely important across the Sahel but beyond the scope of this paper.

3.1 Improved field crop varieties

These technologies relate to four cereal crops with unrealized potential in the Sahel: millet, sorghum, maize, and wheat.

3.1.1 Improved millet

Pearl millet (*Pennisetum glaucum*) is the staple cereal in the harshest of the world's major farming areas: the arid and semiarid region extending between Senegal to Somalia. Withstanding hot, dry, sandy soils, it is adapted toward survival under harsh conditions [20]. It is amazingly drought-tolerant and able to germinate at high soil temperatures and in crusted soil, it withstands "sand blasting" and grows under low soil fertility, and it resists pests and diseases such as downy mildew, stem borer, and parasitic striga. It also grows well in both acidic and saline soils. But its most rugged land races are characteristically low yielding and may not respond well to inputs, and for this reason there is need for improved varieties and their accompanying seed systems. Breeding efforts have led to increased micronutrients (e.g. iron and zinc), and some improved "sugary" types can be harvested at the milk stage, and roasted and consumed like sweet corn. The International Crops Research Institute for the Semi-Arid Tropics (ICRISAT) is responsible within TAAT for millet improvement, offering many new varieties for testing by national systems or release to development efforts.

3.1.2 Improved sorghum

Sorghum (*Sorghum bicolor*) is a physiological marvel; it is extremely drought tolerant and light efficient, with one of the highest dry matter accumulation rates among cultivated crops [20]. It is versatile in its use with some types boiled like rice, others cracked like oats, others malted for brewing, and some milled and baked. The whole plant may be used as forage or hay. ICRISAT is also responsible for sorghum improvement, including in the Sahel. Currently available improved varieties and land races have several favorable characteristics including good seedling emergence and rapid early root development, rapid tillering leading to multiple heads, and long growing cycles to make the best of favorable rains. It can be manufactured into a wide variety of foods and used to substitute for imported grains. These properties combined with sorghum's use as an animal feed suggest that national planners are well advised to regard sorghum as more than a drought-hardy subsistence food.

3.1.3 Drought-tolerant maize

Considerable gains in maize (*Zea mays*) improvement have been achieved in the area of drought tolerance that now make this crop less risky in the southern reaches of the Sahel. Drought tolerant maize varieties have a 20–35% larger grain harvest under moderate drought conditions but may not respond as favorably to occasional years of excellent rains due to their shorter maturity times [23]. Hybrid varieties are marketed under commercial license, while open pollinating varieties can be multiplied and sold free of royalty by farmers and community-based producers. The African Agricultural Technology Foundation has sublicensed 22 seed companies to produce Drought TEGO™ for commercial distribution, and more will follow [18]; but these hybrids have been slow to reach West Africa.

3.1.4 Heat-tolerant wheat

The trait of heat tolerance is now incorporated into improved varieties of wheat (*Triticum* spp). Heat stress and drought are among the most predominant constraints affecting wheat across Africa [24], especially at the reproductive stage during flowering and grain filling, leading to low grain yield or even crop failure [25]. Wheat production has increased significantly in the Sahel over the past several years due to the rapid increase of area planted to these newly released heat-tolerant varieties. Varieties that can withstand temperatures up to 4°C greater than previous lines are available. As a result, farmers are achieving higher and more stable yields, reaping up to 6 t ha^{-1}. The success also has policy implications by convincing country decision-makers that domestic wheat production is a solution to reduce the massive dependence upon wheat imports.

3.2 Improved water management

These technologies relate to different forms of water management, including the design of small-scale irrigation systems.

3.2.1 Combined soil and water conservation

Bunds refer to a micro-catchment technique where low raised walls are arranged in specific patterns on farmlands to collect and conserve water and to reduce soil erosion and gully formation [26]. Bund walls are constructed with soil and/or rock, either by hand or tractor. Designs of bund walls are adjusted to local conditions and sociocultural contexts, but the two main types are contour bunds (or contour ridges) and semicircular bunds (or half-moons). Contour bunds are suitable for uniformly sloping terrains with even runoff, and the retaining walls can stretch hundreds of meters across landscapes. Semicircular bunds operate in a more localized manner [21]. Installing contour bunds can increase grain yields of sorghum by 80% and maize by 300% compared to traditional land management without micro-catchment. Community works that stabilize slopes and better harness seasonal rainfall by constructing and reinforcing bunds are an important element of agricultural development projects in the Sahel.

3.2.2 Water harvesting with zaï pits

Micro-catchment approaches to water harvesting in the Sahel include planting pits, locally known as zaï [15]. Zaï pits also rehabilitate crusted and degraded lands. These structures are made by digging shallow basins of 20–40 cm diameter and 10–20 cm deep into the soil. The pits are prepared during the dry season by farmers allowing the shallow holes to collect water, wind-driven soil particles, and plant debris [5]. Moisture becomes collected inside and below the pits that also serve as localized targets for soil fertility improvement. The technique can improve millet and sorghum production by

60–90% depending on precipitation and soil fertility. When properly managed, these pits become a permanent feature of the field that collects off-season or early rainfall.

3.2.3 Spate management of seasonal water

Exploiting water from rivers and streams during the rainy season to fill channels and direct them to adjacent fields by construction of spates is a strategic small-scale irrigation system. Spate is an ancient approach but under some circumstances, it remains relevant today [5]. This system diverts water from normally dry riverbeds at the onset of seasonal rains and directs it to croplands, converting them into seasonal flood plains. Community consensus assures equitable distribution of these floodwaters, including those further downstream that also rely upon the same water. Managing floodwater is inherently difficult because of the power they hold, but the rewards to managing these waters in arid and semiarid areas are great, and for this reason, the opportunity exists in public support of spate irrigation as a localized civil engineering challenge.

3.2.4 Small-scale irrigation schemes

Irrigation assures that the water requirements of crops are met and the development of community-based irrigation schemes is an essential component of agricultural development in the Sahel [5]. Irrigation consists of two phases, the first where water is diverted from its source and delivered to the vicinity of croplands, and the second where it is applied to fields in a scheduled and calculated manner. Application strategies vary with the volumes, quality, and pressure of water delivery and may be grouped into flood, furrow, sprinkler, and drip irrigation. Irrigation presents a key solution to addressing present and future crop production constrains due to the effects of climate change on weather patterns. Within the context of practical rural development, a focus upon small-scale irrigation schemes in addition to larger, more centralized schemes should be considered.

3.3 Improved soil management

These technologies relate to more efficient use of mineral fertilizers, maximizing symbiotic biological nitrogen fixation and improved use of farmer-available organic resources.

3.3.1 Fertilizer micro-dosing

Fertilizer micro-dosing is based on the application of small amounts of mineral fertilizer in a shallow hole about 5 cm away from the crop stem [15]. Micro-dosing is as simple as applying one bottle cap filled with 3–5 g of fertilizer to each planting hole and is best combined with the addition of organic materials, particularly composts and manures. The total amount of fertilizer used in micro-dosing can vary significantly depending on the planting density, ranging from 50 to 100 kg of fertilizer per ha. This addition results in healthier crops that are better able to counteract mid- and late-season drought as a means to adapt to increased climate variability. A well-timed dose of fertilizer results in increased crop yields ranging from 40% to 120%, providing high returns to modest investment. The micro-dosing technique significantly increases the use efficiency of nutrients and water, particularly when combined with other climate-smart practices such as zaï pits [5].

3.3.2 Better timed nitrogen application

The key to achieving high crop yields and maintaining soil fertility is to apply the right fertilizers at the correct rate and time. Too often, timing is ill considered, particularly in relation to nitrogen (N) topdressing of field crops. Typically, N fertilizer is added to soils once or twice over the season, first as a pre-plant addition and second as a single topdressing, but more frequent and smaller doses are more efficient [27]. The basic principle of this approach is to apply a small quantity of N at planting and progressively add moderate amounts as topdressing during periods with sufficient rainfall when plant nutrient demand is largest. Farmers can top-dress N using readily accessible types of fertilizers such as urea and calcium ammonium nitrate, and the total application rate is based on yield targets and regional recommendations [5]. In some cases, N can be added just prior to, and worked into the soil during weeding, resulting in more efficient combined field operations.

3.3.3 Nitrogen fixation from field legumes

Legumes are very important to the rainfed cropping systems of the Sahel, particularly cowpea (*Vigna unguiculata*) and groundnut (*Arachis hypogaea*) [20]. Intercropping is best practiced by farmers during years of favorable rainfall by growing understory grain legumes between cereal rows at very low densities. More common is crop rotation of cereal and legumes, with a few (e.g. two–four) cycles of cereals punctuated by legumes [15]. Legumes access atmospheric nitrogen through symbiosis with rhizobia, a process that provides both additional protein to the household and residual nitrogen to the land [28]. The rhizobia needed for biological nitrogen fixation of these crops are often native, but their populations may be suppressed in hot, dry soils [14]. When well nodulated, nitrogen fixation is sufficient to secure a grain legume harvest and contribute about 50 kg or so organic nitrogen to the following crop. Unfortunately, legume inoculants containing elite strains of rhizobia are not widely available across the Sahel, so need exists to develop the capacity to manufacture and distribute them through commercial channels [5].

3.3.4 Organic resource management

A majority of soils in the Sahel are characterized by low water holding capacity and limited availability of plant nutrients because of their low clay and high sand content [15]. Farmers across these cereal-based drylands must better manage organic resources in ways that optimize limited rainfall and costly inputs of mineral fertilizer [13]. The maintenance of soil organic matter and carbon stocks is strongly determined by the amount of crop residues available for addition to soils and the competing need for livestock feed and stalks as cooking fuel and building material. Mulches that cover soil surfaces greatly reduce soil erosion, runoff, and evaporation, leading to about 70% increased cereal harvest. Incorporating fresh plant materials or animal manure is another option to compensate for unfavorable soil physical properties. At the same time, mineral fertilizers applied in conjunction with organic resources have greater nutrient use efficiencies. These examples of Integrated Soil Fertility Management illustrate the need for farmers to make best and balanced use of crop residues and other available organic resources [14].

3.4 Systems-level improvements

Several systems improvements result in more resilient agricultural landscapes and are best implemented at the community or landscape levels including the control of insect invasions, elimination of parasitic striga, introduction of trees to open croplands, improvement to rotationally grazed lands, and the local production of biogas.

3.4.1 Controlling insect invasion

The Sahel is characterized by major invasions of insect pests such as the yellow desert locust (*Schistocerca gregaria*) and fall armyworm (*Spodoptera frugiperda*). These outbreaks pose a major threat for farm households and undermine larger efforts to strengthen food systems [29]. Locusts are notoriously difficult to control once large swarms accumulate and spread over expansive areas. Following favorable rains, vegetation is sufficient for multiple generations of locust to spread across agricultural landscapes, devouring everything in their path. Early warning and preventative control are keys to stopping locust populations from reaching epidemic proportions. Spraying with chemical insecticides controls desert locust but to be most effective, insecticides must be applied directly onto migrating swarms. Spraying interventions for smaller areas can be performed by teams on foot with knapsacks, whereas for larger areas there is need for vehicle mounted nebulizers or specialized spray planes.

The invasion of fall armyworm across cereal croplands throughout Africa, including the Sahel, also represents a major threat to food security [30]. TAAT offers a rapid response kit consisting of a custom-built cargo tuktuk, power sprayers, safety equipment, commercially recommended pesticides, farmer information, and communication materials [5]. Early control of armyworm is also achieved through maize seed treatment with Syngenta's FORTENZA DUO, offering protection to maize crops up to 4 weeks after germination. Authorities in countries worst affected by fall armyworm are encouraging all maize seed producers to treat their seed with this product.

3.4.2 Overcoming striga infestation

Striga is a parasitic weed-attacking cereal and other grass and invading cropland of the Sahel. The damage inflicted by striga begins underground where its roots enter the host, feeding on its nutrients and moisture and releasing toxins into the plant causing twisted, discolored, and stunted growth [31]. After feeding below ground for 4–5 weeks, a fast-maturing shoot emerges that produces attractive spikes of violet (*Striga hermonthica*) or red (*Striga asiatica*) flowers that mature into capsules containing abundant, tiny, long-lived seeds. Parasitism greatly reduces crop yields. Striga attacks millet and sorghum, but these crops show some tolerance to its effects; maize is more severely affected. Farmers respond to striga by hand weeding and, less often, burning affected fields, but the efficacy of these practices remains questionable considering the large numbers of tiny seed that a single, mature plant produces and returns to the soil.

The agricultural community has responded by developing several new approaches to striga control. These approaches involve crop resistance to systemic herbicides, striga-tolerant cereal varieties, and striga suppression by nonhosts and trap cropping [32]. Farmers must become aware that striga infestation is a solvable problem and gain experience in the use of breakthrough technologies. Local and national authorities must fully recognize the threat posed by striga and prioritize efforts to overcome it within rural development agendas. By attacking this plant parasite through a combination of approaches, it is now a solvable problem and offers an important element of comprehensive rural development packages wherever this parasitic weed occurs.

3.4.3 Transition to agroforesty

Great potential for agricultural transformation exists through the conversion of open-field cropping to agroforestry parkland [33]. These parklands appear as well-spaced trees that protect the soil and contribute to soil fertility renewal. Because of these benefits, the crops that grow near or below these trees often perform better than those in an open field. Parklands also sequester significantly greater carbon stocks than open croplands in a way that mitigates emissions of greenhouse gasses. These increased carbon stocks may be 20 or 40 MT C per ha greater than that retained by open cropland and hold potential to sequester carbon into deeper soil horizons [34]. The agroforestry parklands that appear in the cultivated drylands are often the result of clearing trees rather than planting them, and this creates difficulty in carbon accounting, but when open cropland is purposefully transitioned to agroforestry parkland, the carbon gains are clear and attributable to the efforts from tree planting and protection [5].

Afforestation of open croplands is best practiced at the community level because of the demand for quality tree seedlings, the need to plant them at scale, and the collective responsibility to protect them until these trees are well established. Transitioning from degrading open cropland to productive agroforestry parkland should be considered within agricultural development efforts as sound from both the food security and climate action perspectives, noting that success also involves capacity development at the community and extension advisory levels.

3.4.4 Improved range management

Raising livestock is a critical enterprise across the Sahel but overgrazing has resulted in extensive land degradation [35]. Cattle, sheep, and goats are regarded as assets among pastoralists living in areas too dry for reliable farming, and strategies are available to improve the grazing and forages that these lands provide. Water harvesting technologies presented in this paper may be practiced on noncultivated lands planted with improved grasses and browse species, particularly near watering holes where animals are likely to concentrate during the dry season. Stover and stubble of cereal fields are grazed following the harvest of millet, sorghum, and maize, and these lands are then fertilized by the manure that is deposited. While this system is robust as long as rotational intervals are of sufficient length, these systems begin to degrade if cropping becomes to frequent. One means to strengthen the crop-livestock system is to improve these rotational pastures using either annual or perennial grasses. These grasses not only provide feed for livestock, but they provide ground cover that resists wind and water erosion.

Improved rangeland management falls into four general categories that are best applied in packages. Agronomic measures are associated with annual crops in a rotational sequence and are impermanent and of short duration. Vegetative measures involve the use of perennial grasses, shrubs, or trees and are of longer-term duration. Structural measures reduce erosion and capture water and may result in a permanent change in landscape. Management measures involve a fundamental change in land use and may be directed through policy intervention [35]. Improved rangeland management is best conducted at the community level where lands are collectively managed. This participation reduces the risks of conflicts between farming and livestock that often lead to larger social misunderstandings.

3.4.5 Local production and use of biogas

This technology refers to the production of combustible gas within small-scale digesters at the household level. It is based on the utilization of plant and animal residues as organic wastes that are decomposed in anaerobic tanks, forming methane and a digested slurry byproduct useful as an organic fertilizer and soil amendment [36]. Gasses rise and collect through an outlet for burning as cooking fuel and the sediments sink into sludge for later collection. Gasses may be produced in a variety of vessels located above- or belowground. These reactors may be fashioned from metal tanks, built from concrete, or purchased as complete units. Attraction to this technology is growing across the Sahel because of its socioeconomic and environmental benefits, and it has a proven ability to improve the lives of rural households that would otherwise burn wood and charcoal, or cook using purchased kerosene [5]. The diversification of energy supply creates economic opportunity to those who build and equip these digesters, and it reduces local air pollution and deforestation due to firewood collection and charcoal making, and increases sequestration of carbon into soils amended with the digested organic sludge. Carbon sequestration is also achieved by the substitution of renewable energy production from methane as compared to reliance upon fossil fuels. Biogas generation is best considered among a suite of rural development options that are designed to educate stakeholders and supply the hardware and infrastructure it requires [37].

4. Impacts from technology deployment

Table 2 presents findings for millet and sorghum from the TAAT Program in seven countries of the Sahel [18]. ICRISAT coordinated this effort based on the delivery of "technology toolkits" through national programs. Millet and sorghum yields were improved by 133% and 140%, respectively, and reached nearly 84,000 households

Parameter	Millet	Sorghum	Units
Average increased productivity	1.00	1.75	MT dw increase ha^{-1}
Yield improvement over baseline	133%	140%	MT increase/MT baseline
Number of technology adopters	12,403	71,217	Households adopting technologies
Innovation coverage	23,765	100,098	Total ha
Total increased production	23,765	175,172	MT on harvest weight basis
Total increase value	4,515,361	37,662,005	Value of increased production in US$
Average adoption area	1.92	1.41	ha $household^{-1}$
Increased food supply	1.92	2.46	MT hh^{-1} yr^{-1}
Increased revenue per household	$364	$529	Total return US$ hh^{-1} yr^{-1}

Table 2.
Benefits from adopting improved technologies for millet and sorghum in the Sahel between 2018 and 2020.

managing about 124,000 ha and leading to the increased production of 199,000 MT of grain worth US $42 million. Individual households greatly benefited in terms of food security, and the average increase income from participating in the technology delivery effort was about US $504 (calculated as a weighted average from **Table 2**). Activities involved 16 partnerships and delivered 1391 MT of improved certified seed. The right technologies taken to scale can deliver benefits to partnering farming communities that rely upon millet and sorghum as a staple crop.

Investment in TAAT technologies results in economic gain across a wider selection of commodities as well. **Table 3** provides information on the increased yields of five cereal crops (rice, wheat, maize, sorghum, and millet), increased cost of production and economic returns to that investment. The average increased productivity was 1.3 MT ha^{-1} worth US $333 resulting from $136 increased investment, mostly as fertilizers. This results in an average increased value of US $197, ranging from $85 (for millet) and $299 for rice. Note that except for rice, these crops were grown under rainfed conditions. The partial benefit to cost ratio ranges between 1.8:1 (for millet) and 3.2:1 (for maize), suggesting that economic returns are solid but not spectacular.

Table 4 shows projections of carbon sequestration resulting from TAAT interventions to cereal production including system gains, values, and household contributions. These projections are based on reports of increased yield, coverage, numbers of adopters (see **Table 2**), and assumptions concerning biomass, moisture content, Harvest Index, crop carbon content, CO_2e:crop C ratio, planning horizons, and the price of CO_2e. This approximation allows for the estimation of realizable gains of CO_2e associated with increased biomass and residual benefits in terms of CO_2e gain per ha and as total average gain per project-year and household [18]. Realizable gains were achieved based on increased focus upon climate-smart field practices and products within the technology toolkits employed by participating farmers and development projects. This approach results in estimated CO_2e gains averaging 4.4 MT ha^{-1} across these five cereals and a total of 2.1 million MT of CO_2e per year worth about US $65 million. When the number of adopters is considered, this amounts to per capita emissions reductions of 1.5 MT CO_2e per household per year, similar to the targets established by Branca et al. [38] and Lipper et al. [39]. This analysis is incomplete, as it does not take into account carbon losses from other farming practices; rather it focuses on peak seasonal increases.

The feasibility of organizing small-scale African farmers into a force devoted to carbon sequestration is an exciting opportunity, but one that does not greatly benefit individual climate-smart practitioners from the standpoint of direct financial benefit as their gains are worth only $16 household per year at current prices of CO_2e.

Commodity	Units	Rice	Wheat	Maize	Sorghum	Millet	Mean (± SEM)
Increased productivity	MT ha^{-1}	1.2	1.9	0.7	1.7	1.0	1.31 ± 0.19
Increased fertilizer cost	US $ ha^{-1}	91	200	69	146	103	122 ± 20
Total increased cost	US $ ha^{-1}	153	203	73	148	105	136 ± 19
Increased crop value	US $ ha^{-1}	452	418	231	376	190	333 ± 44
Increased partial net return	US $ ha^{-1}	299	215	158	229	85	197 ± 30
Partial benefit: cost ratio	US $ US $$^{-1}$	2.95	2.06	3.17	2.55	1.81	2.51 ± 0.22

Table 3.
Economic returns to technology investment in cereals based on TAAT toolkit packages (2018–2020).

TAAT commodity compact	Increased system CO_2e	Annual increased CO_2e[1]	Value of annual increase in CO_2e[2]	Annual reduction per adopter[3]
	($MT\ ha^{-1}$)	$MT\ y^{-1}$	$\$ y^{-1} \times 10^6$	$MT\ CO_2e\ y^{-1}$
Rice	1.06	186,882	4.1	0.08
Wheat	2.92	1,753,606	38.6	1.9
Maize	3.1	869.284	19.1	1.5
Millet	5.71	22,611	0.5	1.8
Sorghum	9.27	154,668	3.4	2.2
Total (*mean*)	*4.41*	2,118,636	65.7	*1.50*

[1]*Mean weighted by coverage from* **Table 2**, ± *standard error of the mean.*
[2]*Based on US $22 per MT CO_2e.*
[3]*Based on the annual increase of CO_2e and overall mean weighted by beneficiary households from* **Table 2**.

Table 4.
Estimated carbon offsets from the adoption of TAAT technologies by African cereal producers (based on [18]).

The benefits of climate-smart technologies are perhaps better advanced in terms of improved livelihood and agricultural resource quality and then factored in terms of realizing national commitments at the landscape level; rather than presented to farmers as an income generating opportunity.

5. A transformational model

The TAAT Clearinghouse is developing a conceptual and mathematical model useful in understanding and managing agricultural transformation in Africa. This model has both qualitative and quantitative features.

5.1 Transformational realms

These realms are based on the roles and responsibilities of three interacting driving sectors: policy, markets, and farmers. It assumes that policies drive public works and rural development programs, markets determine the scope and appeal of commercial products and related investments, and farmers undertake individual and local collective actions. When these roles are depicted along three triangular coordinates, a conceptual model emerges that contains different transformational realms, many of them widely recognized. Grassroots actions occur where farmers dominate adoption processes (**Figure 1**), commerce is conducted where businesses buy and sell agricultural technologies, and government-led parastatal operations exist where government controls agricultural opportunities and trade. Other familiar blended realms exist including agricultural extension, public-private partnerships, and farmer-commercial alliances (e.g. out-grower networks). At the center of these activities, we identify complex alliances, where all three drivers meet on equal terms to pioneer progressive change. Each of these seven realms is briefly described.

5.1.1 Grassroots actions

Grassroots actions (*upper center*) are localized in scope and conducted by farmers and their communities as opposed to being guided by those in more centralized

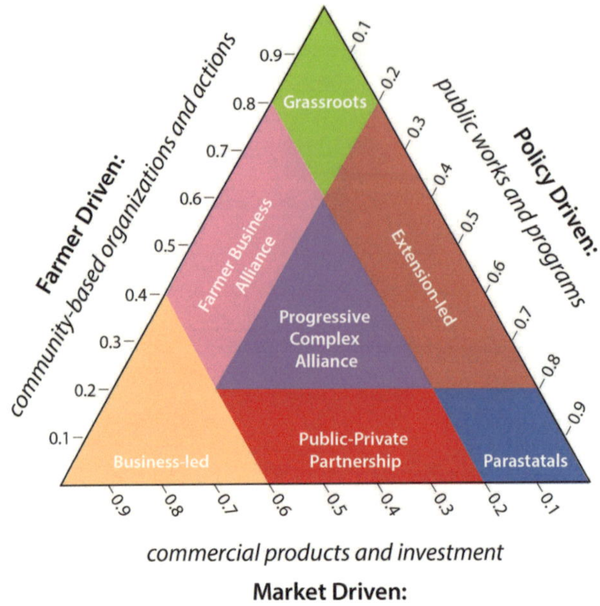

Figure 1.
Rural development realms resulting from policy-, market-, and farmer-driven interests.

positions of power. Farmers belonging to grassroots organizations rely on individual and collective action to effect desired local change and often receive guidance from local agrodealers and extensionists.

5.1.2 Business-led development

Business-led development (*lower left*) incorporates a range of strategies aiming to establish markets and provide economic opportunities that drive rural growth and employment opportunities. In more advanced settings, the private sector plays the lead role in research and development as well, translating breakthrough technologies into useful products and services.

5.1.3 Farmer-business alliances

Farmer-business alliances (*center left*) allow small-scale producers to transition into commercial agriculture by providing information, inputs, and markets, usually based on a focus commodity. This alliance can operate as out-grower schemes and is further advanced through digital services and e-commerce platforms. The "farm to fork" approach relies upon such alliances.

5.1.4 Public-private partnership

Public-private partnership (*center bottom*) is an agreement between the public and private sectors for the purpose of accelerated delivery of products or services beyond the reach of either. In some cases, it increases the efficiency of public services, and in others it is intended as an accelerated pathway to privatization. It may be based on contracts where government assigns some of its responsibilities to a private partner and often involves joint investment under terms attractive to business.

5.1.5 Agricultural extension

Agricultural extension (*center right*) applies new knowledge to agricultural practices through farmer education and advisory services, leading into increased productivity and improved livelihood. It relies on farm visits, group interactions (e.g. demonstrations and field days), and mass information campaigns and is increasingly reliant on digital devices and linkage to education systems. The effectiveness of current extension systems in Africa is often questioned.

5.1.6 Parastatals

Parastatals (*lower right*) are organizations operating under political authority, often as a state-capitalistic form of agricultural production. They are often criticized for being inefficient, corrupt, and for underpaying producers but at the same time have a proven ability to transfer modern farming techniques and new commodities to small-scale producers. Parastatals are increasingly targeted for privatization through public-private partnership.

5.1.7 Progressive complex alliances

Progressive complex alliances (*center triangle*) represent a difficult to achieve form of stakeholder partnership that effectively balances the interests of rural communities and the private and public sectors. In many cases, the loans from development banks are focused on combined actions involving these stakeholders through their combined participation and investment, although formula for success remains ambiguous as it involves complex, knowledge-rich problem-solving across competing interests and site-specific settings.

5.2 Sector interactions

Successful partnership within rural development programs striving for agricultural transformation, particularly within the realm of progressive complex alliances, requires effective communications between sectors (**Figure 2**). Between farming communities and the public sector, these communications involve advocacy on behalf of agricultural producers and their workers, and effective response from agricultural extension services. This dual mechanism ensures that public investment in advisory services is demand-driven. Unfortunately, rural communities often find it difficult to express their needs, and those that do so on their behalf may behave opportunistically. At the same time, public agricultural extension services are too often understaffed and under-resourced, yet it is this communication that can lead to more efficient performance by extension specialists and project designers.

Communication between farming communities and the private sector is more direct. Businesses stream input products through agrodealer networks to farming communities and later purchase their surpluses through buyers. Accompanying these input products is information about them that is intended to achieve or maintain various competitive advantages. Farmer feedback on the availability, efficacy, and affordability of these input products is mainly felt in terms of seasonal purchases. At the same time, businesses seek direct feedback from potential customers to guide their selection of product lines and advertising campaigns. One difficulty in this dual mechanism is the inability of poorer farmers to purchase the full suite of

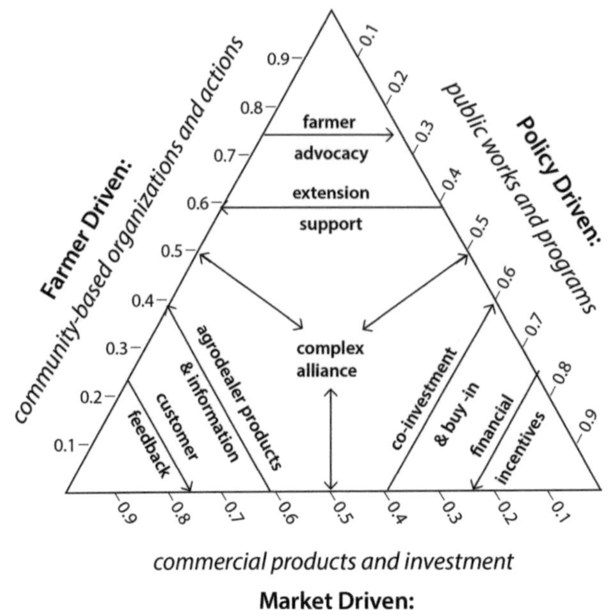

Figure 2.
Key interactions between the public, private, and farming sectors that relate to the design and implementation of rural development projects.

recommended input products proven to maximize their production. There is also the risk that unless accompanying technologies are properly bundled, the returns to any one technology may be disappointing. This communication mechanism can lead to alliances between farmers and businesses in terms of bulk purchase of production inputs and better coordinated marketing of produce.

Interactions between the private and public sector are focused on regulatory approval of products and steering financial incentives, often in ways designed to maximize profits or taxation and that often bypass farming communities. Nonetheless, the opportunities for co-investment into modernizing technologies through these dealings are enormous and can lead to the formulation of needed public-private partnerships that indirectly benefit farmers. One risk of this dialog, however, is where haphazard or opportunistic privatization may result in parastatal inefficiencies being replaced with private sector excesses.

Clearly, the optimal situation is where tripartite communication leads to the design and successful implementation of rural development projects that engage and benefit all three parties: rural communities, the private sector, and government (**Figure 2**). These complex alliances require problem-solving with clear agreement of which difficulties exist, how to merge possible solutions within everyone's best interests, and how different options most appealing to those different interests may be blended or pursued simultaneously. From the programmatic perspective, it is also important to establish how resulting activities may be accurately and continuously monitored within the context of contingencies and corrective adjustment. This level of communication as it relates to the deployment of modernizing agricultural technologies in Africa has proven to be no easy matter.

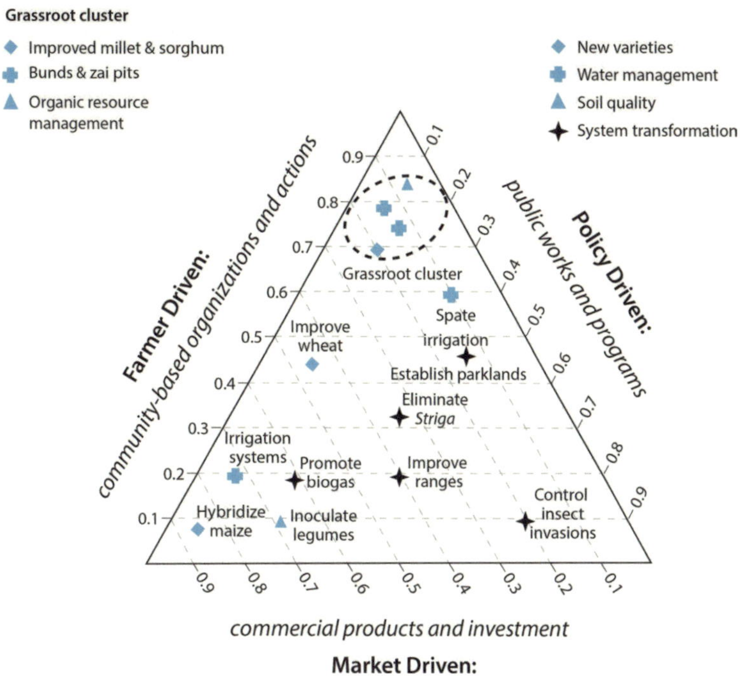

Figure 3.
Selected climate-smart technologies important to the Sahel as positioned within the Agricultural Transformation Triangle.

5.3 An example from the Sahel

Technologies may be positioned within the agricultural transformation triangle assuming that the relative importance of the three different drivers can be assigned (**Figure 3**). This positioning is based on the relative importance of each driver in the deployment of technologies and development outcomes, recognizing that all of them must ultimately be acceptable to rural households to become widely adopted, whether as technology customers or management practitioners. This approach, applied to the 17 technologies appearing in **Table 1**, results in clusters of technologies including those that are mainly achieved through grassroots efforts (upper center), or by private sector investment (lower left). Note that the positioning of new cereal varieties depends largely on whether they are hybridized or open pollinated, as the latter allows for community-based and farmers-own seed production. Also note that systems-level changes (e.g. containment of insect invasions, elimination of *Striga*, agroforestry parkland establishment) require greater involvement of the public sector. One advantage of this approach is that technologies appearing in different clusters and within realms (see **Figure 1**) can be considered mutual objectives within a program's operational framework.

6. Investment volumes

Substantial if not ample investment in the agriculture of the Sahel occurs (**Table 5**). Researchers at the Policy Analysis and Research Group at of Evans

IFI annualized investment	Sahel	All sub-Saharan Africa	Percentage (%)
In smallholder agriculture	$590,213,920	$3,408,245,800	17.3
In rural infrastructure and commercialization	$92,362,308	$2,834,768,443	3.3
In agriculture as a sector	$682,576,228	$6,243,014,242	10.9
Into all development	$3,393,963,264	$28,331,013,684	12.0
Population (2019)	111,121,173	1,045,204,638	10.6
Per capita investment	$30.54	$27.11	112.7
Cultivated lands (ha)	41,883,700	226,540,000	18.5
Per ha investment	$16.30	$27.56	59.1

[1]*Evans School of Public Policy and Governance (EPAR), University of Washington (Project #411).*

Table 5.
Annual investment in African agriculture and natural resource management by three major International Financial Institutions: The African Development Bank, The World Bank, and the International Fund for Agricultural Development (based on EPAR[1]).

School of Public Policy and Governance (University of Washington) recently compiled data from three major International Financial Institutions (The World Bank, the African Development Bank, and the International Fund for Agricultural Development) to provide insights into the "Investment Landscape" in Africa [40]. The database contains all investments in 46 sub-Saharan African countries from the three IFIs as of May 2021 and includes "active" or "implementation" projects, loans, grants, or other financial investments [40]. To make funding by country more comparable, investments were annualized by dividing the total financial commitment per project by the number of years of implementation. Codes were applied that allowed summation for Sahelian countries including Burkina Faso, Chad, Mali, Mauritania, Niger, Senegal, and South Sudan, but not those with a small portion falling within the Sahel (e.g. Benin, Cameroon, and Nigeria). Annual investment in agricultural development across all of sub-Saharan Africa totaled US $6.24 billion in 2019, with 11% of it (=$0.68 billion) directed to the Sahel. This amount is proportionate in terms of population (±0.3%) and represents 20.1% of total IFI investment. Considering the importance of agriculture in the Sahel, this percentage seems somewhat low.

Overall, the per capita annual investment from the three IFS in the Sahel zone is about $30. What can be done with this resource and how may it best be leveraged toward greater benefit? **Table 3** suggests that the cost of modernizing Sahelian farming is about $136 per ha, so these funds are only sufficient for improved production on only 0.22 ha on a household basis. This intervention results in an additional 288 kg food production and revenues worth $73. These modest gains can lead to substantial improvement in lives. If 50% of the funds earmarked to smallholder agriculture in the Sahel (about $295 million, calculated from **Table 5**) was directed to the delivery of TAAT cereal technologies, this is sufficient to "jump start" improved production across 2.17 million ha (calculated from **Tables 3** and **5**) resulting in 2.8 million additional tons of cereal and profits of over $560 million per year from improved agriculture. A similar analysis may be performed based on funds directed to cultivated lands rather than households (**Table 5**). About $16.30 per ha is invested by IFIs in the Sahel, considerably less than the average across sub-Saharan Africa. This level of investment is sufficient to modernize production on 0.34 ha, producing about 445 kg of additional cereal, leading to a huge improvement in food security (calculated from **Tables 3** and 5).

These same gains would lead to an estimated additional 3.3 million MT of sequestered CO_2e across the Sahel worth $71 million (calculated from **Table 4**), assuming that buyers for that offset due to climate adaptation can be found. One complication, however, is that the costs of directly quantifying carbon offsets on a smallholder farm may well be greater than the value of those offsets themselves ($33 calculated from **Table 4**). Clearly, potential exists for combined agricultural development and climate action given the current level of development investment, and the challenge is to better realize these gains so that even more investment will follow.

7. Conclusions

Modernizing technologies literally bring scientific breakthroughs to life in ways that reduce risks and better manage cause-to-effect relationships. Technology transfer determines how this modernization occurs as a process involving a wide assortment of stakeholders from government, the private sector, financial institutions, and research, civil, and educational institutions [41]. This process intends to work on behalf of both the holders of technologies and those who stand to benefit from them most. In the case of climate action through the deployment of agricultural technologies, these users are primarily land managers directed toward larger global needs through practical self-interest, mainly acquisition of more secure harvests and greater protection of farm resources. Policies may set the stage for change, but ultimately environmental gains are achieved through combinations of purchased inputs and improved management practice, with each category representing a different type of technology holder. Input delivery is largely the concern of the private sector in terms of commercial distribution; and management practices are influenced by agricultural service providers, including public extension. Change is quickest when the two work in conjunction, and this forms both a challenge and opportunity to the design of rural development projects.

Two large regional programs of the African Development Bank are well positioned to benefit from the technologies and deployment approaches described in this chapter, The Programme for Integrated Development and Adaptation to Climate Change in the Niger Basin (PIDACC [6]) and The Horn of Africa Project. PIDACC is funded through the Niger River Authority and TAAT is one of its funded partners. It covers nine countries in the Niger River Basin: Benin, Burkina Faso, Cameroon, Chad, Cote D'Ivoire, Guinea, Mali, Niger, and Nigeria. Its activities include climate-smart technologies related to rice, maize, wheat, as well as soil and water management applied at the field, household, and landscape levels. It operates under the premise that farmers who adopt and exchange improved crop varieties, proactively manage pest outbreaks, better utilize water resources, and maintain soil fertility are in a much stronger position to secure food and income for their families and protect their agricultural resource base.

Horn of Africa is an AfDB regional project at an advanced stage of preparation. Its partner countries will deploy proven, climate-smart agriculture technologies across Djibouti, Ethiopia, Kenya, Somalia, South Sudan, and Sudan from 2022 to 2028. The objective of the project is to build resilient food and nutrition insecurity and climate change response, engage women and youth, and reinforce peace and security across the Horn of Africa. Specifically, it aims to (1) improve agro-sylvo-pastoral productivity, (2) increase incomes from that production, and (3) enhance the adaptive capacity of the populations to better prepare for and manage climate risks. Clearly, the right technologies, including those featured in this chapter, are required to achieve these goals. AfDB is also leveraging co-financing from major climate funds in ways that can impact upon UNFCCC Nationally Determined Commitments.

There is a strong relationship between dryland soil and water management technologies available to small-scale farmers and the need for climate action in the Sahel and elsewhere [19]. Within the context of risk reduction, many of the technologies appearing in this chapter are intended to adapt to climate extremes, particularly higher temperatures, moderate drought, and erratic and intense rainfall. These adaptive technologies are particularly important at the field and household level. Farmers that better capture rainfall or protect their cropland soils from wind and water erosion are better able to feed their families. The same is true for communities that adopt and exchange improved seed of open pollinated cereals such as millet and sorghum. In this way, adaptation to climate extremes offers a "drawdown" of greenhouse gasses that are accumulating in the atmosphere.

The most direct mitigative effects are to increase standing biomass and to manage that biomass in ways that become sequestered into soil organic matter and woody biomass. This is readily feasible using improved soil and water management practices across large areas of land over sufficient times to realize these gains. In general, about 50% of increased productivity is carbon and a small proportion of that enters the soil as residues for longer-term retention. One means to greatly increase standing biomass is to move from rainfed to irrigated agriculture, and another is to rehabilitate lands that are degraded and overgrazed. It is possible to combine adaptive and mitigative technologies as when bunds intended to capture water and reduce erosion are planted with perennial vegetation. Also, the same contour structures used to protect croplands may be constructed in adjacent rangeland to assist in the re-establishment of native vegetation. At the same time, carbon gains in rangelands must be weighed against the increased livestock carrying capacity and the methane they release through digestion.

Substantial opportunity for carbon gains across landscapes exists through the steady transition from open-field cultivation to managed parklands, often through the introduction of economically useful trees. The agroforestry techniques to achieve this transition are well described. Re-vegetation has a transnational dimension through the ambitious Great Green Wall for the Sahel and Sahara Initiative to act as a barrier to further desertification [42]. Another proactive mitigation response occurs through bio-digestion in terms of fossil fuel replacement. One huge advantage of mitigation over adaptation is that quantified carbon gains may then be offered for sale and traded with polluters as a condition of their continued emissions. Another is that they can be applied to the Nationally Determined Contributions of countries within climate agreements [43]. Ultimately, rural development projects and climate actions must be viewed as one and the same.

Acknowledgements

Information on these technologies described in this paper was provided by TAAT Compact and Enabler Leaders: Dougbedji Fatondji from ICRISAT for millet and sorghum, Zewdie Bishaw from ICARDA for wheat, Jonga Munyaradzi from AATF for maize, Sander Zwart from IWMI for water management, and Jean Ekwe Dossa from IFDC. Olanrewaju Eniola Olamide graciously provided assistance in formatting this document. The TAAT Clearinghouse is supported through a project of the Bill and Melinda Gates Foundation, and the accompanying TAAT Program is funded by the African Development Fund of the African Development Bank. The "Investment Landscape" database of the University of Washington described in Section 6 is also a funded development of the Bill and Melinda Gates Foundation.

Author details

Paul L. Woomer[1]*, Dries Roobroeck[1] and Didier Yelognisse Alia[2]

1 TAAT Technologies for African Agricultural Transformation Clearinghouse, International Institute of Tropical Agriculture, Nairobi, Kenya

2 Daniel J. Evans School of Public Policy and Governance, University of Washington, United States

*Address all correspondence to: plwoomer@gmail.com

References

[1] IPCC. Summary for policymakers. In: Field CB, Barros VR, Dokken DJ, Mach KJ, Mastrandrea MD, Bilir TE, et al., editors. Climate Change 2014: Impacts, Adaptation, and Vulnerability. Part A: Global and Sectoral Aspects. Contribution of Working Group II to the Fifth Assessment Report of the Intergovernmental Panel on Climate Change. Cambridge: Cambridge University Press; 2014. pp. 1-32

[2] Niles MT, Brown ME. A multi-country assessment of factors related to smallholder food security in varying rainfall conditions. Scientific Reports. 2017;7:16277. DOI: 10.1038/s41598-017-16282-9

[3] Grace K, Davenport F. Climate variability and health in extremely vulnerable communities: Investigating variations in surface water conditions and food security in the West African Sahel. Population and Environment. 2021;**42**:553-577. DOI: 10.1007/s11111-021-00375-9

[4] Sultan B, Roudier P, Quirion P, et al. Assessing climate change impacts on sorghum and millet yields in the Sudanian and Sahelian savannas of West Africa. Environmental Research Letters. 2013;**8**:014040 [Internet]. DOI: 10.1088/1748-9326/8/1/014040

[5] Clearinghouse TAAT. Climate-smart agriculture technologies for the Sahel and Horn of Africa. In: Clearinghouse Technical Report Series 009, Technologies for African Agricultural Transformation. IITA, Nairobi, Kenya: Clearinghouse Office; 2021. p. 32

[6] AfDB. Multinational Programme for Integrated Development and Adaptation to Climate Change in the Niger Basin (PIDACC): Appraisal Report. Cote d' Ivoire: African Development Bank Group, Abidjan (AfDB); 2018. p. 24. plus annexes

[7] Cooper R, Price RA. International Donor Activity on Climate Change Mitigation and Adaptation in the G5 Sahel Countries. K4D Helpdesk Report 587. Brighton, UK: Institute of Development Studies; 2019. Available from: https://opendocs.ids.ac.uk/opendocs/handle/20.500.12413/14585

[8] Ouédraogo M, Partey ST, Zougmoré RB, Nyuor AB, Zakari S, Traoré KB. Uptake of Climate-Smart Agriculture in West Africa: What can we learn from Climate-Smart Villages of Ghana, Mali and Niger? Info Note, CGIAR: CCFAS. 2018. Available from: https://ccafs.cgiar.org/publications/uptake-climate-smart-agriculture-west-africa-what-can-we-learnclimate-smart-villages#.XMhhyChKg2w

[9] Barrett CB, Benton TG, Fanzo J, Herrero M, Nelson RJ, Bageant E, et al. Socio-technical Innovation Bundles for Agri-food Systems Transformation, Report of the International Expert Panel on Innovations to Build Sustainable, Equitable, Inclusive Food Value Chains. Ithaca, NY, and London: Cornell Atkinson Center for Sustainability and Springer Nature; 2020. p. 172. Available from: https://hdl.handle.net/10568/110864

[10] AfDB. Feed Africa: Strategy for Agricultural Transformation in Africa 2016-2025. Cote d' Ivoire: African Development Bank Group (AfDB), Abidjan; 2016. p. 79

[11] Doso S. Land degradation and agriculture in the Sahel of Africa:

Causes, impacts and recommendations. Journal of Agricultural Sciences and Applications. 2014;**3**:67-73

[12] Clearinghouse TAAT. Wheat Technology Toolkit Catalogue. Clearinghouse Technical Report Series 011, Technologies for African Agricultural Transformation. IITA, Nairobi, Kenya: Clearinghouse Office; 2021. p. 36

[13] Schlecht E, Buerkert A, Tielkes E, Bationo A. A critical analysis of challenges and opportunities for soil fertility restoration in Sudano-Sahelian West Africa. Nutrient Cycling in Agroecosystems. 2006;**76**:109-136

[14] Sanginga N, Woomer PL, editors. Integrated Soil Fertility Management in Africa: Principles, Practices and Developmental Process. Nairobi, Kenya: Tropical Soil Biology and Fertility Institute of the International Centre for Tropical Agriculture; 2009. p. 263

[15] Fatondji D, Martius C, Vlek PLG, Bielders CL, Bationo A. Effect of Zai soil and water conservation technique on water balance and the fate of nitrate from organic amendments applied: A case of degraded crusted soils in Niger. In: Innovations as Key to the Green Revolution in Africa. Dordrecht, The Netherlands: Springer; 2011. pp. 1125-1135

[16] AfDB. Technologies for African Agricultural Transformation: Framework program in support of Feed Africa. Cote d' Ivoire: African Development Bank Group (AfDB), Abidjan; 2017. p. 26

[17] TAAT Clearinghouse. TAAT Technology Toolkits and their Strategic Deployment. Clearinghouse Technical Report Series 001, Technologies for African Agricultural Transformation. Cotonou, Benin: Clearinghouse Office; 2018. p. 18

[18] Woomer PL, Mulei WM, Zozo R. A new paradigm in the delivery of modernizing agricultural technologies across Africa. In: Technology in Agriculture. London: Intech Open; 2021. p. 23. DOI: 10.5772/intechopen.98940

[19] World Bank Group. World Bank Group Climate Change Action Plan 2021-2025: Supporting Green, Resilient, and Inclusive Development. Washington, DC: World Bank; 2021. Available from: https://openknowledge.worldbank.org/handle/10986/35799

[20] USAID. Agricultural Adaptation to Climate Change in the Sahel: A Review of Fifteen Crops Cultivated in the Sahel. Washington DC, USA: USAID; 2014. p. 101

[21] Tabor JA. Improving crop yields in the Sahel by means of water-harvesting. Journal of Arid Environments. 1995;**30**:83-106

[22] Rockström J, Karlberg L, Wani SP, Barron J, Hatibu N, Oweis T, et al. Managing water in rainfed agriculture: The need for a paradigm shift. Agricultural Water Management. 2009;**97**:543-50. DOI: 10.1016/j.agwat.2009.09.009

[23] TAAT Clearinghouse. Maize Technology Toolkit Catalogue. Clearinghouse Technical Report Series 008, Technologies for African Agricultural Transformation. Clearinghouse Office: IITA, Cotonou, Benin; 2021. p. 32

[24] Negassa ABS, Jawoo Koo K, Sonder M, Smale HJ, Braun SG, Zhe Guo D, et al. The Potential for Wheat Production in Africa: Analysis of Biophysical Suitability and Economic Profitability. Mexico, D.F: CIMMYT; 2013. Available from: https://repository.cimmyt.org/handle/10883/4015

[25] Poudel PB, Poudel MR. Heat stress effects and tolerance in wheat: A review. Journal of Biology and Today's World. 2020;**9**(4):217

[26] Barry B, Olaleye AO, Zougmoré R, Fatondji D. Rainwater Harvesting Technologies in the Sahelian Zone of West Africa and the Potential for Outscaling. Colombo, Sri Lanka: International Water Management Institute; 2008. p. 40. (IWMI Working Paper 126)

[27] Piha MI. Optimising fertilizer use and practical rainfall capture in a semi-arid environment with variable rainfall. Experimental Agriculture. 1993;**29**:405-415

[28] Giller KE. Nitrogen Fixation in Tropical Cropping Systems. CAB International; 2001. p. 423

[29] Sánchez-Zapata J, Donázar J, Huertas DA, Forero M, Ceballos O, Hiraldo F. Desert locust outbreaks in the Sahel: Resource competition, predation and ecological effects of pest control. Journal of Applied Ecology. 2007;**44**:323-329. DOI: 10.1111/j.1365-2664.2007.01279.x

[30] FAO and CABI. Fall Armyworm Field Handbook: Identification and Management. Rome, Italy: 1st ed, Food and Agriculture Organization (FAO); 2019. p. 36

[31] Awadallah B. Biology and physiology of witchweed (*Striga* spp.): A review. International Journal of Academic Multidisciplinary Research. 2019;**3**(10):42-51. Available from: https://www.researchgate.net/publication/336927326_Biology_and_Physiology_of_Witchweed_Striga_spp_A_Review

[32] Woomer PL, Bokanga M, Odhiambo GD. Striga management and the African Farmer. Outlook on Agriculture. 2008;**37**(4):277-282

[33] Bayala J, Sanou J, Teklehaimanot Z, Kalinganire A, Ouedraogo SJ. Parklands for buffering climate risk and sustaining agricultural production in the Sahel of West Africa. Current Opinion in Environmental Sustainability. 2014;**6**:28-34

[34] Nair PK, Nair VD, Kumar BM, Haile SG. Soil carbon sequestration in tropical agroforestry systems: A feasibility appraisal. Environmental Science and Policy. 2009;**12**:1099-1111. DOI: 10.1016/j.envsci.2009.01.010

[35] Liniger HP, Mekdaschi Studer R. Sustainable Rangeland Management in Sub-Saharan Africa—Guidelines to Good Practice. World Bank: Washington, DC, USA; 2019. p. 408

[36] Rabezandrina R. Biogas: Evolution of Actions and Prospects for the Rural Environment in Africa. Ambio. 1990;**19**(8):424-426

[37] GTZ. Feasibility Study for a National Domestic Biogas Programme in Burkina Faso. Eschborn: Deutsche Gesellschaft fur Technische Zusammenarbeit (GTZ); 2007. p. 214

[38] Branca G, Lipper L, McCarthy N, Jolejole MC. Food security, climate change, and sustainable land management. A review. Agronomy for Sustainable Development. 2013;**33**:635-650. DOI: 10.1007/s13593-013-0133-1

[39] Lipper L, Thornton P, Campbell BM, Baedeker T, Braimoh A, Bwalya M, et al. Climate-smart agriculture for food security. Nature Climate Change. 2014;**13**:1068-1072. DOI: 10.1038/nclimate2437

[40] Figone K, Porton A, Kiel S, Hariri B, Kaminsky M, Alia D, et al. Summary of Three International Financial Institution (IFI) Investments in Sub-Saharan Africa. EPAR Technical Report #411. University of Washington.

Available from: https://epar.evans.uw.edu/research/tracking-investment-landscape-summary-three-international-financial-institutions-ifis: Evans School of Public Policy & Governance; 2021 [Accessed: December 2, 2021]

[41] Özçatalbaş O. Technology transfer and change management. In: Leal FW, Azul A, Brandli L, Özuyar P, Wall T, editors. Zero Hunger. Encyclopedia of the UN Sustainable Development Goals. Cham: Springer; 2020. DOI: 10.1007/978-3-319-69626-3_53-1

[42] AUC. Harmonized Regional Strategy for Implementation of the "Great Green Wall Initiative of the Sahara and the Sahel". Addis Ababa, Ethiopia. p. 33. The African Union Commission (AUC); 2016. Available from: http://www.fao.org/fileadmin/templates/europeanunion/pdf/harmonized_strategy_GGWSSI-EN_.pdf

[43] United Nations. Transforming Our World: The 2030 Agenda for Sustainable Development [Internet]. 2015. Available from: https://sustainabledevelopment.un.org/content/documents/21252030%20Agenda%20for%20Sustainable%20Development%20web.pdf

Chapter 3

The Linear and Nonlinear Relationship between Infrastructure and FDI in India

Nenavath Sreenu and Kondru Sunda Sekhar Rao

Abstract

The study examines the linear and nonlinear relationship between Infrastructure and FDI, to understand whether there is a significant difference or not concerning the FDI equity inflows to infrastructure projects. The ARDL and Granger causality methods to cointegration; propose the existence of long-run function in two-directional causalities between foreign direct investment and infrastructure, whereas the nonlinear autoregressive distributed lag (ARDL) validates the asymmetries in the relationship between FDI and Infrastructure. The outcomes of the study are that foreign direct investment inflows are significant to improve the infrastructure projects in various sectors, in the short-run and long run. As enlightening infrastructure is dynamic to attract FDI, outcomes will be predominantly valuable to policymakers and related to the emerging markets.

Keywords: infrastructure, FDI, GII, ARDL and market

1. Introduction

The secondary information has extensively recognized the key role of infrastructure growth in fascinating FDI inflow. A sound developed infrastructure strategy boosts markets integration and entices FDI inflow in any nation [1], whereas the deficiency of comprehensive infrastructure interrupts markets relationship and herewith slowdown the foreign direct investment (FDI) inflow in particularly developing countries [2]. The obtainability of advanced infrastructure principally decreases the cost of trade, boosts the ease of doing business and invites the foreign direct investment (FDI) inflow. Infrastructure tool is used in this paper as an engine for economic growth and facilitate a comparative advantage to a developing nation in terms of foreign direct investment inflow [3]. Additionally, the secondary data has shown evidence that the nation with good infrastructure engrossed more foreign direct investment inflow [4], whereas the nation's deficient with infrastructure development are stereotypically unsuccessful to attract the FDI inflow [5] and those nation economies also shown that the poor condition [6]. Moreover, it also determined that the impact of the infrastructure development on FDI is positive and significant in a growing economy, preceding research [7] assessed that a deficiency of Infrastructure castigates FDI inflow. The significance of the infrastructure plays a vital role in the promotion of foreign direct investment inflow. The research data extensively explore the query of how the nonexistence of infrastructure can affect foreign direct investment inflow along with the different results of foreign direct investment in the host nation. Though the literature review has given little concentration to examine the role of FDI in improving the obtainability and quality of infrastructure in developing nations' economies. According to Pradhan et al. [8] illustrated that the foreign organizations carried progressive technology and skills to the host nation herewith encouraged new technological dissemination in the nation along with investment. FDI inflow also facilitates home organizations with an unintended opportunity to learn from the foreign firms by studying and permeating an intelligence of plenteous needed competition [9]: which develops a cumulative output of the ant nation economy. Foreign direct investment plays a significant role in economic growth [10], the positive impact of FDI inflow is not only inadequate to the transfer of better technology, in circumstance, it also needs any nation to develop the quality infrastructure [11]. As a developing nation economy like Indian have enough resources but do not advance technology to effective utilization of the resources, to improve the infrastructure facilities in India, advanced technology is required, it is required the support of overseas capital to improve their infrastructure facilities. Foreign companies cooperate in R&D in enhancing innovative technology and development of any nation, specifically in infrastructure development to bond up with various markets in the different nations. The literature review broadly examined the various determinants factors of FDI like population, political stability and institutional quality etc. [12]. As this critical point lacks in review, this paper aims to examine the causal two functional relationships between total FDI and total infrastructure and enhance the literature review on this vital singularity. Furthermore, the literature discloses that prevailing research articles on the subject matter hurt from numerous data limitations [13]. The original paper on infrastructure focuses on variables representative of infrastructure for a large nation sample during 1990–2018 but it does not formulate an index of cumulative infrastructure. Likewise, extensively the review on infrastructure projects in different sectors depend on a very inadequate description of infrastructure while examining its stimulus on different economic indicators are investment, trade, and growth. Gnangnon (2018) assessed the impact of the telecommunication infrastructure on economic growth in a developing nation like India. Chakraborty and Nunnenkamp [14] uses ITC (international telephone circuits), the inclusive road infrastructure length and the number of airways as a proxy for the infrastructure development to examine the relationship between public infrastructure and foreign capital. Hall et al. [15] investigated broader insight and apprehension of the various infrastructure components to assess the association between transportation cost and infrastructure growth.

As for the given information limitations, this research paper employs a recently established inclusive using global infrastructure index 2020 which comprehends numerous infrastructure extents for India to overcome statistics limitations in secondary information. Predominantly the global infrastructure index 2020 is grounded on an annual wide-ranging of minimum 15 indicators datasheet of the obtainability and quality of infrastructure during 1995–2018 formulated by Khan et al. [16, 17]. In this research paper, the following infrastructure parameters (like power sector, construction, transportation, telecommunication, health, finance and energy) are used. UCM (Unobserved Components Model) is employed expedient infrastructure from the sub-parameters of infrastructure development. Additionally, the paper highlights some important points According to the Reserve Bank of India, infrastructure covers the following sectors also Power, Telecommunications, Railways, Roads including bridges, Seaport and airport, Industrial parks, Urban infrastructure, Mining, exploration and refining, and Cold storage and cold room facility, including for farm level pre-cooling for the preservation or storage of agricultural and allied produce, marine products and meat.

2. Infrastructure and FDI inflow in India

India, among the worldwide five major countries in the emerging economy, has significant potential to improve rapidly and thus made up to be an appropriate end-point for FDI inflow. Despite all difficulties, India has attracted reasonable FDI inflow comparatively with other developing economic developed issues. In the 20th century's, policy reforms, Indian performed better and received comparatively greater FDI inflow [18]. The FDI inflow for the financial year 2009–2010 was 37,745 US million dollars, and over the ensuing period of 10 years, the FDI inflows have recovered stable growth in each year. For the year 2019–2020, the inflow was $27.1 billion higher than the annual inflows in previous years from 2010 to date. The FDI inflows reduced during the year 2011–2012 and 2013–2014 where they fell by 8 per cent and 26 per cent respectively due to some reason like slowdown of economic development issues in India. **Table 1** has explained the input of specific sector FDI out of total FDI inflow in the Indian sector. The table illustrates the evocative variations in the infrastructure sectoral composite of FDI inflow in India from the last 20 years. An inclusive investigation of sector-wise FDI discloses that external investors favored in Manufacturing sectors throughout 2000–2020. The manufacturing sector has accounted for more than US$ 89.40 billion from April 2000 to March 2020. During 2020 the Government of India increased FDI in manufacturing under the automatic route from 49–74%.

Similarly, the FDI inflow has significantly contributed to the above-mentioned three sectors and has shown the reforms of sectoral-FDI also account for significant variation for the time-to-time period. The distribution of the service sector out of total FDI inflow has increasingly and expansively improved in the same period. FDI equity inflow amount for services sector India FY 2015–2020. In the financial year 2020, the foreign direct investment equity inflow in the services sector in India was worth approximately 7.86 billion U.S. dollars. The foreign investment inflows have been consistently increasing over the last five years in this sector. To assess the post- and pre-reform of sector-wise FDI performance in the economy of India, this paper calculated the FDI performance index sector level that indicates the share of FDI sector-wise, comparative to its influence to aggregates of the India GDP. A value higher than 1 presents that the specific sector has recognized additional FDI inflow than its comparative economic size while a value less than 1 suggests that the specific sector received less FDI- inflow than its relative contribution; the below method was also used by Shah et al. [19].

Years	Service sector	Manufacturing sector	Primary sector
2000–2005	0.086	0.983	27.341
2005–2010	0.193	0.671	18.037
2010–2015	1.386	1.472	24.726
2015–2020	1.047	1.057	9.163

Source: Calculations of authors'.

Table 1.
Index of sector-level FDI performance.

2.1 Rural infrastructure

Rural infrastructure in the country is crucial for agriculture, agro-industries and poverty alleviation in the rural areas. Rural infrastructure provides essential production conditions which are required for social and economic growth and for promoting the quality of life in rural areas. As per the government statistics clean tap water is available to only 24% rural households. About 56% of rural households had electricity connections. Centre and state government have over all estimated a total capital expenditure of Rs. 7,73,915 crore between fiscals 2020 and 2025 on rural infrastructure development in India.

According to the Department for Promotion of Industry and Internal Trade (DPIIT), the Indian food processing industry in rural has cumulatively attracted Foreign Direct Investment (FDI) equity inflow of about US$ 10.24 billion between April 2000 and December 2020.

In the year 2021 infrastructure activities accounted for 13% share of the total FDI inflows of US$ 81.72 billion. The government invested US$ 1.4 trillion in infrastructure development as of July 2021.

Department of Drinking Water and Sanitation will be implementing the Jal Jeevan Mission to provide functional household tap connection to every rural household i.e., "Har Ghar Nal se Jal" by 2024. The program will be implemented at an estimated total capex of Rs. 3,60,000 crore shared between states and center as follows: Rs. 2,48,626 crore would be invested in rural housing under PMAY Gramin and about Rs 162,329 crore would be invested to improve rural roads under PMGSY. Improving the rural road connectivity by providing all-weather roads to connect eligible habitations in rural areas. As on December 31, 2019, road length worth Rs. 2.9 lakh crore had been sanctioned and expenditure of Rs. 2.17 lakh crore incurred. World Bank sanctioned about INR 2462 billion (US$ 37 billion) through its Country Assistance Strategy committed to a series of loans/credits to support "Pradhan Mantri Gram Sadak Yojana (PMGSY) to complete 165,411 Road projects in rural areas. The total projected rural infrastructure investment from 2020 to 2025 is given in the Table below.

From the table given above it can be understood that, the rural infrastructure investment is 7% in the total infrastructure investment in India. The projected cumulative investment from 2020 to 2025 is 773,915 million rupees.

Sector-wise break-up of capital expenditure of Rs. 111 lakh crore during fiscals 2020–2025.

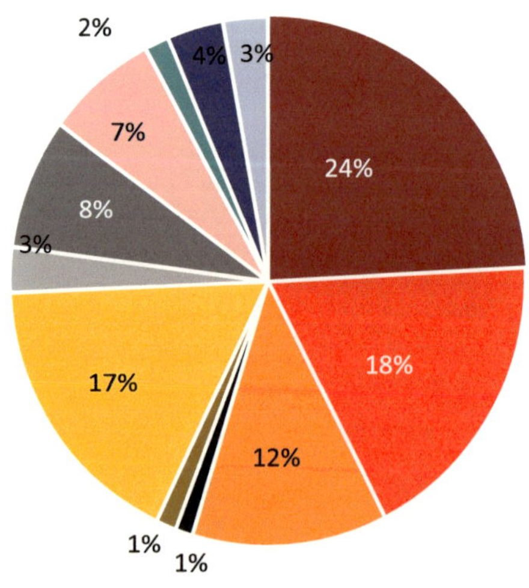

Department	FY20	FY21	FY22	FY23	FY24	FY25	No phasing	FY20-FY25
Rural infrastructure	103,555	116,306	109,930	27,055	27,055	27,055	0	410,955
Water and sanitation	36,758	60,497	100,881	84,822	80,002	0	0	362,960
Total rural infrastructure	140,313	176,803	210,811	111,877	107,057	27,055	0	773,915
Total infrastructure	1,442,131	2,153,779	2,132,274	1,647,122	1,540,813	1,315,091	899,218	11,130,428

Table 2.
Table shows the projected investment in rural infrastructure in India from 2020 to 2025. (rupees in crores).

from the above diagram it can be understood that, energy sector 24%, roads sector 18%, railways 12%, ports 1%, Airports 1%, urban infrastructure 17%, digital infrastructure 3%, irrigation 8%, rural infrastructure 7%, agriculture & food processing 2%, social infrastructure 4% and industrial infrastructure 3%. Hence, it is concluded that, the total share of the rural infrastructure in total FDI is 7%.

$$\text{Present FDI inflow} = \frac{FDI_t/FDI_i}{GDP_t/GDP_i}$$

From the above equation used for the determination of present FDI inflow, whereas, FDI_i inflow in the infrastructure sector I; FDI_t is cumulative FDI inflow, $GDP_{i\&t}$ indicates GDP of the infrastructure sector I and overall value of GDP is t.

Table 2 has shown the variance between the infrastructure project performance during pre-and post-reforms periods of sector-wise FDI growth and better performance indices. The performance indexes illustrate that during the pre-and post-reforms era, the major and important sectors are gas, oil, power sector, transportation, construction and mining sectors, which attracted FDI inflow and contributed to GDP growth. In the present situation, the Indian industries have overcome the shortage of electricity and the deficiency of proper infrastructure facilities. Both private and public manufacturing sectors are facing low-level problems against the lack of infrastructure issue, it looks like the latter is winning. Based on the literature review, the paper has tested the following two hypotheses.

Null Hypothesis (H0): There is no significant difference in FDI equity inflows to Infrastructure projects.

Alternative Hypothesis (H1): There is a significant difference in FDI equity inflows to Infrastructure projects.

3. Data and methodology

3.1 Explanation of variables and data gathering

To assess the relationship between infrastructure and FDI inflow from 2000 to 2020, the paper depends on the global infrastructure index 2020 (GII-2020) a compound index and also sub-sector of infrastructures such as transportation (TI), telecommunication (CI), power sector (PI) and energy sector (EI) and financial sector (FI) recognized on data collected from various sources (RBI, world bank and Global infrastructure index and CMIE reports).

The global infrastructure index 2020 (GII-2020) encompasses different quality and quantity magnitudes of infrastructure for India. The GII-2020 is created every year on a comprehensive range of infrastructure development parameter datasets of the accessibility and quality of infrastructure throughout 2000–2020. Besides, the paper used the institutional quality component, trade openness and human capital factors as control variables.

3.2 Research methodology

The present research investigates the two functional short and long-run causal dynamic relationships between infrastructure and FDI inflow, employing granger causality, ARDL (autoregressive distributed lag), and NARDL (Nonlinear autoregressive distributed lag) estimators to cointegration. This method is recognized in the case when the carefully chosen indicator is integrated either at the 1(0) level or the first difference I (1).

Moreover, from the simple linear transformation, the error ECM (−1) correction model easily may originate [16, 17]. To calculate the relationship between FDI and infrastructure the autoregressive distributed lag model assesses the following unlimited error correction model:

$$\begin{aligned}\Delta FDI_t = \alpha_0 fdi &+ \sum_{i=t}^{p} \partial_{fdii} FDI\_inf_{t-i} + \sum_{i=1}^{p} \in_{fdii} \Delta HC\_inf_{t-1} + \sum_{i=1}^{p} \beta_{fdii} \Delta TO\_inf_{i=t} \\ &+ \sum_{i=1}^{p} \gamma_{fdii} \Delta IQ\_inf_{t=i} + \sum_{i=1}^{p} \varnothing_{fdii} \Delta GI\_inf_{I=1} + \mu_{1fdii} FDI\_inf_{t=i} + \mu_{2fdii} \Delta HC\_inf_{t-1} \\ &+ \mu_{3fdii} \Delta TO\_inf_{i=t} + \mu_{4fdii} \Delta IQ\_inf_{t=i} + \mu_{5fdii} \Delta GI\_inf_{I=1} + D_t + \forall_{11}\end{aligned} \tag{1}$$

$$\begin{aligned}\Delta HC\_inf_t = \alpha_0 fdi &+ \sum_{i=1}^{p} \in_{fdii} \Delta HC\_inf_{t-1} + \sum_{i=t}^{p} \partial_{fdii} FDI\_inf_{t-i} + \sum_{i=1}^{p} \beta_{fdii} \Delta TO\_inf_{i=t} \\ &+ \sum_{i=1}^{p} \gamma_{fdii} \Delta IQ\_inf_{t=i} + \sum_{i=1}^{p} \varnothing_{fdii} \Delta GI\_inf_{I=1} + \mu_{1fdii} FDI\_inf_{t=i} + \mu_{2fdii} \Delta HC\_inf_{t-1} \\ &+ \mu_{3fdii} \Delta TO\_inf_{i=t} + \mu_{4fdii} \Delta IQ\_inf_{t=i} + \mu_{5fdii} \Delta GI\_inf_{I=1} + D_t + \forall_{1t}\end{aligned} \tag{2}$$

The measuring the long-run relationship between FDI and infrastructure this paper employs the bound testing techniques. The process of bound testing technique analysis of the hypothesis of no cointegration between the chosen indicator and the existence of cointegration between the indicators of study interest. The lower and upper bound critical values are significant role-plays as a determinant for the cointegration test [20]. If the calculated F-statistic value is higher than the upper bound critical value, then the H0 (Null hypothesis) is rejected. If the F-statistic value is lower than the lower bound critical value,

The Granger causality model using the I(I) of variables all over a VAR may cause uncertainty in the results in the existence of cointegration among selected variables. Hence, an advanced form of traditional Granger causality model relating the error correction method (ECM) is articulated in VECM as follow:

$$\Delta FDI_t = \alpha_0 fdi + \sum_{i=t}^{p} \partial_{fdii} FDI\_inf_{t-i} + \sum_{i=1}^{p} \in_{fdii} \Delta HC\_inf_{t-1} + \sum_{i=1}^{p} \beta_{fdii} \Delta TO\_inf_{i=t} + \sum_{i=1}^{p} \gamma_{fdii} \Delta IQ\_inf_{t=i} + \sum_{i=1}^{p} \varnothing_{fdii} \Delta GI\_inf_{I=1} + +\Omega ECM_{t-1} + D_t + \mu_{3t} \quad (3)$$

$$\Delta HC\_inf_t = \alpha_0 fdi + \sum_{i=1}^{p} \in_{fdii} \Delta HC\_inf_{t-1} + \sum_{i=t}^{p} \partial_{fdii} FDI\_inf_{t-i} + \sum_{i=1}^{p} \beta_{fdii} \Delta TO\_inf_{i=t} + \sum_{i=1}^{p} \gamma_{fdii} \Delta IQ\_inf_{t=i} + \sum_{i=1}^{p} \varnothing_{fdii} \Delta GI\_inf_{I=1} + +\Omega ECM_{t-1} + D_t + \mu_{3t} \quad (4)$$

3.3 The non-linear auto-regressive distributive lag model (NARDLM)

According to Pesaran et al. [20] the cointegration test makes available proof of a linear relationship among the chosen variables. The current research paper also uses the NARDL [19] model to examine the existence of an association between FDI inflow and infrastructure in India. The non-linear auto-regressive distributive lag model [21] is a nonlinear extended form of the autoregressive distributive lag model for consistent impeding both short and long-run irregularity in the autoregressive distributive lag model.

The non-linear auto-regressive distributive lag model is calculated in the current paper that determines the short run and long run of the positive and negative partial sums. Thus, the non-linear auto-regressive distributive lag model contemplates the form of the resulting equation:

$$\Delta FDI_t = \alpha_0 fdi + \sum_{i=t}^{p} \partial_{fdii} GII\_inf_{t-i} + \sum_{i=1}^{p} \in_{fdii} \Delta HC\_inf_{t-1} + \sum_{i=1}^{p} \beta_{fdii} \Delta TO\_inf_{i=t} + \sum_{i=1}^{p} \gamma_{fdii} \Delta IQ\_inf_{t=i} + \sum_{i=1}^{p} \varnothing_{fdii} \Delta GI\_inf_{I=1} + \mu_{1fdii} FDI\_inf_{t=i} + \mu_{2fdii} \Delta GII\_inf_{t-1} + \mu_{3fdii} \Delta TO\_inf_{i=t} + \mu_{5fdii} \Delta HC_{I=1} + D_t + \forall_{1t} \quad (5)$$

$$\Delta GII_t = \alpha_0 fdi + \sum\nolimits_{i=t}^{p} \partial_{fdii} GII\_inf_{t-i} + \sum_{i=1}^{p} \in_{fdii} \Delta FDI_{t-1} + \sum_{i=1}^{p} \beta_{fdii} \Delta FDI_{i=t} + \sum_{i=1}^{p} \gamma_{fdii} \Delta IQ\_inf_{t=i} + \sum_{i=1}^{p} \varnothing_{fdii} \Delta HC_{I=1} + \sum_{i=1}^{p} \beta_{fdii} \Delta TO\_inf_{i=t} + \mu_{2fdii} \Delta GII\_inf_{t-1} + \mu_{3fdii} \Delta TO\_inf_{i=t} + \mu_{5fdii} \Delta HC_{I=1} + D_t + \forall_{1t} \quad (6)$$

4. Empirical outcomes and argument

4.1 Descriptive statistics and unit-root testing

Table 3 explains descriptive statistics value, this table helps to highlight the how data descriptive statistics like. Descriptive statistics values comprise of several observations with determined values, mean, minimum, maximum, central value and standard deviation point with corresponding variables to transportation infrastructure (T_inf), telecommunication infrastructure (Te_inf), energy sector infrastructure (E_inf), Financial sector infrastructure (F_inf), global infrastructure index 2020 (G_inf-2020), quality in institutional approach (IQ_inf), the primary sector of FDI inflow (PFDII), FDI inflow service sector (FDIIS), export and import to GDP or trade openness in infrastructure (T_inf) and human capital (H-inf).

Ouattara (2004) illustrated that the level of stationary among all the chosen variables of the study was of interest to observe the probable variables of FDI inflow in infrastructure sectors wise during the long run and short run. Due to the circumstance that if the factors of the study interest are stationary at I (2) the estimated F-test value will not be significant. In the current paper, use the two types of tests are structural break analysis which is (1) and (1) contemplate the structural break in the given timer series data to examine the order of integration among the selected variables.

Table 4 explore that each variable is integrated either at I(1) OR I(0) order and none of the indicators is stationary at I(2) order, According to (1) in this condition, the auto-regressive distributive lag model is suitable moderately another cointegration process. To assess the existence of a long-run relationship among chosen variables,

Years	α	β	∅	μ	∂	γ
2000–2005	17.845	19.462	7.374	7.562	6.461	2.479
2005–2010	9.351	24.522	23.538	22.218	9.704	7.483
2010–2015	27.361	21.580	17.650	31.492	16.739	14.695
2015–2020	39.720	31.301	6.361	38.572	29.537	27.968

Source: calculations of authors. The data has been collected from RBI.

Table 3.
Shares' of different economic groups in % of cumulative FDI inflow in India.

Variable	N	Mean	Std.dev	Min	Max
T_inf	37	−.765	.026	−.967	−.644
Te_inf	37	−.564	.531	−.931	−.854
E_inf	37	−.786	.201	−.2.872	−.117
F_inf	37	.043	.746	−.797	.708
G_inf-2020	37	−0.069	2.836	−1.417	−.835
IQ_inf	37	213.962	21.690	−2.630	2.648
PFDII	37	231.067	214.495	141.759	241.640
FDIIS	37	146.947	732.571	−.138.503	271.492
FDI	37	214.057	261.837	127.395	382.708
TO-inf	37	0.056	0.073	.352	.893
HC-inf	37	31.672	2.873	41.153	32.163

Source: calculations of authors.

Table 4.
Descriptive statistics value.

this paper used auto-regressive distributive lag model and error correction model techniques to cointegrate by using equations no 1&2. The study calculates the regressions techniques that FDI is substituted by sectoral FDI inflow (like by FDI in the primary sector, FDI in the service sector, and FDI in manufacturing and trading) to evaluate the probable long-run association regarding FDI inflow and cumulative infrastructure. As this paper investigates the two functional causalities between infrastructure and FDI inflow sector (like by FDI in the primary sector, FDI in the service sector, and FDI in manufacturing and trading), so for this determination of inverse impact, the study also take infrastructure as a dependent variable and then substitute the infrastructure into sub-indices of infrastructure (such as (T_inf), (Te_inf), (E_inf), (F_inf), (G_inf-2020), (IQ_inf), (PFDII), (FDIIS), (T_inf) (H-inf)). The optimum lag length is grounded on AIC for measuring the present models of interest.

5. Linear cointegration outcomes (autoregressive distributed lag-ARDL)

From **Table 5**, the significant level calculated with help of the F-statistics values, the Ho (Null hypothesis) rejected the there is no cointegration and **Table 5** suggested that there is a possible relationship between the FDI inflow, Infrastructure and all other control variables (such as Trade openness (export and import), Infrastructure intuitional quality and Human capital in infrastructure projects) exist in long-run. From equation no 1, the calculated F-statistics values are the upper bound critical value at 5% and 1% significant level with control variables. The outcomes are in the same order with preceding research studies [22, 23]. Furthermore, it was also examined in **Table 4**, the probable association exit in the long run between FDI, Infrastructure and control variables. From equation no 2, the calculated F-statistics value is higher than the upper bound critical factor value at 5% and 1% significant level

Augmented dickey fuller test (ADF)				DF-GLS Test	Zivot-Andrews			
Variables	**I (0)**	**I (1)**	**I (0)**	**I (1)**	**I (1)**	**Break**	**I (0)**	**Break**
T_inf	−1.847	−3.849***	−2.401	−4.842**	−1.372	2003	−2.390**	2010
Te_inf	−2.673***	−4.283***	−3.207***	−2.194***	−3.408	2005	−3.490***	2008
E_inf	−3.670***	−6.381***	−1.869	−5.784***	−4.784***	2010	−1.539***	2012
F_inf	−1.934	−5.298**	−2.301***	−4.483***	−4.389	2013	−4.382***	2007
G_inf-2020	−2.873	−2.602***	−1.403	−6.492***	−1.950**	2016	−2.367***	2014
IQ_inf	−3.438***	−4.391***	−2.672*	−3.502***	−3.041	2014	−5.361***	2015
PFDII	−5.785***	−2.428***	−1.069	−5.302***	−3.361***	2017	−6.351***	2017
FDIIS	2.480	−4.406***	−2.301	−3.401***	−2.701	2018	−4.287***	2006
FDI	3.561***	−3.371***	−1.285**	−4.295***	−1.361	2020	−6.351**	2008
T0-inf	1.015	−5.103**	−3.40***	−2.491***	−2.730	2012	−3.537***	2008
HC-inf	2.638**	−2.502**	−1.289	−3.089***	−1.308	2008	−1.628**	2009

****, **, and * indicates significance level at per cent of "10%", "5%" and "1%" correspondingly. The "critical values" of intercept are −2.701, −2.730, −1.950 significant level at 1%, 5% and 10% correspondingly, where the "critical values" for Zivot-Andrews are −3.490, −6.351, −6.351 significant level at 1%, 5% and 10% 'correspondingly.*

Table 5.
Unit root test outcomes.

without considering the control variables of human capital and quality in infrastructure institutional body). The experiential outcomes of the certain test for equation no 1&2, the H1 (alternative hypothesis) accepted, the existence of cointegration between the chosen variables and Ho (Null hypothesis) rejected, due to no cointegration between the selected variables, according to Asiedu [24]. The study also checked the robustness for the determination of the long-run relationship between FDI and infrastructure projects. The study also used the depended-on variables with and without control variables, indicates in **Table 6** which gives constant outputs in both the cases (with control variables and without control variables).

Table 6 illustrated that the association among the aggregates infrastructure, manufacturing sector and FDI inflow in the long run. Based on **Table 6**, the outcomes show the expected positive relationship among the chosen variables in the long run. The calculated F-statistics values are lower than higher bound critical factor value and significant at 5 per cent level with control variables are infrastructure institutional quality, human capital and export and import trade openness, while on the other side (i.e., opposite causality) the pragmatic outcomes of the bound test have advised robust relationship between total infrastructure and FDI inflow in the manufacturing sector (column 5–6). The predictable F-statistic value is higher than the upper bounds critical factor value at 5% and 1% correspondingly. Thus, the described outcomes disclose the existence of two functional associations between total infrastructure and manufacturing FDI. In this connection, the null hypothesis was rejected because there is no positive relationship between the selected variables. The current paper showed that two functional associations between total infrastructure FDI inflow in the primary sector in the long run in the column no 4 and 8. The empirical outcomes

Cumulative Global Infrastructure Index 2020 to cumulative and disaggregate FDI inflow								
Variables	**FDI_I**		**FDI_P**		**FDI_S**		**FDI_M**	
	F-Sta	**ECM**	**F-Sta**	**ECM**	**F-Sta**	**ECM**	**F-Sta**	**ECM**
FDI/GII	2.65*	−3.10	3.51	−3.01***	6.84***	−1.84	6.56**	−3.98
FDI/GII/IQI	2.04*	−4.31**	3.47	−4.31	6.56***	−3.05	4.62	−379
FDI/GII/IQI/TO	4.35	−5.71**	4.23***	−4.28	3.43	−5.40**	3.69	−4.37
FDI/GII/IQI/TO/HC	7.69***	−2.82***	7.06	−5.14***	8.35***	−6.57***	6.24**	−6.45***
FDI/GII/IQI/TO/HC/	3.71***	−4.52***	10.75***	−9.56***	5.40	−5.83**	7.62***	−747***
Aggregate and disaggregate FDI Inflow to Cumulative Global Infrastructure Index 2020								
Variables	**FDI_I**		**FDI_P**		**FDI_S**		**FDI_M**	
	F-Sta	**ECM**	**F-Sta**	**ECM**	**F-Sta**	**ECM**	**F-Sta**	**ECM**
FDI/GII	7.13***	−3.76	6.74***	−4.17***	2.93	−2.14	2.98	−4.91
FDI/GII/IQI	6.36***	−3.42	3.68***	−4.38***	3.52	−2.21	3.53	−4.13
FDI/GII/IQI/TO	3.61	−2.28	4.35**	−3.57	3.21	−3.75**	3.89	−5.52**
FDI/GII/IQI/TO/HC	4.39	−7.01***	3.78*	−2.18	4.89**	−4.27**	4.27**	−5.63**
FDI/GII/IQI/TO/HC/	5.31**	−5.39**	4.05**	−5.37***	4.14	−3.19	3.52	−5.53**

****, **, and * Indicates significance level at "10%", "5%" and "1%" correspondingly.*

Table 6.
Cointegration outcomes (ARDL constraints test and error correction model result).

explored the positive association among the infrastructure, FDI in the primary sector in the long run and all the selected variables of the paper. The F-statistics is higher than the upper bound critical factor value at a 1 per cent significance level with consideration of with and without control variables.

Table 6 indicates the two-function association between FDI inflow in the service sector and infrastructure in columns 4 and 8. The outcomes show the existence of two functional relationships between FDI and infrastructure development in India. The F-statistics values are greater than higher bound critical factor values in the case of with and without control variables of column 4. On the other hand, the Aggregate infrastructure to FDI services the values of assessed F-statistics are lesser than the greater bound critical factor values in case of without control variables and experiential greater than higher bound critical factor values at 10% significance level.

6. Granger causality method

The granger causality test determines the long-run relationship between the condition for causality and the selected variables of the study according to Morley [22, 23]. The Confirmation of long-run existence between the variables indicators shows that there should be the minimum non-functional causality between the selected variables for this research [25]. Henceforth, VECM is employed to calculate the function of the long run and short-run causality relationship between total infrastructure and FDI inflow with the consistency of the service sector, manufacturing sector and the primary sector.

Table 7 illustrates the long and short-run granger causality relationship from FDI inflow in different sectors and total infrastructure in Indian, the results show that the coefficient of error correction in long run period and the CEC term is strongly significant when cumulative FDI inflow FDI in the manufacturing sector, FDI in the primary sector and FDI in the service sector are used as dependent variables. Whereas the F-statistics value does not indicate a significant impact on the selected variables during the short run, i.e., FDI inflow in selected sectors to total infrastructure. The empirical output of the Granger causality method estimator promoted the long-run causality exists from the study variables (FDI_I, FDI_T, FDI_M, FDI_S) to total infrastructure which advises the infrastructure play a significant role to attract FDI inflow in service sectors, primary sectors, and manufacturing sectors of India (1). However, there is no causality existing in the short-run (from FDI_I, FDI_T, FDI_M, FDI_S to total infrastructure) which discloses that the total infrastructure does not affect the ability to attract FDI inflow in sectors wise in the short run. The results in **Table 7** also indicates the short and long-run causality from total infrastructure to total FDI inflow. The output indicates that the long and short-run causality in current and significant level at 5 per cent. It means that in Indian total FDI inflow affects the availability of infrastructure and quality level.

Whereas the transportation infrastructure, telecommunication infrastructure, energy infrastructure, infrastructure in the power sector and financial infrastructure variables are used in this paper as dependent variables. This empirical output shows that total FDI inflow causes aggregates and sub-indices of infrastructure in the long run. The finding of the study reveals that inverse causality in FDI inflows indicates positive and significant effects on overall infrastructure sub-indices in the long-run period. Furthermore, the outcomes show that the sectors are FDI_P, FDI_S and FDI_M are used as descriptive variables. The output demonstrated that the error correction model is significant at the level of 5 per cent while FDI_P, FDI_S and FDI_M are used as independent variables. The outcomes also indicate that in the long

	F-Statistics (Short-run)				**ECM_{t-1} (Long run)**				**A/R**
	FDI inflow to total Infrastructure								
ΔFDI_I	ΔGII	ΔIQ	ΔTOE&I	ΔHC	ΔGII	ΔIQ	ΔTOE&I	ΔHC	H1:A
	1.62	0.23	6.694	0.64***	−0.31***	−0.63***	−0.34***	−0.52***	
	FDI inflow in Primary sector to total Infrastructure								
ΔFDI_P	ΔGII	ΔIQ	ΔTOE&I	ΔHC	ΔGII	ΔIQ	ΔTOE&I	ΔHC	H1:A
	0.56	0.023	3.75	7.05***	−0.65***	−0.23***	−0.76***	−0.78***	
	FDI inflow in the Manufacturing sector to total Infrastructure								
ΔFDI_M	ΔGII	ΔIQ	ΔTOE&I	ΔHC***	ΔGII	ΔIQ	ΔTOE&I	ΔHC	H1:A
	2.01	3.04	3.05	1.72	−0.32**	−0.15**	−0.37**	−0.28***	
	FDI inflow in Services sector to total Infrastructure								
ΔFDI_S	ΔGII	ΔIQ	ΔTOE&I	ΔHC	ΔGII	ΔIQ	ΔTOE&I	ΔHC	H1:A
	2.49	0.67	2.25	0.72	−0.67***	−0.15***	−0.71***	−0.19***	
	Total infrastructure and FDI inflow (Opposite causality)								
ΔGII	ΔFDI_I	ΔIQ	ΔTOE&I	ΔHC	ΔFDI_I	ΔIQ	ΔTOE&I	ΔHC	H1:A
	4.78***	0.84	2.46	0.04**	−0.51***	−0.42**	−0.71***	−0.91***	

****, **, and * Indicates significance level at "10%", "5%" and "1%" correspondingly.*
Source: calculations of author.

Table 7.
Granger causality test output.

run extension of FDI inflow, it can grow infrastructure quality and availability (1). Can grow infrastructure quality and availability (1).

Null hypothesis: The Diagnostic tests are not affected by the mention Econometric problem. Alternative: The Diagnostic tests are affected by the mention Econometric problem.

Table 8 illustrates the determined causal relationship between GII and FDIP in the long run, in the same order where FDI_S is used as a dependent variable. The fact that FDI_ S is used as a dependent variable. This indicates that the impact of total

Econometric problem	**F-statistics**	**P-values**	**Test**	**Hypothesis A/R**	**Support equation**
Heteroscedasticity	0.0471	0.0378	Breusch-Pagan-Godfrey	Accepted	Equation no 1
Specification	2.4914	0.0121	Ramsey RESET	Accepted	
Serial Correlation	0.4891	0.3861	Breusch-Godfrey LM	Accepted	
Normality	1.0382	0.0137	Jarque-Bera		
Heteroscedasticity	1.0381	0.0027	Breusch-Pagan-Godfrey	Accepted	Equation no −2
Specification	3.2037	0.0271	Ramsey RESET	Accepted	
Serial Correlation	0.4891	0.0461	Breusch-Godfrey LM	Accepted	
Normality	0.8325	0.0294	Jarque-Bera	Accepted	

Table 8.
Diagnostic tests.

infrastructure is positive but insignificant in the long run without considering the control variables, while significant considering the control variable. The empirical outcomes indicate that the spill-over effect of FDI inflow is more than infrastructure in the long run in Indian. The model's constancy is established by recursive estimation. They recommend that statistically valid inference can be drawn from the selected models. The rest of the diagnostic tests are indicated in **Table 8**.

Variables	FDI_I	FDI_P	FDI_S	FDI_M
		SR dynamic		
ΔFDI_I	0.481**	0.361	0.732	0.301
ΔGII_PS	0.701	1.332*	−5.952	2.639
ΔGII_PS_{t-1}	0.427	2.549	−7.603	12.812
ΔGII_NG	−2.481*	0.367	3.225	−9.374
ΔGII_NG_{t-1}	−1.302**	−1.837	6.361	−11.371
ΔH_C	9.326***	−7.418	−37.291	−26.385*
ΔH_C_{t-1}	7.589***	−11.720**	−16.320*	−35.679**
ΔI_Q	0.375	0.211*	−0.391**	0.581
ΔI_Q_{t-1}	0.017	−0.417	−2.438	0.793
ΔT_O	−2.940	−3.491	−4.210	3.482
ΔT_O_{t-1}	−0.013	0.036	−7.364	−1.596
ΔDM_GII	1.058*	2.972	−4.647	4.795
ΔDM_GII_{t-1}	0.381	3.285	−2.431	3.061
		LR dynamics		
GII-PS	0.503	−0.602	2.442	−0.640
GII_NG	−0.186	−3.673	8.927	−6.582
H_C	12.036***	9.027	23.183***	36.284**
I_Q	2.748**	5.327***	−0.947	4.473
T_O	6.274**	−0.390**	5.963	9.739
DM_GII	−2.491	−0.283	13.327	−3.406
PSS F-Stat	8.384**	4.372***	0.113**	1.728***
ECM_{t-1}	−3.374**	−1.361**	0.341***	−3.273**
Constant	−107.849	−52.286***	−103.325***	−178.957***
N	57	57	57	57
R^2	0.749	0.702	0.821	0.384
Adj. R^2	0.853	0.731	0.873	0.648
SR asymmetries	3.593	0.668	0.703**	0.478**
LR asymmetries	0.472***	0.004***	3.251	0.561

*Note: * DU_FDI is time dummy variable confirmed for operational break in FDI_I, FDI_S, FDI_P, and FDI_M. ***$p < 0.01$, **$p < 0.05$, *$p < 0.1$.*

Table 9.
Nonlinear effect of global infrastructure index on aggregate and disaggregate FDI inflow in India.

7. Nonlinear cointegration test

There may be a nonlinear relationship exist of time series variables, thus, after the newest methodology proposed by Shin et al. [21], this paper tested the cointegration method by exempting the linear relationship restriction. The outcomes are described in **Table 9**, which authorizes the cointegration relationship (attained negative and

Variables	FDI_I	FDI_P	FDI_S	FDI_M
		SR dynamic		
ΔGII_I	0.052	0.034	−0.081	−0.342
ΔFDI-PS	−0.512	0.164*	0.051	0.892
$\Delta FDII\_PS_{t-1}$	−0.901*	0.522	0.073	0.036
ΔFDI_NG	−0.541	−0.701	0.067	0.307
ΔFDI_NG_{t-1}	−0.062	0.381	0.097	0.431
ΔH_C	21.983*	0.821	−6.842	−8.031
ΔH_C_{t-1}	2.092	−5.057	0.462	−0.582
ΔI_Q	−0.037*	−0.371	2.092	0.482
ΔI_Q_{t-1}	−0.361*	−0.781	0.879	1.462
ΔT_O	3.267	4.381	0.956	1.549
ΔT_O_{t-1}	2.471	4.381	0.472	5.391
ΔDM_FDI	0.302	−0.613	0.462	−0.945
ΔDM_FDI_{t-1}	−0.461	−0.203	0.034	−0.126
		LR dynamics		
FDI-_PS	0.705***	−0.027	−0.231	−0.479
FDI_NG	0.523***	0.916	−0.362	−0.253
H-C	−16.538	−7.527	16.437	27.481
I_Q	0.567	0.937	0,371	0,738
T_O	−3.601	−1.385	−0.482	−2.481
DM-FDI	0.471**	0.681	−0.385	0.462
PSS F-Stat	1.372	0.375*	−0.471***	0.463
ECM_{t-1}	−0.472*	−0.638	−0.617**	−0.739**
Constant	97.153**	32.230**	−17.926	−27.631
N	57	57	57	57
R^2	0.471	0.746	0.857	0.431
Adj. R^2	0.521	0.648	0.427	0.172
SR asymmetries	2.461***	0.374	1.046	0.597
LR asymmetries	0.031	0.204	0.381	0.046

*Note: * DU_FDI is time dummy variable confirmed for operational break in FDI_I, FDI_S, FDI_P, and FDI_M.* *$^{***}p < 0.01$, $^{**}p < 0.05$, $^{*}p < 0.1$.*

Table 10.
Nonlinear effect of aggregate and disaggregate foreign direct investment inflow on global infrastructure index.

significant statistics of Error correction model) among the FDI_I, FDI_S, FDI_P, FDI_M and G_I_INF. Though, the co-movement of FDI_I, FDI_S, and G_I_INF is maintained by a significant PSS test. The feature that differentiates the non-linear auto-regressive distributive lag mode [25] from the traditional autoregressive distributive lag mode is the asymmetries testing. Fascinatingly, the outcomes illustrate that in the case of equation no 1 there is an indication of SR asymmetries, and in equation no 3 there exist LR asymmetries.

Likewise, **Table 10** explores the dependent variable is exchanged with an independent variable and the non-linear auto-regressive distributive lag model is assessed. The outcome of the paper is that cointegration exists when FDI inflow and FDI in the services sector are taken as descriptive variables, while unpredictably, the PSS F-Test does not sustain to Error correction term or model. Concerning the asymmetric relationship, only equation no 1 shows the existence of SR asymmetries, which is confirmation of the outcomes stated in **Table 9**. Thus, the paper infers that in the relationship of FDI and G_I_INF, traditional auto-regressive distributive lag may not be acceptable to rely upon and to articulate effective strategies, as it proceeds from asymmetric circumstances, which may lead to unsuitable policy measures. Hence, it is suggested to contemplate the non-linearities that may exist while testing linear modeling between the variables.

8. Conclusion

The current paper determined to examine the linear and nonlinear cointegration between FDI inflow and total infrastructure, together with various sub-indices of infrastructure and sectoral FDI inflows of India. To accomplish this objective, the paper used Granger causality to determine the causal relationship between FDI inflow and infrastructure, while linear and nonlinear situations are used to find the cointegration relationship. The observation of the findings confirms the existence of the linear and nonlinear cointegration between the cumulative as well as sub-indices of infrastructure and aggregated and disaggregated FDI inflow. Additionally, the findings of asymmetric testing are motivating, which article mix outcomes in terms of the existence of SR and LR asymmetries in the appropriate manner as stated in **Tables 9** and **10**. So, we infer that in the fitting together of FDI and Global infrastructure index, traditional ARDL may not be acceptable to depend on and to articulate effective policies, as it proceeds from asymmetric circumstances, which may lead to a weedy policy assertion. Hence, it is suggested to study the non-linearities that may exist while testing linear modeling. Furthermore, the conclusions elaborate that to make the economy attract more FDI, the government shall further expand the system of infrastructure in education, quality of the institution and to promote the exports. Furthermore, the empirical results advise that improved quality and availability of infrastructure stocks are the most to attract high FDI inflow in the primary sector, services sector and manufacturing sector of India's economy in the long run. Hence, the emphasis of policies should be to progress both infrastructure facilities and to make available a conducive atmosphere for global investors to obtain high FDI because FDI inflows indicate to improve the quality and availability of infrastructure. This research concludes that, the current study offers a worthwhile understanding to policymakers and supervisors to consider the sectoral level FDI inflow in India, as an alternative of planning policies exclusively based upon aggregate FDI. Likewise, the study brings into the argument an exceptional measure of infrastructure index

(G_I_I) that incorporates the broader aspects together with telecommunication, energy, transportation and financial infrastructure. On the other hand, the previous secondary information deeply depends on only the telecommunication and IT infrastructure which may not be acceptable to represent the sundry dimensions of the infrastructure, reported by the Global Infrastructure Index 2020.

Author details

Nenavath Sreenu[1*] and Kondru Sunda Sekhar Rao[2]

1 Department of Management Studies, NIT Bhopal, MP, India

2 School of Management, Presidency University, Bangalore, India

*Address all correspondence to: srinunaikphd@gmail.com; sri_cbm@yahoo.com and sreenu@manit.ac.in

References

[1] Abdulahi ME, Shu Y, Khan MA. Resource rents, economic growth, and the role of institutional quality: A panel threshold analysis. Resources Policy. 2019;**61**:293-303

[2] Huang Y, Tang H. FDI policies in China and India: Evidence from firm surveys. The World Economy. 2012;**35**(1):91-105

[3] Shan J. A VAR approach to the economics of FDI in China. Applied Economics. 2002;**34**(7):885-893

[4] Ahmad MH, Ahmed QM. Does the institutional quality matter to attract the foreign direct investment? An empirical investigation for Pakistan. South Asia Economic Journal. 2014;**15**:55-700

[5] Bhattarai K. FDI and growth. Advances in Management and Applied Economics. 2016;**6**(2):1-23

[6] Chang SC. The interactions among foreign direct investment, economic growth, degree of openness and unemployment in Taiwan. Applied Economics. 2007;**39**(13):1647-1661

[7] Athukorala PC. Trends and patterns of foreign direct investments in Asia: A comparative perspective. Margin. 2009;**3**(4):365-408

[8] Pradhan RP, Bagchi TP. Effect of transportation infrastructure on economic growth in India: The VECM Approach. Research in Transportation Economics. 2013;**38**:139-148

[9] Donaubauer J, Meyer BE, Nunnenkamp P. A new global index of infrastructure: Construction, rankings and applications. World Economics. 2016;**29**(2):236-259

[10] Dash RK, Parida PC. FDI, services trade and economic growth in India: Empirical evidence on causal links. Empirical Economics. 2013;**45**(1):217-238

[11] Gurbuz A, Aybars A. The impact of foreign ownership on firm performance, evidence from an emerging market. American Journal of Economics and Business Administration. 2010;**2**(4):350-359

[12] Economou F. Economic freedom and asymmetric crisis effects on FDI inflows: The case of four South European economies. Research in International Business and Finance. 2019;**49**:114-126

[13] Elheddad MM. What determines FDI inflow to MENA countries? Empirical study on Gulf countries: Sectoral level analysis. Research in International Business and Finance. 2018;**44**(C):332-339

[14] Chakraborty C, Nunnenkamp P. Economic reforms, FDI and economic growth in India: A sector level analysis. World Development. 2008;**36**(7):1192-1212

[15] Hall A, Banerjee A, Dolado JJ, Galbraith JW. Co-integration, error correction, and the econometric analysis of non-stationary data. The Econometrics Journal. 1996;**106**(439)

[16] Khan MA, Khan MA, Abdulahi ME, Liaqat I, Shah SSH. Institutional quality and financial development: The United States perspective. Journal of Multinational Financial Management. 2019a;**49**:67-80

[17] Khan MA, Kong D, Xiang J, Zhang J. Impact of institutional quality on financial development: Cross-country

evidence based on emerging and growth leading economies. Emerging Markets Finance and Trade. 2019b;**56**:1-17

[18] Kumar N. Infrastructure availability, foreign direct investment inflows and their export-orientation: A cross-country exploration. Indian Economic Journal. 2006;**54**(1):125-144

[19] Shah SH, Ahmad MH, Ahmed QM. The nexus between sectoral FDI and institutional quality: empirical evidence from Pakistan. Applied Economics. 2016;**48**:1591-1601

[20] Pesaran MH, Shin Y, Smith RJ. Bounds testing approaches to the analysis of level relationships. Journal of Applied Econometrics. 2001;**16**(3):289-326

[21] Shin Y, Yu B, Greenwood-Nimmo M. Modelling Asymmetric Cointegration and Dynamic Multipliers in a Nonlinear ARDL Framework. Festschrift in Honor of Peter Schmidt. USA: Springer; 2014. pp. 281-314

[22] Morley B. Causality between economic growth and immigration: an ARDL bounds testing approach. Economics Letters. 2006a;**90**(1):72-76

[23] Morley B. Causality between economic growth and immigration: an ARDL bounds testing approach. Economics Letters. 2006b;**90**(1):72-76

[24] Asiedu E. On the determinants of foreign direct investment to developing countries: Is Africa different? World Development. 2002;**30**(1):107-119

[25] Narayan PK, Smyth R. Electricity consumption, employment and real income in Australia evidence from multivariate Granger causality tests. Energy Policy. 2005;**33**(9):1109-1116

Chapter 4

No Sustainability without Planning for It: Scope and Dimensions for Sustainable Rural Planning in Zimbabwe

Innocent Chirisa, Jeofrey Matai and Tafadzwa Mutambisi

Abstract

This chapter explores and discusses the instrumentality of planning in the pursuit of rural sustainability using Zimbabwe as a case study. This is against the background that there is a growing bias towards urban areas in terms of planning for sustainability even though there are still more people living in rural places than in urban areas and that urban areas largely depend on rural areas for resources. The chapter argues that sustainability in rural areas is a function of deliberate planning. It engages the qualitative research approach, making use of document reviews and key informant interviews as data collection instruments. It is revealed that sustainability in rural Zimbabwe can remain a pie in the sky if the current approaches to planning are not revisited. Sustainability being plural and multidimensional concept, planning for it requires a multi-stakeholder approach, transcending rural and urban areas, the public and private sector, civic organisations and the rural communities as well as strong institutional arrangements that provide for transparent governance.

Keywords: sustainability, planning, economic, social, strategy, rural, Zimbabwe

1. Introduction

Development trends point to the fact that the global population is urbanising [1] and that more than 50% of the world's population is now urban [1, 2]. Whilst this is the global reality, in Africa, more people are leaving in the rural areas [3]. These people are exposed to several challenges, most of which are a result of the urbanisation process. For example, they compete for food, land, water and other resources [4]. The rural areas are also the recipients of waste that is produced by the cities. Climate change, on the other hand, is making farming which is the major rural economic activity, difficult as yields are continuously reducing owing to extreme weather patterns. This exposes the rural communities to high levels of poverty and distress and also to exploit the natural resources in an unsustainable manner. In light of the aforementioned issues, and the need to meet global goals of eradicating hunger and poverty, achieve food security, make human settlements safe, inclusive, resilient and sustainable [5], there is need to rethink approaches to rural development; a paradigm shift from the old paradigm guided by modernisation to a new paradigm that is driven by the quality of life and sustainability [6]. The United Nations Development Programme (UNDP) [7] recommends that such an approach requires the integration of the various actors and stakeholders who are involved in the development of rural places; planning is one tool that can integrate multiple actors and stakeholders towards the attainment of sustainable rural development [8].

Several studies have been conducted on rural development and planning in Zimbabwe [9–11]. These studies are critical in the literature of rural planning and development. However, they are silent on the relationship between sustainable rural development and planning. For example, PlanAfric [11] is more evaluative on the rural development approaches in Zimbabwe. Davies [9] analyses the trends in rural planning pre and post-independence in Zimbabwe whilst Hahlani [10] looks into factors that hinder integrated rural planning. This chapter emphasises the instrumentality of planning in sustainable rural development and argues that sustainability in rural planning is a product of deliberate planning. It further argues that without planning, rural areas will not be able to cope with the prevalent issues of climate change particularly considering that most rural areas in Zimbabwe are agro-based. The chapter has eight sections. Following this section is the background and overview. This is followed by a section on the theories that inform the study, literature review and methodology. The results are presented in the proceeding section from which a discussion and recommendations section follows. The last section is the conclusion of the chapter.

2. Background and overview

Rural development has been at the centre of development for most countries across the globe. However, the assumed model of developing rural places was inspired or guided by the modernist philosophy whose focus was mainly hinged on, among other processes, capital investment, the application of science to production and urbanisation [6]. This approach was also materialistic in its conceptualisation of development; hence, little consideration was paid to qualitative aspects of development such as improved quality of life and access to basic services. The modernist approach to the development of rural places resulted in the bias towards urban areas as it was generally assumed that urbanisation of rural places was an indicator of development, thus, as put forward by Rao [12], more attention was put to industrialisation than the agricultural sector, hence the rural suffered. Development was also top-down in approach [6]. The resultant effect of this approach to development was increased poverty in rural places [13, 14], unsustainable exploitation of raw materials from the rural areas as inputs for industrialisation in urban areas, environmental degradation and the consequential effects such as climate change.

After observing that the modernist approach to development was associated with several problems, there was a paradigm shift to a sustainable centred approach to development. According to [6], this approach paid more attention to agricultural development and a participatory approach to development programmes. Following a series of global goals for sustainable development, for example, the World Commission on Environment and Development [15], the Millennium Development Goals and the Sustainable Development Goals, the need to use resources within the regenerative capacity of the environment was put at the centre of developmental goals. In addition, the need to combat the effects of climate change and eradicate poverty and hunger has made it necessary to revisit the approaches to development planning, rural areas included. The UNDP [7] has observed and placed planning at the centre of sustainable development because of its coordinative and integrative capacity.

In Zimbabwe, rural development has been the centre of development policy since independence. According to Davies [9], it has been politically motivated; the aim being that of redressing disparities caused by colonial policies. Several policies, for example, the Growth with Equity and the Growth Point Policy, were put in place to ensure that rural development was realised. The model to rural development, however, reflected the modernist approach as key among the policies, was the development of growth points that were to lead the industrialisation of rural places. However, the effectiveness of the policies in addressing the problems of disparities is debatable as little progress has been made [16]. Like most rural places across the globe, Zimbabwe's rural places are experiencing a myriad of challenges that range from social, economic and environmental [11, 17]. Food insecurity, poverty, income inequalities and the lack of employment opportunities reinforce each other in a vicious cycle by eroding human capital and decreasing labour productivity, thereby perpetuating poverty and social inequalities [18].

Despite having multiple interested stakeholders in rural planning, the approaches that have been adopted to guide and encourage rural development are disintegrated and are characterised by gaps and overlaps [10, 11], necessitating an approach that integrates the various activities and programmes of the multiple actors in rural planning and development [6, 16]. Planning is important in achieving sustainability as it helps to cover the social, political, environmental and physical issues that affect the prosperity of rural areas [7, 19]. Against this background, the chapter examines the instrumentality of planning in enhancing sustainability in rural development and planning in Zimbabwe and shows that without planning for it, sustainability in rural development will not be achieved.

3. Theories informing the study

This study is informed by the sustainability theory and the style/models of planning theory. These models seek to highlight how planning is done and its procedures. This section also seeks to highlight the multi-dimensions of planning as well as the importance of planning in different scenarios and in promoting sustainable development.

4. Sustainability theory

The theory of sustainability is derived from the concept of sustainable development which is premised on the notion of maintaining or improving the state and availability of resources or materials over the long term [20]. It entails meeting fundamental human needs while conserving the life support systems of the earth. This implies that chasing the sustainability goal should aim at maintaining, improving or sustaining conditions that are beneficial to people, their communities and the environment [21]. Sustainability is preoccupied with the long-term treatment of the biophysical environment and the resources therein, social systems and people in ways that are consistent with human wellbeing and the stability of a dynamic system [20]. However, Bossel [22] outlined that the sustainability of human society is now at stake because the change of the environment is so fast that there is inadequate time for adaptive response. This overwhelming rate of change affects the ability of the ecological system to respond, consequently making the system to lose viability and sustainability. In this study, the theory of sustainability plays a critical role in justifying the need to plan for sustainability in rural areas. The reasons, according to the sustainability, include the desire to improve maintain and improve the social, economic and environmental conditions of the rural areas within the development process of rural places. The rational is that, by planning for sustainable rural development, both rural and urban areas become sustainable because of the rural-urban relations [1, 2, 14].

5. Literature review

This section reviews the literature on the key aspects of rural sustainability, planning for rural sustainability, the actors in the sustainability of rural places as well as their role. It also looks at the factors that enhance or hinder planning for sustainability in rural areas. Rural sustainability is defined by Bryant and Granjon [23] as the development strategies that maintain and produce healthy rural communities. Such communities will have compatible economic, political, socio-cultural and environmental values that respond to any requirements in these dimensions in the long run. Rural sustainability is dynamic as information, conceptualisation and values about the different dimensions of sustainability change with time [23, 24]. As such Bryant and Granjon [23] recommend that in planning for sustainability there is a need to take account of the fact that what is appropriate at a given place in time maybe inappropriate in the same place as changes occur. For example, the nature and extent of rural-urban links have changed [25, 26]. This significantly affects the approaches to rural sustainability planning. Similarly, what is appropriate in a place at a given time may not be appropriate in another place at the same time. Rural places that are closer to cities have a different set of needs compared to rural places that are further away from cities.

Sustainability, whether urban or rural, maintains the same characteristics, goals and dimensions. Bryant and Granjon [23] highlight those previous conceptions about sustainability in rural places focused on the ecological imperative, giving less attention to socio-cultural, economic and political issues. However, with a continued evolution in the concept of sustainability, and changes in social and economic aspects in both rural and urban places, sustainability is now fundamentally the same, with differences only seen in the environments, population density and activities between the rural and urban places. This entails that certain characteristics of the rural places influence sustainability planning in rural contexts. Some of these characteristics are linked to the biophysical environment which is considerably important for rural communities who depend largely on its resources for livelihood activities. Thus, for most people in the rural places who are mostly small holder peasant farmers, availability of arable land, fertile soils, sufficient water, good climatic conditions and the presence of flora and fauna are important. Land availability and favourable climatic conditions for example, are significant biophysical factors for successful farming [27].

The natural resources that are entrenched in the biophysical environment are an important source of many livelihoods in rural places. Bryant and Granjon [23] observe that besides farming-related activities, there are other livelihood activities such as sand mining and gravel extraction as well as forestry and woodland management. Rural communities are also characterised by small communities that can be concentrated into small villages, dispersed or both. However, rural places that are close to cities and towns have certain characteristics that differentiate them from those that are distant from the cities. Resources in such places are critical for the well-being of the population in cities as well.

Planning for sustainability should therefore take account of the characteristics of the rural places and the relationship between the rural places and the urban areas since these are interlinked systems [8, 23, 25]. However, focusing on rural development and sustainability tend to treat the rural places as homogeneous places [28] and as independent from urban areas [29]. A comprehensive approach is required for the realisation of sustainability as both a process and a goal, hence, the United Nations, [30] and Marsden [24] argue that planning is key if sustainability is to be attained.

Scholarly and policy works have identified several factors that affect planning for sustainability [9, 20, 23, 24, 31]. One of the factors that impact on planning for sustainability and its subsequent realisation is state policy [31]. His argument is informed by Lipton's [14] urban bias theory which, among other issues, argue that states act to protect the interests of cities and to discriminate against the interests of the rural places. In this instance, financial, human and productive resources are deliberately moved out of the rural places to benefit urban dwellers. Douglass [32] argues that, on top of providing food and natural resources to cities, rural places have pronounced poverty and hunger with relatively low income; they continue to experience out-migration and environmental degradation. State policies in this instance are discriminatory and lead to the extraction of resources from the rural places. This incentivise urban areas resulting in a reduced supply of labour in rural places. This in turn, endangers the sustainability of rural places [31, 32].

The continued separation of rural and urban places in terms of policy and planning is another factor that affects rural sustainability planning [33]. This continued separation of focus between the rural and urban places in the face of changing rural and urban relations [26] has a negative influence in the planning and realisation of sustainability [23]. To address the problems associated with the separation, [33] recommends that plans that transcend rural and urban places whilst protecting the natural and urban activities are required.

Planning for sustainability involves multiple actors which range from the public institutions, private organisations, civic organisations, non-governmental organisations and the local people who participate in the planning processes and activities. The public sector usually provides a framework for sustainable development through the various arms and tiers of the government. The private sector is usually involved in the implementation of sustainability programmes through investments while the civic organisations and non-governmental organisations play a variety of activities that ranges from advocacy to supporting and mobilising the local people. With such a range of actors, [33] recommends that there is a need for strong institutional frameworks that coordinates the plans and activities of the various actors that are involved in the planning of sustainability. Failure to coordinate the plans and activities of the various actors and the rural and urban places results in chaotic and complex spaces that make sustainability difficult to plan for and achieve [34]. Thus, without planning for it, sustainability will remain a pipe line dream.

Several approaches have been put in place to address sustainability challenges across the globe. At the global level, ratifications have been put in place and international goals set to ensure that sustainability. The Brundtland Commission of 1987 marks the dedicated commitment towards sustainable development. The Millennium Development Goals and the current Sustainable Development Goals are some of the international efforts towards sustainability. These frameworks have been adopted by different countries in a localised manner to plan for sustainability. The adopted approaches and strategies are not uniform across the globe [35]. The approaches are linked to the main challenges to sustainable development that places face, hence, priority also differs among and between nations. The approaches include strategies earmarked to deal with poverty, environmental deterioration, marginalisation, water and sanitation, safe, resilient and sustainable human settlements as well as climate change issues among others [5, 28, 30, 35]. To address the challenges, some countries have developed sector-specific programmes and projects. States have developed policies that aim to trigger planning towards sustainability in different contexts, for example, climate change policies, housing policies and economic blueprints. However, scholarly work shows that most of these approaches are planned and implemented in silos, making planning for sustainability insufficient [11, 29, 34].

In summary, scholarly work shows that the sustainability of rural places is key to the attainment of sustainability in both rural and urban areas. There is also clear evidence from scholarly work that the approaches to sustainability planning generally lack comprehensiveness, making the efforts disintegrated and less effective. The question that remains to be answered is, what should planning-for-sustainability, particularly in the rural context, take?.

6. Research methodology

Research methodology is the specific procedures or techniques used to identify, select, process and analyse information about a topic [36]. The research methodology allows for a critical evaluation of a study's overall validity and reliability. The study made use of the case study approach, which is an approach that is used to generate an in-depth, multifaceted understanding of a complex issue in its real-life context [37, 38]. Planning for sustainability is being discussed and analysed looking specifically at rural areas in Zimbabwe. Binga, Mbire and Runde rural districts. The study also made use of document reviews. Document review is a systematic collection, documentation, analysis and interpretation of data as a data collection method in research. Documents review include, e-books, journal articles, policy papers as well as relevant planning statutes. Key informant interviews were conducted in the study to get a clear assessment of sustainable rural development and planning in Zimbabwe. Key respondents included key stakeholders in Non-Governmental Organisation (NGOs) and rural district councils.

7. Results

According to the population census of 2012, the rurality of the population is measured based on three criteria. It classifies any settlement with more than 50% of the population engaged in agriculture and less than 2500 inhabitants as rural. Rural areas in Zimbabwe generally have poor economic and social conditions. Food insecurity, poverty, income inequalities and the lack of employment opportunities reinforce each other in a vicious cycle by eroding human capital and decreasing labour productivity, thereby perpetuating poverty and social inequalities across generations in rural Zimbabwe [18].

Rural planning may usefully be seen as comprising three crucial elements. The first concerns the content of rural planning that is the strategies and policies that underlie what rural planning seeks to achieve. The second may be termed the institutional framework within which rural planning operates, especially the agencies and actors involved and how they interact. The third aspect can be called the approach to rural planning and is often seen in terms of the polarities of a top-down, blueprint approach or a bottom-up process approach [39].

8. Composition of the rural population and livelihoods in Zimbabwe

According to the 2016 ZimVac Rural Livelihoods Assessment (RLA) the 2015/16 and 2016/17 are projected to be the two poorest consumption years since 2009. During the first quarter for the 2016/17 consumption year, 987,000 people could not meet their annual food requirements which was an increase from last year when 151,000 people were estimated to be food insecure during the same period. A large share of the rural poor still depends on low-productivity subsistence farming for their livelihoods. The poorest rural households lack access to productive assets and often rely on income from wage employment. Of the 300–500 million wage workers in agriculture [18]. Concern among state legislators about rural development and rural land use is not new. In many states, agriculture remains an important feature of the economic, cultural, and political landscape [29].

Adopted from: NFP Co-ordination Unit [4].

From **Box 1**, it can be noted that development is rural areas is mostly centred on land use, environmental and physical planning. It has been stated previously that

Physical planning in rural areas

- Activity was largely undertaken by the Department of Physical Planning (DPP) within the MLGNH.
- DPP acts for local authorities but hopes to devolve more planning functions to them.
- DPP provides technical support to provinces and RDCs which are the lead planning agencies for development planning at the district level.

Environmental planning in rural areas

- Undertaken by the Department of Natural Resources (DNR) in the Ministry of Mines, Energy and Tourism (MMET) in partnership with RDCs. District environmental action plans (DEAPs) are funded by UNDP.
- The relationship of DEAPs to other planning systems is not clear.

Land-use planning in rural areas

- The lead agency for land-use planning is the Department of Agriculture and Extension (AGRITEX) situated within the Ministry of Lands and Agriculture. AGRITEX works closely with the RDC and traditional leaders.
- The water catchment process is led by the Department of Water Development in the Ministry of Rural Resources and Water Development (MRRWD). This will pass to the Zimbabwe National Water Authority (ZINWA) when it is established. The programme has strong donor inputs from GTZ, Dutch aid and DFID.
- National park plans are led by the Department of National Parks and Wildlife Management (DNPWLM) and it has established a stakeholder consultation group including RDCs, representatives from CAMPFIRE (Communal Areas Management Programme for Indigenous Resources) and others.
- A planning unit has been established within DNPWLM supported by World Bank, JICA and UNDP.
- Forest action plans are led by the Forestry Commission and other stakeholders (especially rural district councils, RDCs) and supported by GTZ and DFID.

Box 1.
Actors in rural planning.

rural livelihoods mostly come from agricultural produce in Zimbabwe. Therefore, proper land use planning is the corner stone of sustainable social and economic development in rural areas. Physical factors mostly threaten the sustainability of rural livelihoods and way of life. Therefore, many factors threaten rural well-being such as urbanisation or urban growth. Urbanisation has been the major cause of rural to urban migration which has led to the loss of human resources for rural areas. More so, the relative decline of farm prices, for example, has eroded farm profitability and led to bankruptcy. Soil erosion degrades soil productivity and escalates the need for chemical additives. Nutrient and pesticide runoff degrades waterways and groundwater aquifers. Declining rural populations have closed rural schools, hospitals, and other public facilities. In short, eroding farm profitability, soil erosion, environmental degradation, and loss of public services represent far greater threats to rural health than urban growth [29]. High inequality in the rural sector, in particular in the distribution of assets such as land, water, capital, education and health is an obstacle that needs to be addressed to enhance food security. The underlying issue of discrimination in the rural sector, including against women and children is also another social aspect in need of planning to make rural areas sustainable communities.

Achieving sustainability in rural areas means entailing the integration of four pillars: social development, economic development, environmental management, and urban governance [40]. Therefore, planning is never a unilateral venture by one actor but involves many stakeholders, leaving no stone unturned on matters that matter (**Figure 1**).

Rural development aims to improve livelihoods by implementing comprehensive development for rural areas where a majority of people in poverty live. Pillars of sustainability are an accurate presentation of comprehensiveness in planning. Comprehensive plans have become common for many regions because development is complex, development decisions are interrelated, and the development process could be improved through careful analysis, foresight, and planning. These conditions now hold in rural areas, too. Successful rural development requires that agricultural infrastructure and property markets facilitate farming at its most efficient scale; that safeguards are in place to protect the integrity of environmental systems; and that critical education, health care, and other social services remain fiscally viable. Equally important is system interdependence. Large scale farms, for example, pose greater

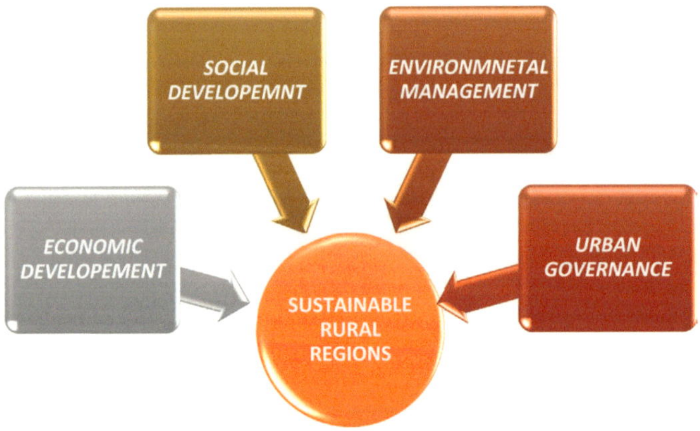

Figure 1.
Pillars of sustainable rural planning. Adapted from: The Department of Economic and Social Affairs [40].

threats to environmental systems, require more non-resident, high skilled labour, and place greater demands on social service and health care systems [29].

9. Planning approaches in rural areas

Planning and sustainability are for the people'; therefore, many development programmes in rural areas have been seen to be participatory. That is planning by the government as well as NGOs have implemented projects using the bottom up approach. This approach as implemented after the realisation that the top don approach in rare cases met the demands of the rural people [41]. Therefore, many project plans in rural districts use the public participatory approach. When policies are evidence-based there is a greater chance of success [42].

10. The REDD+ project in Mbire district

Planning is multi-dimensional and lateral. Therefore, Planning policies even though they are specific they should also consider the dimensions of society. This is so as the REDD+ plus project is an environmental project plan that was meant to help the people in Mbire District to conserve the environment, build their economic fabric through carbon sequestration as well as to give them a path to growing their knowledge in plant and biodiversity preservation. The REDD+ project is a bottom-up approach that allows local people to use their Indigenous Knowledge formulate ways to minimise deforestation and enable carbon sequestration to minimise the rate at which emissions affect the ozone layer.

The REDD+ Conservation Project I aimed at providing sustainable livelihood opportunities for poor communities. This is aimed and reducing forest deforestation, land degradation, poverty and drought. This strategic plan involves the education of techniques that increase their productivity substantially and reduce the need to cut down more trees. This initiative involved a multiple stakeholder approach as it is community based. It used the participatory approach as it worked there was community participation from the local people, the rural district council, the government as well as the carbon asset management. The multi stakeholder approach in planning is critical in promoting sustainability as it allows many stakeholders to have a multi-faceted view of social, economic and physical issues of a specific problem or solution. Carbon credit s are an economic solution to an environmental problem and they boost the social life of the community. Carbon credits from the REDD+ program also help to physically develop the Mbire District. Therefore, planning is everything in sustainable rural development.

11. WASH programmes in Runde district and Chivi district

Water is essential for life. Without water, humans can survive only for days. Water comprises 75% body weight in infants to 55% in elderly and is essential for cellular homeostasis and life (World Health Organisation 2013) Water represents a critical nutrient whose absence can be lethal to humans in days [43]. Water is important in framing sustainability in rural areas as it also has a direct impact on women and children as they are responsible for etching water in rural areas as well the health of the community. The importance of incorporating a gender-based perspective in WASH programmes is vital based on the fact that women and girls are the primary users of water in their households and are guardians of household hygiene. The government and NGOs such as the World Division have been implementing various wash projects. This has been seen through the construction of boreholes and wells. Areas such as Chivi experience dry spells more than other rural districts. Therefore, supplying water assists in eradicating the issues of dehydration and promotes good health and sanitation. Irrigation schemes have also been introduced in these districts. Most of the water comes from the Runde River. This River has helped supply rural districts with water that has helped them increase their agricultural produce. They have booted economic independence of various villages in these districts as it gives them food and a source of income.

An improvement in rural livelihoods has a direct impact on the increase of the Gross Domestic Product Of the country. These sentiments are in line with those of [43] when he states that The establishment of new water point sources has improved the livelihoods of the Runde as they can now access a basic right to clean water. In addition, the availability of water for productive purposes has increased the incomes of households that rely on agricultural production as a form of livelihood. It can be seen that development programmes are important in promoting sustainability in rural areas. WASH programmes in Zimbabwe are important as most rural areas are facing frequent drought spells due to the ever-changing climate change. This is also evidence that the linkages and causal relationships between the pillars of sustainability show that planning is important in achieving sustainability it allows the planning for all dimensions of rural livelihoods. Taking the case of rural communities' access to water resolves gender inequality issues, economic livelihood, and health as well as sanitation problems.

12. Discussion and recommendations

The rural economy holds significant potential for creating decent and productive jobs and contributing to sustainable development and economic growth. This is why the government adopted the 2030 Agenda for Sustainable Development embraces the three dimensions of sustainability economic, social and environmental [18]. However, much remains to be done to adequately adjust the British-modelled urban and rural planning system to suit local conditions. Rural planning should be iterative with a balance of top-down and bottom-up approaches with clear national policy guidelines coming down and local needs and priorities filtering up [44]. A broad-based rural development strategy has to include infrastructural investments to better connect producers to output markets, including in rural-urban linkages and the improvement of distribution systems and storage facilities. Social protection mechanisms, including social safety nets, must also be part of a broader rural development strategy to facilitate access of low-income households to food [40]. They are also required to better connect producers and smallholders to output markets, including rural-urban linkages. A broader rural development strategy should also include social protection mechanisms, including safety nets, to facilitate access of lower-income groups to food, in particular during economic shocks. Agricultural extension services need to be further supported, with a particular focus on growing crops appropriate to the land, and on sustainable environmental management [45].

13. Conclusion

In conclusion, the key aspect in the evaluation of multiple planning scenarios and for the definition of sustainable transition pathways is key in rural development. Sustainable development must inevitably be based on long-term planning and must consider relations among other actors. Rural development is critical for an integrated approach to sustainability and for reducing poverty. Ensuring wider and inclusive access to public services can reduce rural/urban inequalities, disaster risk and food insecurity, as well as strengthen networks between cities and villages.

Author details

Innocent Chirisa[1*], Jeofrey Matai[2] and Tafadzwa Mutambisi[3]

1 Department of Demography Settlement and Development, University of Zimbabwe, Harare, Zimbabwe and Department of Urban and Regional Planning, University of the Free State, Bloemfontein, South Africa

2 Department of Architecture and Real Estate, University of Zimbabwe, Harare, Zimbabwe

3 Department of Civil Engineering, Harare Polytechnic College, Harare, Zimbabwe

*Address all correspondence to: innocent.chirisa@gmail.com

References

[1] McGranahan G, Satterthwaite D. Urbanisation Concepts and Trends. London: International Institute for Environment and Development; 2014

[2] Turok I. Urbanisation and Development: Reinforcing the Foundations. Routledge: The Routledge Companion to Planning in the Global South; 2017

[3] Saghir J. Urbanization in Sub-Saharan Africa. In: Meeting Challenges by Bridging Stakeholders. Washington DC: Centre for Strategic and International Studies; 2018. p. 7

[4] NFP Co-ordination Unit, editor. Malawi's National Forestry Programme. Lilongwe, Malawi: Govt. of Malawi, Dept. of Forestry, NFP Co-ordination Unit; 2001

[5] Sachs J et al. Sustainable Development Report 2020: The Sustainable Development Goals and Covid-19 Includes the SDG Index and Dashboards. Cambridge: Cambridge University Press; 2021. DOI: 10.1017/9781108992411

[6] Shepherd A. Sustainable Rural Development. Basingstoke, UK: Macmillan International Higher Education; 1998

[7] UNDP. Handbook on Planning Monitoring and Evaluating for Development Result. New York: UNDP; 2009

[8] Stead D, Nadin V. Spatial Planning. Key Instrument for Development and Effective Governance with Special Reference to Countries in Transition. Geneva: United Nations; 2008

[9] Davies DH. Zimbabwe's experience in rural development. Development Southern Africa. 1990;**7**(Sup.1):467-482. DOI: 10.1080/03768359008439557

[10] Hahlani CD. Bottlenecks to integrated rural development planning in Zimbabwe: A focus on the Midlands province. The Dyke. 2012;**6**(2):136-158

[11] PlanAfric. Evolution of Rural Planning in Zimbabwe. London: International Institute for Environment and Development; 2000

[12] Rao VKRV. Review article: Urban bias and rural development. Indian Economic Review. 1980;**15**(1):75-83

[13] Bezemer D, Headey D. Agriculture, development, and urban bias. World Development. 2008;**36**(8):1342-1364. DOI: 10.1016/j.worlddev.2007.07.001

[14] Lipton M. Why Poor People Stay Poor : A Study of Urban Bias in World Development. Temple Smith: Australian National University Press; 1977

[15] World Commission on Environment and Development. Our Common Future. Oxford: Oxford University Press; 1990

[16] Mushuku A. Growth Points or Ghost Towns? Post-Independence Experiences of the Industrialisation Process At Nemamwa Growth Points In Zimbabwe. International Journal of Politics and Good Governance. 2013;**4**(4):27

[17] Chirisa I, Nel V. Resilience and climate change in rural areas: A review of infrastructure policies across global regions. Sustainable and Resilient Infrastructure. 2021;**0**(0):1-11. DOI: 10.1080/23789689.2020.1871538

[18] Sibanda S. Promoting decent work in the rural economy Lessons

from Zimbabwe (2008-2018). Harare: International Labour Organization; 2019. p. 48

[19] Williams A. Strategic planning in the executive office of the UN secretary-general. Global Governance. 2010;**16**(4):435-449

[20] Harrington LMB. Sustainability theory and conceptual considerations: A review of key ideas for sustainability, and the rural context. Papers in Applied Geography. 2016;**2**(4):365-382. DOI: 10.1080/23754931.2016.1239222

[21] Baker S et al. The Politics of Sustainable Development: Theory, Policy and Practice Within the European Union. East Sussex, UK: Psychology Press; 1997

[22] Bossel H. Indicators for Sustainable Development: Theory, Method, Applications ; A Report to the Balaton Group. Winnipeg: IISD; 1999

[23] Bryant C, Granjon D. Rural sustainability. Rural Sustainability. 2009;**2**:158-167

[24] Marsden T. The Condition of Rural Sustainability. Assen, Netherlands: Uitgeverij Van Gorcum; 2003

[25] Bah M et al. Changing rural-urban linkages in Mali, Nigeria and Tanzania. Environment and Urbanization. 2003;**15**(1):13-24. DOI: 10.1177/095624780301500104

[26] Tacoli C. The links between urban and rural development. Environment and Urbanization. 2003;**15**(1):3-12. DOI: 10.1177/095624780301500111

[27] Knowledge Portal. Smallholder Agriculture. 2018. Available from: https://www.stars-project.org/en/knowledgeportal/magazine/table-of-contents.html [Accessed: November 22, 2021]

[28] Leal Filho W et al., editors. Sustainable Cities and Communities. Cham: Springer International Publishing (Encyclopedia of the UN Sustainable Development Goals); 2020. DOI: 10.1007/978-3-319-95717-3

[29] Knaap G, Chakraborty A. Comprehensive planning for sustainable rural development. Journal of Regional Analysis and Policy. 2007;**37**(1):18-20

[30] United Nations, editor. Trends in Sustainable Development: Africa Report. New York: United Nations; 2008

[31] Wegren SK. The quest for rural sustainability in Russia. Sustainability. 2016;(7):8, 602. DOI: 10.3390/su8070602

[32] Douglass M. Rural habitat - spatial development planning for improving rural livelihood. Agriculture + Rural Development. 2000;7(2):38-41

[33] Mylott E. Urban-Rural Connections: A Review of the Literature. 2009. Available from: https://www.academia.edu/27817208/Urban-rural_connections_a_review_of_the_literature

[34] Scott AJ et al. Disintegrated development at the rural–urban fringe: Re-connecting spatial planning theory and practice. Progress in Planning. 2013;**83**:1-52. DOI: 10.1016/j.progress.2012.09.001

[35] OECD. Sustainable Development Strategies: What are They and How Can Development Co-Operation Agencies Support Them?. 2001. Available from: https://www.oecd.org/dac/environment-development/1899857.pdf [Accessed: November 23, 2021]

[36] Bashir M, Afzal MT, Azeem M. Reliability and validity of qualitative and operational research paradigm. Pakistan

Journal of Statistics and Operation Research. 2008;**4**(1):35. DOI: 10.18187/pjsor.v4i1.59

[37] Creswell JW. Research Design Qualitative, Quantitative, and Mixed Methods Approaches. United States of America: SAGE Publications; 2014

[38] Fidel R. The case study method: A case study. Library and Information Science. Research. 1984;**6**(3):273-288

[39] PlanAfric. Rural District Planning in Zimbabwe: A Case Study. Bulawayo Zimbabwe: Environmental Planning Issues; 2000

[40] United Nations Department for Economic and Social Affairs (UN-DESA). World Economic and Social Survey 2013: Sustainable Development Challenges. New York: UNDESA; 2013. Available from: https://www.un.org/en/development/desa/publications/world-economic-and-social-survey-2013-sustainable-development-challenges.html

[41] Siambombe A. Rural communities and policy participation: The case of economic policies in Zimbabwe. African Journal of Social Work. 2015;**5**(2):87-107

[42] Manjengwa J, Feresu S, Chimhowu A, editors. Understanding Poverty, Promoting Wellbeing and Sustainable Development. Harae: Sable Press; 2012

[43] Ngwira M. Sustainable Development Goals and the Impact of Water Sanitation and Hygiene (WASH) Programmes in Rural Zimbabwe: Case of Runde District Zimbabwe. Bachelor Thesis, Midlands State University; 2018

[44] United Nations, editor. Socio-Economic Impacts of Ebola on Africa. Addis Ababa: United Nations Economic Commission for Africa; 2015

[45] Save The Children. A1 Resettlement Areas & Mutorashanga Informal Mining Communities Zvimba District, Mashonaland West Zimbabwe. Zimbabwe: Save the Children (UK); 2003

Chapter 5

A Framework for Facilitating Holistic Interventions for Building Community Resilience to Climate Change for Sustainable Community Development

Precious Tirivanhu

Abstract

The realities of climate change in Africa have led to a growing need for innovative approaches to livelihoods programming that promote resilience among rural communities for sustainable community development. Although several community resilience frameworks are emerging there is a need for practice modalities. This paper proposes a programming framework grounded in soft systems thinking that brings an understanding of the multi-dimensional and integrated nature of resilience programming. The author utilizes experiential knowledge from over a decade of rural development facilitation in Zimbabwe coupled with secondary reviews to address two key research questions: What are the critical components of a systemic programming framework for community resilience? And, how is such a framework facilitated in practice? The paper concludes by giving critical components of the systemic programming framework and recommends that the framework should be tested empirically for its components to be integrated into resilience programming in Zimbabwe.

Keywords: systemic programming framework, soft systems methodology, resilience, climate change, sustainable community development, Zimbabwe

1. Introduction

Zimbabwe, like most southern African countries, currently faces erratic weather patterns amidst a rural population that is highly dependent on rain-fed agriculture for livelihood. The impacts of climate change have become a reality and this scenario calls for urgency in improving rural livelihoods and sustainable community development. There is growing recognition amongst development practitioners, academics, policymakers, non-governmental organisations (NGOs), and funding agencies for innovative approaches to the design and implementation of livelihoods programs that promote resilience among rural communities. Several proposed frameworks are emerging mostly based on improving the adaptive, absorptive, and transformational capacities of vulnerable communities. A key question remains on how such frameworks can be developed into practical programming models for rural development practitioners. Research on the impacts of climate change on agriculture and food systems is not new. In fact, a review of the literature shows that such studies in Zimbabwe and most parts of southern Africa can be traced to the 1980s.

Considerable research has been conducted on the impacts and potential impacts of climate change on Zimbabwean rural households, see, for example, Masiyiwa et al. [1]; Brown et al. [2]; Gwimbi [3]; Mutekwa [4]; Nhemachena and Hassan [5]; Kinuthia [6]; Buckland [7]; Matarira et al. [8]; Downing [9]. Some of the key documented climate change impacts include food insecurity; malnutrition; increases in incidence of drought, extreme temperature fluctuations, unpredictable seasons, reduced run-off necessary to sustain the country's hydro-electric power supply, damage and destruction of infrastructure. There has been an increase in both minimum and maximum temperatures in Zimbabwe. In addition, it is documented that the most adverse impacts of climate change are in the developing world because of geographic exposure, reliance on climate-sensitive sectors, low incomes, and weak adaptive capacity. According to Heltberg et al. [10], most vulnerable households are those with assets and livelihoods exposed and sensitive to climatic risks and who are most dependent or rain-fed agriculture. This paper makes a deliberate departure from the discourse on climate change impacts (which has been given much attention) and focuses on how programs or projects on building community resilience against the impacts of climate change can be facilitated in practice since there is a dearth of such studies.

2. Climate change interventions amidst evolving rural development approaches

Approaches to rural development programming are always in transition. Such dynamism can be attributed to several factors including the continuous shift in development paradigms or thinking; the need for donor effectiveness; continuous learning and knowledge generation from rural development practice; and continuous shifts in global and local socio-economic, socio-cultural, and political factors affecting rural livelihoods. A synopsis of the rural development programming trajectory identifies several phases and shifts in development paradigms from the 1950s to the present that influenced praxis. These include development as economic growth and modernization in the 1950s; state intervention in the 1960s; market liberalisation in the 1980s; poverty reduction, participation, and empowerment in the 1990s; environment, climate change concerns, sustainable livelihoods, and millennium development goals (MDGs) in the early 2000 and more recently sustainable development goals (SDGs) and the focus on resilience [11–13]. These development paradigms shape worldviews, beliefs, and perceptions on appropriate programming approaches.

Although a detailed account on the evolution and progression of programming approaches is beyond the scope of this paper, a few approaches are highlighted here to give a context. A review of literature highlights varying nomenclature in classifying rural development programming approaches.[1] Westoby and Dowling [14] identify several of these approaches. These include community driven development (CDD);

[1] The approaches outlined here are not necessarily the most effective but are perceived as commonly applied in the African community development context.

rights-based community development (RBCD); asset-based community development (ABCD); sustainable livelihoods approach (SLA), people-centred capacity building approach (PCCBA); comprehensive community initiatives, and most recently, community resilience approaches which are the focus of this paper and are detailed in proceeding sections. CDD is associated with investments by the World Bank although its origins are linked to post-colonial years in India and Bangladesh during the 1940s and 1960s ([15], p. 27). Such a scenario has been linked to huge investments by the World Bank[2] into CCD projects in the last decade. It is an approach that empowers the community by giving control of decision-making and resources. Communities are given the power to plan, execute and monitor projects. It places emphasis on improving governance capacity of the community and local development institutions.

Rights-based community development sets the achievement of human rights as a development objective and utilizes international human accountability to support development [17]. Its tenets are linked to the 1948 Universal Declaration of Human Rights and the 1986 United Nations Declaration of the Rights to Development (UNDRD). According to Cornwall and Nyamu-Musembi [18], the rights-based approach calls for existing resources to be shared more equally and for assisting the marginalised people to assert their rights to those resources. Its origins are highly contested with some scholars arguing linkages with gender and human rights struggles [19]. Other scholars link it with rights of the disabled [20] and civil, political, economic, social, environmental, and cultural rights [21].

ABCD is based on the assertion that communities can organise and drive their own development through the identification and mobilisation of existing resources at their disposal [22]. The SLA links socio-economic and environmental development concerns within communities and focuses on people's strengths. It looks at five types of household assets; natural, social, financial, physical, and human capital and how they sustain livelihoods. The approach is premised on livelihoods, which are regarded as means of gaining a living through capabilities or livelihood strategies (e.g., agricultural intensification, livelihood diversification) and assets (both tangible and intangible). Livelihoods are taken as sustainable through the ability to recover from stress, and shocks to maintain and enhance capabilities and assets without undermining the natural resource base [23]. The approach emphasizes the importance of contexts, institutions, and supportive policies in enhancing livelihoods.

Related to SLA is community capitals framework (CCF) which is a systems approach to analysis of communities for holistic interventions. It emphasises seven different forms of capital; natural, human, social, financial, built, cultural and political, types of assets found in each capital, and how capitals are converted and coordinated. It provides tools for identifying capabilities for change of vulnerability situations. The CCF is related to the SL framework, with five capitals (human, social, financial, natural, and physical) being part of the assets in SLA framework. CCF adds cultural and political capital; the former brings dimensions of values, norms, and world views while the latter caters to influencing power dynamics, laws, policies, and strategies that affect livelihoods. Another approach within the African development discourse is the comprehensive rural development program (CRDP) or integrated rural development program or approach which cuts across all sectors and comprehensive approach whose components include agrarian reform, rural development, and land reform [24].

---

[2] Over the period 2000–2010, the World Bank has invested an average of USD 2 Billion a year for its CDD portfolio. In 2003, CDD represented $7 Billion of the World Bank commitments [16].

In the last two decades, there has been a proliferation of comprehensive community initiatives (CCIs). These are multi-sectoral, multi-stakeholder approaches to rural development [25]. They aim at a system-wide approach to community empowerment. They provide communities with leadership skills, youth, and women empowerment, aim at improving health systems and entrepreneurial skills, and enhance the utilisation of information communication technologies within communities. Comprehensive community initiatives (CCIs) present a shift from project-specific interventions toward a multi-faceted approach that aims at community-wide socio-economic transformation. They cover multiple development sectors (health, social services, leadership development, information communication technologies, youth development, institutional strengthening, women empowerment, and entrepreneurial development). CCIs engage multiple stakeholders including government departments, community-based organisations, private sector companies, research, and academic institutions. They are the shift from developmental approaches that view communities as recipients of aid and empower communities through decision-making and financial control. Communities are viewed as partners in developing community-led local development solutions. The following section explores the concept of community resilience which is at the core of this paper and is currently a core theme in rural development programming in the context of climate change.

3. Building community resilience

Resilience focuses on how a community or individual can deal with disturbance, surprise, and change. It entails framing a sustainable future in an environment of growing risk and uncertainty. The concept was originally coined from ecology but currently borrows from various disciplines including ecosystems stability, complex adaptive systems, engineering infrastructure, psychology, behavioral sciences, and disaster risk management [26, 27]. The concept of resilience does not have a common definition and its building blocks are highly contested. However, it is generally formulated around the continued ability of a person, group, or system to adapt to shocks and stress and continue to function, or quickly recover its ability to function, during and after stress [28].

In the rural development context, it focuses on how communities can recover after a hazard, to their reference state of livelihood status or improve for the batter. It is the ability to withstand (absorb) shocks and stresses, as well as the ability to adapt to dynamic conditions and put in place mechanisms that enable longer-term, systemic responses to the underlying causes of vulnerability [29]. The need for building community resilience to the impacts of climate change has become central to rural development programming [26, 30–32]. Effort has been put into developing theories of change that build/strengthen household and community resilience. This requires helping people cope with current change, adapt their livelihoods, and improve governance systems and ecosystem health so they are better able to avoid problems in the future. It requires an integrated approach and a long-term commitment to improving three critical capacities (absorptive, adaptive, and transformative), which are interconnected, mutually reinforcing, and exist at multiple levels, i.e., individual, household, community, national, and ecosystem levels [26, 32]. Absorptive capacity leads to persistence, adaptive capacity leads to incremental adjustments/changes and adaptation, while transformative capacity leads to transformational responses [26].

According to Frankenberger [29], most NGO work on resilience programming has focused on five fundamental variables in developing theories of change. These variables a focus on shock dynamics, a multidimensional capacity, resilience functions, outcome-indexed capacities, and a multilevel and systems-based approach. Shock dynamics focus on understanding the type of shock(s) and the effects of the shock(s). A multidimensional approach draws on human, social, economic, physical, ecological, and programmatic (for example, safety nets) resources, the optimal configuration of which varies by type of shock, level of aggregation, context, and community. Resilience functions prepare for and respond to a particular type of disturbance or configuration of disturbances. They may require different types of absorptive, adaptive, and transformative capacities. Outcome indexed capacities stipulate that resilience should be indexed to a given well-being outcome. The specific capacities drawn upon may vary depending on the outcome of interest (for example, health, food security, poverty. Multilevel, systems-based approach argues that resilience is observed at a given level (such as household or community) but is understood as a multilevel construct. Interventions should be sensitive to nested dependencies between levels (for instance, households and communities, communities, and regions).

Absorptive capacity is the ability to minimize exposure to shocks and stresses through preventative measures and appropriate coping strategies to recover quickly and avoid permanent, negative impacts [26]. It is built through various incremental changes and adaptations that people undergo to continue functioning in response to a shock or growing stress, without making major qualitative changes to the way they operate. These adjustments can take many forms. In the context of rural households affected by food insecurity, examples include the adoption of new farming techniques, the diversification or adjustment of household's livelihood activities, and the decision of taking out loans or connecting to new social networks. Disaster risk reduction/management (DRR/DRM) supports improved absorptive capacity by helping households and communities reduce risk and absorb the impacts of shocks without permanent, negative impacts on their livelihoods [32].

Adaptive capacity is the ability to make proactive and informed choices about alternative livelihood strategies based on an understanding of changing conditions. It is the capacity to learn, combine experience and knowledge, adjust responses to changing external drivers and internal processes, and continues operating ([33], p. 13). According to Brooks [34]; Smit and Wandel [35] adaptation refers to adjustments in a systems' (household, community, group, sector, region, country) behavior, characteristics, actions, or outcomes that enhance its ability to cope with, manage or adjust to some changing condition, stress, hazard, risk or opportunity in order for the system to improve livelihoods. The rural development discourse derives knowledge of adaptation mostly from studying vulnerability to natural hazards and impacts on food insecurity [36]. Adaptation strategies realise that communities can take concrete steps to minimise net losses from climate change including taking advantage of opportunities for gains. Improved adaptive capacity results from adjustments that include livelihoods diversification, asset accumulation, and improved social and human capital.

Transformative capacity refers to system-level changes that enable more lasting resilience at the household and community levels. In recent years, resilience programming has shifted the balance of effort and resources from short-term humanitarian assistance efforts toward a combination of disaster risk management, climate change adaptation, livelihood diversification, social protection programs, and longer-term institutional development and systemic change [32]. Transformative capacity enables more lasting resilience at the household and community levels through altering permanently and drastically the system's functioning or its structure to ensure the immediate "survival" of the household/system. It encompasses the governance mechanisms, policies/regulations, infrastructure, community networks, and formal and informal social protection mechanisms that constitute the enabling environment necessary for systemic change [32].

4. The need for a systemic programming framework for community resilience

The need for a systemic or holistic programming framework for community resilience and improved livelihoods advocated in this paper is justified by three assertions. Firstly, there is a growing paradigm focused on improving community resilience against the shocks and stresses of climate change. Secondly, non-governmental organisation (NGO) work in Zimbabwe has been highly fragmented, and uncoordinated. Finally, there is a dearth of literature that chronicles good practices in rural development facilitation.

NGO activities in Zimbabwe have been highly fragmented in practice. Social experiments by these organisations have in most instances focused on isolated projects which tend to ignore the holistic nature of community life and make abstract assumptions from reality. In essence, such individual projects should be a part of a bigger puzzle and avoid duplication. Within the context of resilience being addressed in this paper, the components of building resilient communities should not be piecemeal but rather be integrated and have emergent properties. In recent years, it has become increasingly clear that, through encouraging systems-based thinking the concept of resilience has the potential to radically transform the compartmentalised and somewhat fragmented ways of developing vulnerable rural communities [31]. By recognising the complex interplay of the conditions of vulnerability, resilience could provide a means for more holistic understandings of such complexity by shifting attention away from individual project approaches toward addressing the complex milieu of community conditions.

This paper focuses on how resilience programming can be facilitated in practice. Although several NGOs have implemented various programming approaches, there is a dearth of literature on how such processes are conducted. Such a scenario is understandable as programming approaches determine the competitiveness and comparative advantage among NGOs. Thus, it becomes irrational to expose the 'secrets' of their programming successes. However, in recent years, potential approaches to building community resilience against the impacts of climate change have been conceptualised and documented but need to be tested empirically [29, 32]. It is against this scenario that this paper is premised. The key research questions addressed are: What are the critical components of a systemic programming framework for livelihoods and resilience? And how is such a framework facilitated in practice?

5. Conceptual framework

The concepts of soft systems methodology provide conceptual building blocks for the development of a systemic or holistic programming approach for building community resilience. Laslo and Krippner [37] define a system as, "a complex of interacting components together with the relationships among them that permit the identification of a boundary-maintaining entity or process". The underlying principle of systems theory is that the effects or outputs of a system are dependent on the interactions among various components. Studying the components in isolation will not provide an accurate picture of the system [38–40]. Unlike the reductionist research approaches that rely on drawing samples, systems theory does not separate individual components under study but focuses on how these components interact with each other in their entirety [40]. Central to systems theory is the concept of wholeness that aims at bringing together fragmented research findings in a comprehensive view of man, nature, and society. Systems thinking is championed on the premise that there are emergent properties of systems that do not exist when systems are decoupled into smaller parts [41, 42].

There is a diverse array of system thinking methodologies including system dynamics, critical systems thinking, viable systems, and critical systems heuristics, among others [43]. This paper adopts soft systems methodology (SSM) as a conceptual framework. The core of SSM is the construction of models of the system(s) being studied. These models are used to discuss how to bring about organizational/community change. They allow the community to engage in debate and the practitioner/facilitator to elicit multiple perspectives. The learning that takes place leads to purposeful[3] action systems. The models constructed through SSM are regarded as learning systems, instead of incontestable representations of reality. Within the context of this paper, therefore, the term 'system' refers to the process of inquiry, *i.e.*, the analysis of a situation, rather than to an "objective" view of the world that the observer assumes to be capable of managing. SSM is premised on structuring coherent debates and allowing those involved in the problem situation, and those likely to be affected by any solutions to define the problem to be addressed [44]. It allows those with differing perspectives to understand each other sufficiently, so that they may act in the world in a way that all parties can live with ([43], p. 143). This paper adopts the classical SSM[4] implemented following a seven-step process as illustrated in **Figure 1**.

The first stage explores the situation within a framework of the real world. It focuses on the mapping of cultural history, stakeholder analysis, community perspectives and assumptions, historical trends, among other factors. The purpose is not to define the problem but to solicit holistic unstructured parameters of the problem situation through dialogue and debates with all the affected and those capable of bringing relevant choices. Stage 2 expresses the problem situation through development of a rich picture from the unstructured problem in stage 1.

Stage 3 provides root definitions of relevant systems in the problem situation. This is a departure from the real world and provides perceived choices. A root definition is a statement defining what is relevant to the system and who is either affected or affects it. Defining root definitions is guided by a CATWOE analysis (Customers, Actors, Transformation process, *Weltanschauung*, Owner, and Environmental Constraints). Customers are the victims/beneficiaries of the purposeful activities. Actors are responsible for the activities while the transformation process is expressed as inputs, transformation, and outputs. *Weltanschauung* are the worldviews for a meaningful system and owners are community members who can stop the system from functioning. Environmental constraints are taken as 'given' and difficult to influence, affect, and change (**Figure 2**).

---

[3] A system is purposeful if it allows debate and reflection.

[4] SSM has gone through reviews over the last three decades based empirical studies from different discipline. See for example Checkland and Scholes [45].

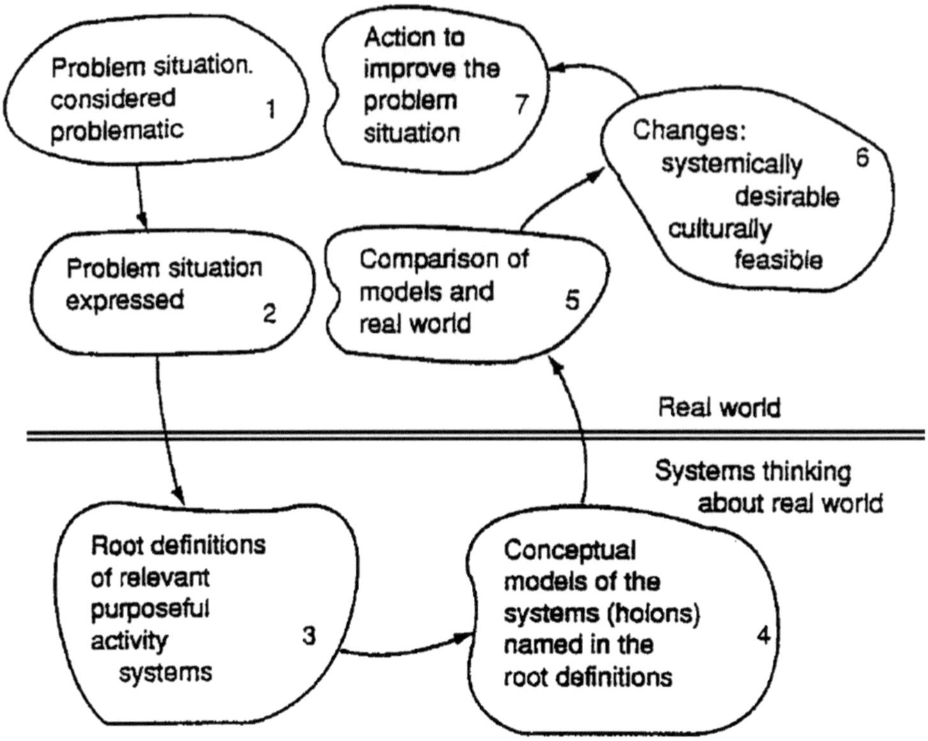

Figure 1.
The seven-step process in classical soft systems methodology adopted from Mingers [46].

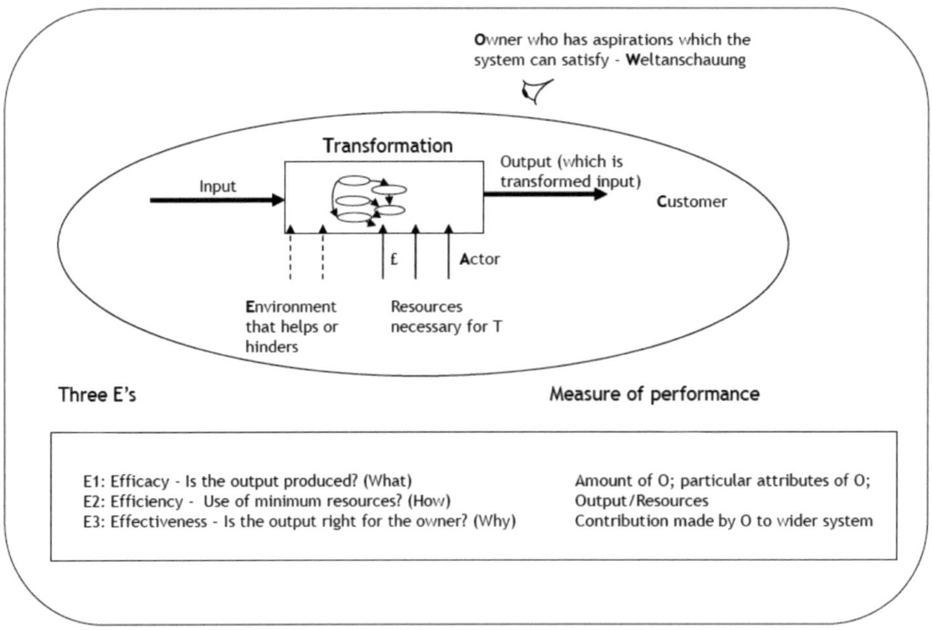

Figure 2.
CATWOE analysis. Adopted from Wang et al. [47].

Stage 4 is the construction of conceptual models that present holistic stakeholder perspectives about the desired system and associated human activities. It prepares for the dialogical process that will take place during the implementation of SSM-based interventions. For each root definition, the analyst makes a conceptual model. The conceptual model is the structured set of activities that logic requires in a system, defined in the root definition. Stage 5 compares the conceptual model with the real-world problem situation. It provides a dialogic process and debate on the perceived situation and an opportunity for stakeholders to critique their assumptions. Stage 6 determines the desirable and feasible systemic changes. Checkland [44] identifies three types of changes: structural, procedural, and attitudes. Within the rural development discourse, structural might refer to community groupings, communication, social capital, and functional responsibilities. Procedural will include community and other stakeholder modes of operation, while attitudes include changing mental models, and practices. Stage 7 is the implementation stage and outlines the implementation strategy, resources, and skills requirements. According to Mingers [46], in practice, these steps are not taken sequentially and some may be omitted and combined.

6. Methodology

This paper utilised (a) experiential knowledge and expert experience from action research by the author based on more than a decade of engagement in rural development facilitation in Zimbabwe and (b) document reviews. The experiential knowledge was acquired through an action research process where the author engaged in action research between 2002 and 2014, under the WK Kellogg foundation programs as a development facilitator in Manicaland Province of Zimbabwe. Experiential knowledge is based on a participative inquiry paradigm and grounded in the belief that experiential encounter with the presence of the world is the ground of our being and knowing [48]. It assumes the creative shaping of a world through the transaction of imaging it, perceptually and in other ways. Experiential knowing thus articulates reality through inner resonance with what there is, and through perceptually enacting its forms of appearing. It further asserts that to experience, anything is to participate in it and to participate in both to mold and to encounter, hence experiential reality is always subjective/objective. Document review included reviewing the literature on systems thinking and rural development facilitation.

7. Results and discussions: developing a systemic programming framework for community resilience

This paper adopts theoretical constructs from soft systems methodology to develop a community resilience programming framework. The proposed framework is illustrated in **Figure 3**. It incorporates principles from SSM and pillars from action research based on experiential knowledge by the author. The framework adopts a project cycle-based typology with two interrelated cycles (**Figure 3**). The outer cycle highlights the key components based on SSM while the inner cycle proposes key pillars for effectiveness of the programming framework. It should be highlighted that these cycles should be integrated and implemented simultaneously. The proposed components of the framework are detailed in the proceeding sections.

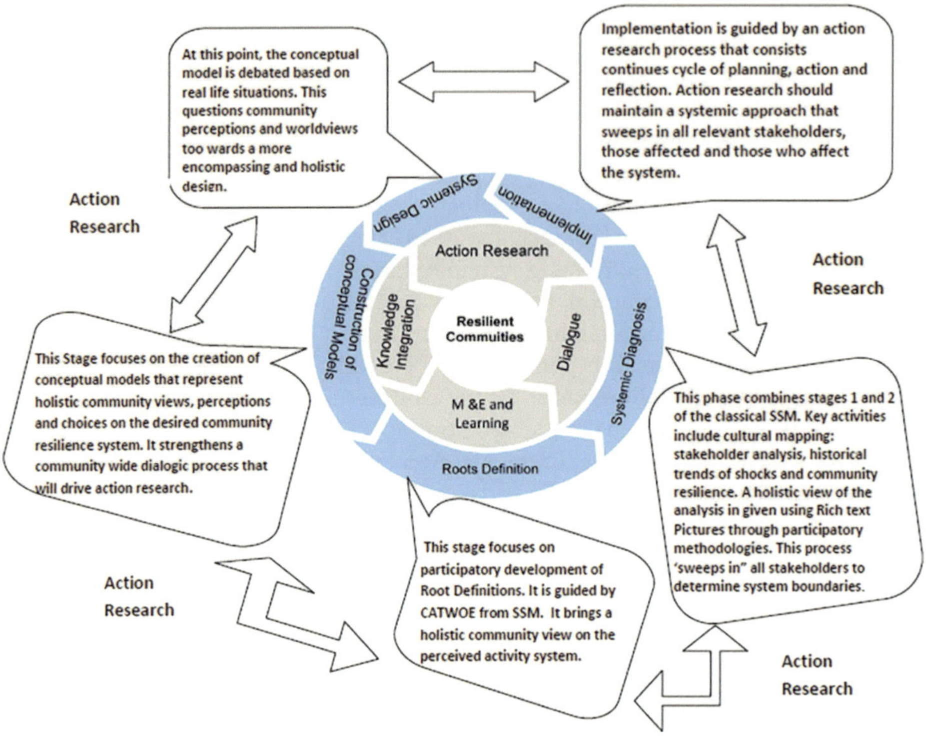

Figure 3.
The proposed systemic programming framework.

7.1 Systemic diagnosis

This phase focuses on a holistic diagnosis of key resilience issues within the community. It is holistic in the sense of 'sweeping in' representatives of all stakeholders and community groupings as well as understanding interactions and synergies among various actors. Unlike most fragmented projects that select specified beneficiaries, system boundaries are stretched to incorporate different worldviews. A number of variables are mapped in the process including cultural history, stakeholders and their roles, community perspectives on resilience, historical trends of major shocks and coping strategies, and the role of social capital, among others. The process aims at capturing a wide range of perspectives and choices. It brings the real world and unstructured community perceptions on adaptive, absorptive, and transformative capacities. Experiences from action research highlight a few critical factors that might negatively affect such a holistic mapping process.

Firstly, most communities in Zimbabwe are polarised due to political defenses and past interactions with non-governmental organisations (NGOs) and other funding partners. Experiences by the author in Chimanimani District for example identified biases in beneficiary selection on the basis of political affiliation. On the other hand, participation in the different programs creates 'camps' between those selected and those left out. Such divisions often affect knowledge sharing and participation in local development activities. Other factors include local leadership wrangles and power dynamics as well as perceptions that participation in NGO activities need to be rewarded. Some cultural beliefs may affect participation by women. Dwindling confidence in NGO work in some sections of rural communities due to previously unsustainable projects may affect commitment by some community members.

The diagnosis process is conducted using rich text pictures (RTPs). An example of a TRP is given in **Figure 4**. The essence is to provide different perceptions, world-views, and proposed actions on resilience within the community. Such diagrams could ideally be done at the village level and later consolidated at the Ward level through dialogue and participatory techniques for a more holistic view. The present is an unstructured description of issues on shocks and resilience within the community.

7.2 Developing root definitions

Root definitions are statements that present an ideal system with regards to relevant stakeholders, community actors, the perceived transformation process, world views, and endogenous and exogenous environmental factors. It should clearly highlight the required community-wide transformation process required to build resilience. It should be guided by inputs from the systemic diagnosis process in the preceding section. Root definitions should incorporate the three capacities of resilience (adaptive, absorptive, and transformative). An example of a root definition could be: "*A community led resilience building process for improved livelihoods*". Experiences from community planning and reflection exercises during action research suggest that such a process should start at the village level. A ward-level approach would complicate the visioning process since villages have; different resource endowments, different visions, and worldviews due to different leadership styles by the traditional leadership and have engaged in different development programs due to heterogeneous engagement by various NGOs leading to different levels of appreciating rural development approaches. Experience has also shown that a lot of communities are "over researched" and have gone through numerous planning activities hence it is important to take cognisance of their existing plans. Clear root definitions should be set incorporating community actors, key stakeholders, possible endogenous and exogenous factors that will impact the transformation process.

7.3 Construction of conceptual models

These models illustrate the relationships among the various elements defined in the root definitions. They define activities that the system must implement to achieve the proposed transformation. It is important to note that conceptual models define the process and not the methods applied. These models must be as holistic as possible to 'sweep in' all relevant stakeholder views. They should show interactions and synergies (both positive and negative) of various elements. Communities in most rural areas in Zimbabwe have gone through different shocks/hazards and have developed their own coping mechanisms. The construction of conceptual models should, as much as possible, elicit tacit knowledge from these situations.

Conceptual models are normally represented in the form of bubble diagrams with activities enclosed in bubbles. The bubbles link each other through arrows that depict dependencies (positive and negative). They should demonstrate an ongoing purpose for improving community resilience, a means of assessing performance, decision-making processes, components that are sub-systems, an environment, continuity, and required resources. **Figure 5** shows examples of conceptual models.

The development of conceptual models should also be done at the village level. Experience from Action research indicates that such planning processes need to be

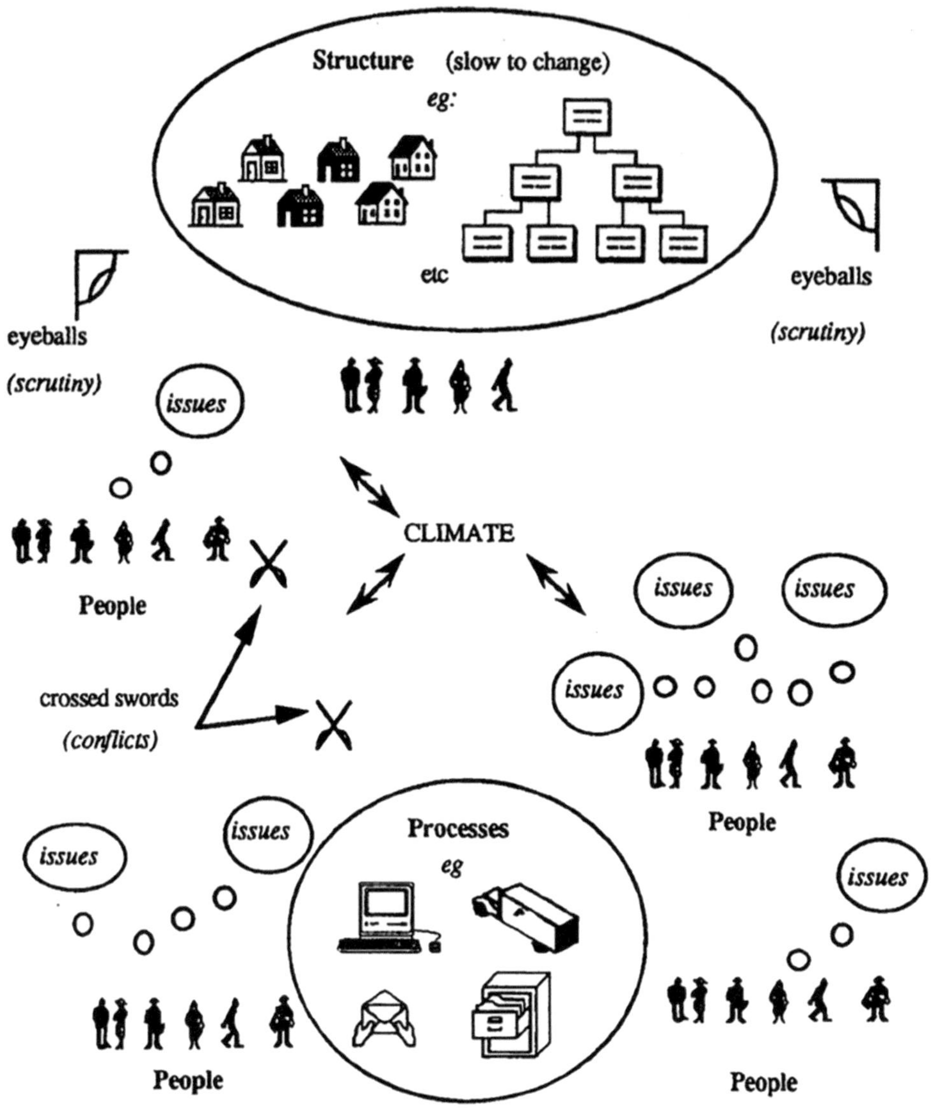

Figure 4.
Example of a rich text picture. Adapted from Patching [49].

conducted properly to manage reticence by some community members. Facilitators need to understand community dynamics that affect effective participation. Women for example may not participate effectively around men. Facilitators must be creative and adopt participatory methodologies that elicit the views of all stakeholders involved in the process. There might be need, for example, to form development groups, have separate inputs, and then conduct a village-wide process of consolidating the conceptual models.

7.4 Systemic design

This is an iterative process that focuses on comparing the conceptual models with the real-life situation from the unstructured problem presented in the rich text pictures. The conceptual and the real situations are compared to come up with realistic interventions. This process should ideally be facilitated at the village level with the participation of all key stakeholders through community-wide dialogue. This process can be replicated in other countries with similar socio-cultural conditions, particularly southern African countries such as Malawi, Mozambique, and Zambia. From action research experience, such a process can be challenging as village-wide dialogue has

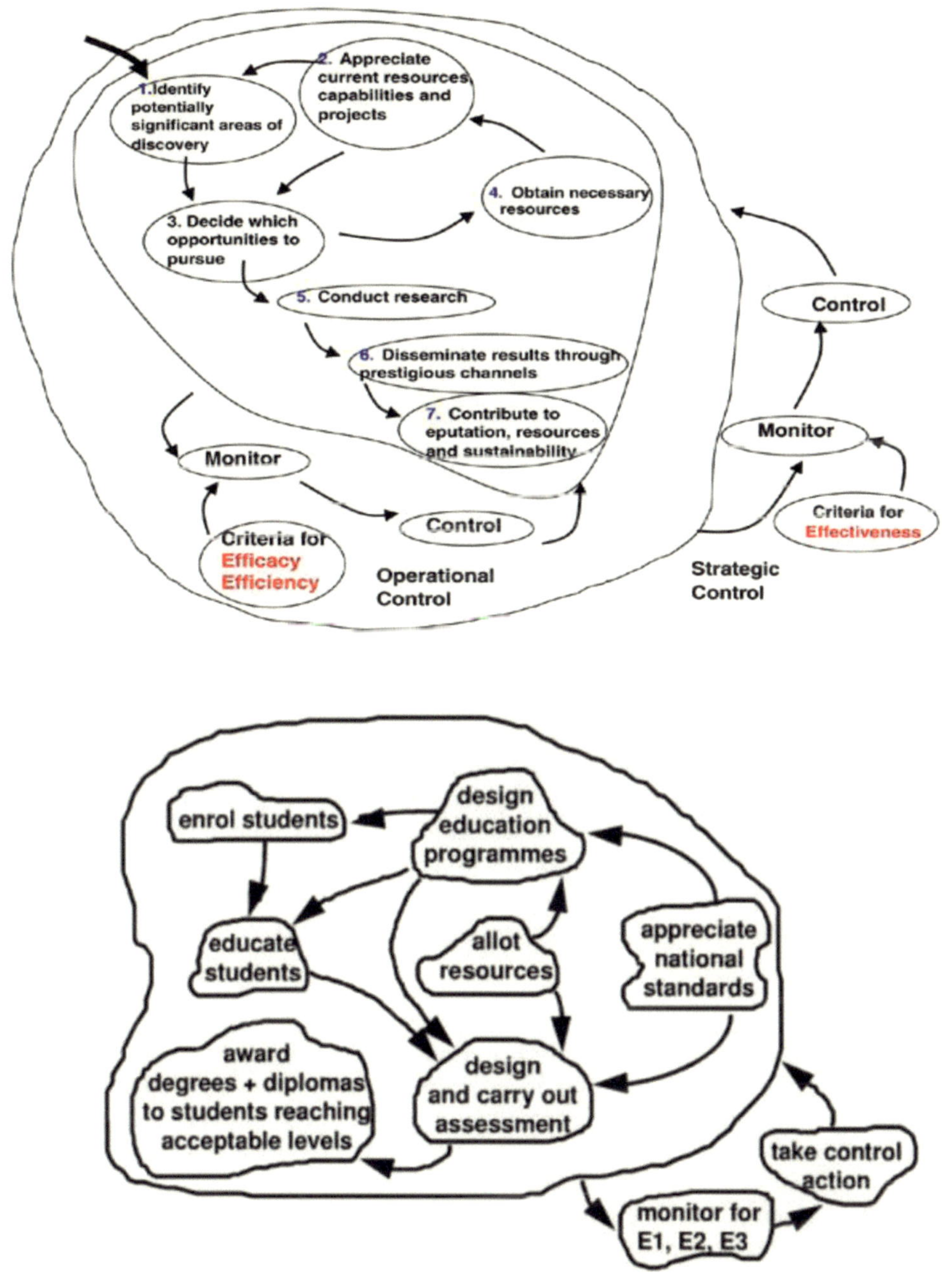

Figure 5.
Example of a conceptual model (http://users.actrix.co.nz/bobwill/ssm.pdf).

a number of challenging factors including breaking the 'conspiracy of silence' where community members won't share knowledge and information as some perceive their tacit knowledge as inferior to technical expertise knowledge; gender and cultural dynamics where in some cases, certain community members' contributions are regarded as inferior; de-politicking community dialogue; and detangling NGO specific forums. The latter normally stems from NGOs competing for space and developing their own forums. The comparison of the conceptual model to the rich text pictures should be done activity by activity in a tabular format. The table will have several columns. These columns might include the following headings: activity; status (indicating if the activity is already being implemented); current challenges and coping mechanisms; measures of performance; recommendations; and comments. The facilitators should draw as much as possible from participatory planning tools such as community based planning (CBP).

7.5 Implementation of sustainable development

The implementation process should be action research-oriented informed by contemporary extension approaches. According to Özçatalbaş [50], such an extension approach should utilize proven, accurate information based on research findings to improve welfare. This process will allow the delivery of information and knowledge to target groups for socioeconomic development. Rather than a pure development practitioner/community member dichotomy, it should be based on a knowledge co-creation agenda. Such knowledge transfer should ideally be based on technology transfer within the current context of the fourth industrial revolution (4IR). Zimbabwean community members have been through numerous shocks; including droughts, and economic transitions among others. Extension and technology transfer through action research ensure that the implementation of the designed social activity systems engages a learning agenda. Experience shows that engaging Zimbabwean communities in action research have a key challenge in facilitating learning. There is a general belief by communities that rural development practitioners are more knowledgeable that communities and communities should learn from them. This poses a challenge for facilitators in changing this mindset to allow a knowledge-sharing agenda. In such instances, a tool such as appreciative inquiry, for example, was found to be effective in Chimanimani District during action research. The approach deviates from the traditional assumptions that community systems have inherent flaws that need to be fixed through problem solving and interventions. Rather than treating communities as problems, it focuses was placed on identifying positive capacities within communities which are the facilitators utilise to drive dialogue.

7.6 The strategic pillars

The proposed programming frame has strategic pillars that enhance its effectiveness (monitoring evaluation and learning, action research, knowledge integration, and dialogue). As outlined in the preceding sections, these should not be treated as a stand but should be integrated and interweaved with the entire SSM cycle (**Figure 3**). Implementation of the strategy should adopt monitoring and evaluation mechanisms that enhance community empowerment and learning. Experience in Chimanimani District through action research shows that tools such as community based monitoring and evaluation (CBME) and community score cards, where communities are given skills enhance knowledge sharing, improve project performance and accountability of local institutions and stakeholders. As described in preceding sections, action research, knowledge integration, and dialogue ought to be integral components of the proposed systemic programming framework.

8. Conclusion

This paper provides a framework for systemic programming for community resilience and sustainable community development in Zimbabwe. It is not a step-by-step programming manual but rather provides key tenets for researchers and development practitioners. The building blocks proposed are neither exhaustive nor prescriptive. Rural development is highly contextual, *thus*, tools and approaches to implementing the framework are not specified. This calls for creativity in the empirical testing of the proposed framework. Two research questions are addressed; what are the critical components of a systemic programming framework for livelihoods and resilience? And, how is such a framework facilitated in practice? The components of the proposed systemic programming framework are systemic diagnosis, roots definition, and construction of conceptual models, systemic design, and implementation. These are supported by four strategic pillars; monitoring, evaluation, and learning; knowledge integration; action research, and dialogue. It is recommended that the systemic framework be tested empirically, and its components become part of resilience programming in Zimbabwe.

Author details

Precious Tirivanhu
Human Sciences Research Council (HSRC), Pretoria, South Africa

*Address all correspondence to: ptirivanhu@gmail.com

References

[1] Musiyiwa K, Leal-Fihlo W, Harris D, Nyamangara J. Implications of climate variability and change for smallholder crop production in different areas of Zimbabwe. Research Journal of Environmental and Earth Sciences. 2014;**6**(8):394-401

[2] Brown D, Chanakira RR, Chatiza K, Dhliwayo M, Dodman D, Masiiwa M, et al. Climate change impacts, vulnerability and adaptation in Zimbabwe. In: IIED Climate Change Working Paper No. 3. London: International Institute for Environment and Development (IIED); 2012

[3] Gwimbi P. Cotton farmers' vulnerability to climate change in Gokwe District (Zimbabwe): Impact and influencing factors. Journal of Disaster Risk Studies. 2009;**2**(1):81-92

[4] Mutekwa VT. Climate change impacts and adaptation in the agricultural sector: The case of smallholder farmers in Zimbabwe. Journal of Sustainable Development in Africa. 2009;**11**(2):237-256

[5] Nhemachena C, Hassan R. Micro-level analysis of farmers' adaptation to climate change in Southern Africa. In: IFPRI Discussion Paper 00714. Washington, DC, USA: International Food Policy Research Institute; 2007

[6] Kinuthia JH. Global warming and climate impacts in Southern Africa: How might things change? Internet Journal of African Studies. 1997;**2**

[7] Buckland RW 1997. Implications of climate variability for food security in the Southern African development Community (SADC) from the lessons learned from the 1991-92 Southern African Drought. In: Glantz, M. editor. Usable Science: Food Security, Early Warning and El NiCΓ́o, Workshop Proceedings. Budapest, Hungary; 25-28 October 1993:185-194

[8] Matarira CH, Makadho JM, Mwamuka FC. Zimbabwe: Climate change impacts on maize production and adaptive measures for the agricultural sector. In: Ramos-Mane C, Benioff R, editors. US Country Studies Program. Interim Report on Climate Change Country Studies. Washington, DC: US Department of Energy; 1995

[9] Downing TE. Climate Change and Vulnerable Places: Global Food Security and Country Studies in Zimbabwe, Kenya, Senegal and Chile. Oxford, UK: Environmental Change Unit, Oxford University; 1992

[10] Heltberg R, Siegel PB, Jorgensen SL. Addressing human vulnerability to climate change: Toward a 'no-regrets' approach. Global Environmental Change. 2009;**19**:89-99

[11] Ashley C, Maxwell S. Rethinking rural development. Development Policy Review. 2001;**19**(4):395-425

[12] Delgado CL. Africa's changing agricultural development strategies: Past and present paradigms as a guide to the future. Brown Journal of World Affairs. 1998;**5**(1):175-214

[13] Ellis F, Briggs S. Evolving themes in rural development 1950s–2000s. Development Policy Review. 2001;**9**(4):437-448

[14] Westoby P, Dowling G. Theory and Practice of Dialogical Community Development: International Perspectives. London and New York: Routledge; 2013

[15] Binswanger-Mkhize HP, Aiyar SSA, De Regt JP, Serrano-Berthet R, Hellinh L, Van Domelen J, et al. Historical roots and evolution of community driven development. In: Binswanger-Mkhize HP, De Regt JP, Spector S, editors. Local and Community Driven Development: Moving to Scale in Theory and Practice. Washington, DC: The World Bank; 2010

[16] Arcand J, Bassole L. Does Community Driven Development Work? Evidence from Senegal. Washington, DC: World Bank; 2007

[17] Manzo K. Africa in the rise of rights-based development. Geoforum. 2003;**34**:437-456

[18] Cornwall A, Nyamu-Musembi C. Putting the 'rights-based approach' to development into perspective. Third World Quarterly. 2004;**25**(8):1415-1437

[19] Boerefijn, I, Brouwer, M and Fakhred-Dine R, 2001. Linking and learning in the field of economic, social and cultural rights. SIM Special No. 27. The Hague: The Netherlands Institute of Human Rights

[20] Hurst R. 'Rights Now!', in Activate. Derby: British Council of Disabled People of Great Britain; 2001

[21] UNHCR. Human Rights in Development: Rights Based Approaches. Geneva: UNHCR; 2001 Available from: www.unhchr.ch/development/approaches.html [Accessed: 15 June 2016]

[22] Mathie A, Cunningham G. From clients to citizens: Asset-based community development as a strategy for community-driven development. Development in Practice. 2003;**13**(5): 474-486

[23] Scoones I. Sustainable rural livelihoods: A framework for analysis. Working Paper 72. Institute of Development Studies; 1998

[24] Nhemachena C, Chakwizira J, Dube S, Maponya G, Rashopola R, Mayindi D. Integrating indigenous knowledge systems (IKS) in improving accessibility and mobility in support of the comprehensive rural development program in South Africa. Memio; 2011

[25] Tirivanhu P, Matondi PB, Groenewald I. Comprehensive community initiative: Evaluation of a transformation system in Mhakwe community in Zimbabwe. Development Southern Africa. 2015;**32**(6):785-800

[26] Béné C, Headey D, Haddad L, von Grebmer K. Is resilience a useful concept in the context of food security and nutrition programmes? Some conceptual and practical considerations. Food Security. 2016;**8**:123-138

[27] Mitchel T, Harris K. Resilience: A Risk Management Approach. London: Overseas Development Institute; 2012

[28] Bahadur A, Doczi J. Unlocking Resilience through Autonomous Innovation. London: Overseas Development Institute; 2016

[29] Frankenberger TR, Constas MA, Nelson S, Starr L. Non governmental organisations approaches to resilience programming. In: 2020 Conference Brief 7. Washington, DC: International Food Policy Research Institute; 2014

[30] Brown K. Rethinking progress in a warming world: Interrogating Climate Resilience Development. In: `Rethinking Development in an Age of Scarcity and Uncertainty' EADI/DSA Conference; September 2011; York. 2011. Submitted paper.

[31] Harris K. Resilience in practice: Operationalising the ten characteristics

of resilience through the case of greening Darfur. In: Strengthening Climate Resilience Discussion Paper 10. Brighton: Institute of Development Studies; 2011

[32] UNDP. Building Resileince in Zimbabwe: Towards Resilience Strategic Framework. Harare: UNDP; 2015

[33] Berkes F, Colding J, Folke C. Navigating Social–Ecological Systems: Building Resilience for Complexity and Change. Cambridge: Cambridge University Press; 2003

[34] Brooks N. Vulnerability, risk and adaptation: A conceptual framework. Working Paper 38. Norwich: Tyndall Centre for Climate Change Research, University of East Anglia; 2003

[35] Smit B, Wandel J. Adaptation, adaptive capacity and vulnerability. Global Environmental Change. 2006;**16**:282-292

[36] Dilley M, Boudreau TE. Coming to terms with vulnerability: A critique of the food security definition. Food Policy. 2001;**26**:229-247

[37] Laslo L, Krippner S. Systems theories: Their origins, foundations, and development. In: Jordan JS, editor. Systems Theories and A Priori Aspects of Perception. Amsterdam: Elsevier; 1998

[38] Cabrera DA. Systems thinking [PhD dissertation]. Faculty of the Graduate School of Cornell University in Partial Fulfilment of the Requirements; 2006

[39] Trochim WM, Cabrera DA, Milstein B, Gallagher RS, Leischow SJ. Practical challenges of systems thinking and modelling in public health. American Journal of Public Health. 2006;**96**(3):538-546

[40] Waldman JD. Thinking systems need systems thinking. Systems Research and Behavioural Science. 2007;**24**:271-284

[41] Mingers J, White L. A review of the recent contribution of systems thinking to operational research and management science. European Journal of Operational Research. 2010;**207**:1147-1161

[42] Rubenstein-Montano B, Liebowitz I, Buchwalter J, McCaw D, Newman B, Rebeck K. 2001. A systems thinking framework for knowledge management. In: Sandrine S. Systemic evaluation methodology: The emergence of social learning from environmental ICT prototypes. Systemic Practice and Action Research. 2004;**17**(5):471-496

[43] Floyd J. Towards an integral renewal of systems methodology for futures studies. Futures. 2008;**40**:138-149

[44] Checkland P. Systems Thinking, Systems Practice. West Sussex: John Wiley & Sons; 2009

[45] Checkland P, Scholes J. Soft Systems Methodology in Action. Chichester: West Sussex; 1990

[46] Mingers J. An idea ahead of its time: The history and development of soft systems methodology. Systemic Practice and Action Research. 2000;**13**(6):733-755

[47] Wang W, Liu W, Mingers J. A systemic method for stakeholder identification using soft systems methodology. Working Paper No. 304. University of Kent; 2014. Available from: file://mcsfs02/users$/2009004223/Desktop/CATWOE%20Analysis.pdf

[48] Heron J, Reason P. A participatory inquiry paradigm. Qualitative Inquiry. 1997;**3**(3):274-294

[49] Patching D. Practical Soft System Analysis. Essex: Prentice Hall; 1990

[50] Özçatalbaş O. Technology transfer and change management. In: Leal Filho W, Azul A, Brandli L, Özuyar P, Wall T, editors. Zero Hunger. Encyclopedia of the UN Sustainable Development Goals. Cham: Springer; 2020. DOI: 10.1007/978-3-319-69626-3_53-1

Chapter 6

Do the Collaboration Dimensions Pay in Manufacturing Reverse Supply Chain? An Empirical Approach

Ifije Ohiomah, Clinton Aigbavboa and Nita Sukdeo

Abstract

The purpose of this paper is to examine empirically the enablers and practices of collaboration in relation to reverse supply chain. The research method used in this research was a quantitative method using a survey approach to empirically test if the following collaboration enables and practices are applicable. The statistical approach was AMOS 26. The findings revealed that, the relationship building and management for implementing collaboration was ranked highest, resource investment and development in reverse supply chain was ranked the next. Furthermore, quick response on returned goods and information sharing with suppliers on the returned products were highest ranked. The research was limited because the study was based in the Gauteng region, which means that a generalised statement cannot be made of the finding, as well there is a need for the study to be industry specific such as electronics, online retailers. The practical implications of the findings are that the enablers and practices are needed for reverse supply practices to achieve its aims. There is lack of research in the reverse collaboration space, this has paper has fulfilled the following gap.

Keywords: collaboration, reverse supply chain, manufacturing, sustainable practices

1. Introduction

In the recent decades, environmental considerations, cost reduction, and consumer pressure have become significant concerns worldwide [1] with the intensifying calls for environmental concerns, namely depletion of resources, landfills exhaustion in many countries, and several legislative measures by governments to take back the end of life products, issues like reverse logistics, product recovery, remanufacturing, and reusing have come across as significant areas for development [2]. For this paper, the discussion is around reverse supply chain (RSC).

Rural development is a very key driver in the bid to achieve sustainable development. Development of economic activities in rural areas plays a significant role not just in South Africa, but every country. This cannot be achieved without the preservation of the environment. Sustainable rural development is a mixed developmental concept which is created by an integral merging of sustainable and rural developments and represents a particular combination of their basic elements. In this study the basic elements is reverse supply chain, as reverse supply chain will drive the rural development in South Africa as the concept of sustainable rural development includes social and economic dimensions in the South African context. The economic dimension of rural development refers to the economic growth through the achievement of vitality, efficiency and effectiveness of the economic activity in the rural areas. The social dimension of rural development refers to the social progress through the improvement of the human potential and the creation of equal opportunities for a living in rural areas.

The reverse SC is known as reverse logistics (RL) in most of the literature; however, reverse SC and RL are used interchangeably in this paper. However, logistics is central to the supply chain. The reverse SC is an emerging business practice that supports sustainable production and consumption. Further, the importance of RSC has gained prominence in recent years, as there are high returns by customers due to the expansion of product choices and shorter product life cycles [3].

Reverse Supply Chain (RSC) is the collaborative responsibility of both the producers and consumers to reduce the waste by recycling, remanufacture, reusing, and adequately to dispose unacceptable products or items to enhance environmental sustainability [4, 5]. Globalisation has driven companies to become more flexible and productive by rapidly developing new products and reducing delivery times to meet customers' needs resulting in supply chains collaborating entirely with supply chain partners to reach short delivery times and shorten product release time. *Collaboration* is a term primarily used by the forward supply chain organisation. Collaboration works in conjunction with different parties to complete tasks and accomplishes mutual goals. It is a mechanism by which several individuals or organisations work collectively. It includes just over the unification of mutual goals, as in cooperative organisations, and a mutual willingness to fulfil a common aim [6].

One of the most often discussed terms in supply chain management has been cooperation in the forward supply chain. Over the past decades, organisations have seen the need to look beyond their organisation for opportunities to work with partners to ensure that the supply chain is efficient and responsive to dynamic market needs [6]. There is little or no research in collaboration in the reverse supply chain since most collaborative studies have been on the forward supply chain.

Supply chain cooperation is a powerful tool for achieving productive and receptive supply chain management (SCM) [7, 8]. Therefore, extended to the reverse supply chain since the literature revealed that the reverse supply chain is inefficient and unpredictable. Reverse supply chain collaboration is a relationship between partners in the reverse supply chain that aim to share information among themselves to jointly improve the performance of the reverse supply chain and improve the profit margin by redesigning business practices [9].

Supply chain collaboration (SCC) has different definitions. This study considers several definitions by authors such as [8, 10, 11]. The definition from Simatupang and Sridharan [12] defined SCC as "two or more independent companies work jointly to plan and execute supply chain operations with greater success than when acting isolation." SCC can take on two different relationships: vertical collaboration, the relationship between a manufacturer and a client, or horizontal collaboration, which is the relationship between companies at the same echelon of a supply chain, including between rivals [13].

This study adopts the approach taken by Barratt [14] and Lafferty and van Fossen [15] to classify Supply Chain Collaboration since it does not cover the same dyadic associations among supply chain partners that fit this aim study. This study concentrates on the organisation's vertical collaboration with its partners, i.e., suppliers, organisations, and customers.

Collaboration in supply chain if well thought out, could be central to successful business operations [16] and brings about competitive advantage [17]. It has become a buzzword in the business sphere in recent times [18]. Nonetheless, a survey was conducted, which revealed that 35 per cent of collaborative initiatives were moderately successful [19]. In practice, however, studies have shown that most attempts to introduce cooperation do not meet the participant's standards and end in failure [20]. Literature has shown that a collaboration record of accomplishment of execution finds all management strategies in the supply chain unsatisfactory [21]. In the light of this awareness, collaboration must be effective in the RSC.

Successfully implementing collaboration in the RSC and recognising the causes of collaboration failure is the product of a limited understanding of the elements required to implement the company's collaborative initiatives [22, 23]. Ho et al. [24] developed steps and processes (as shown below in **Table 1**) necessary to implement and strengthen collaboration for the RSC. These steps and processes are necessary to implement vital collaborative activities effectively in the reverse supply chain. Organisations need to collaborate moving forward as a necessary activity [25, 26], as well as including the prerequisite activities as a backdrop to building the capacity and competence to collaborate in main activities and to help and enhance activities [27] (**Table 1**).

Hence for reverse supply chain to achieve its full potential of been included in the reverse supply chain process as seen in the forward supply chain. It is important that, collaboration is the centre of the reverse supply chain process. Furthermore, studies have shown that there is little, or no research carried out in the reverse supply chain literature, this poses a significant issue in the reverse supply chain domain as there is little or no research to back findings in collaboration as it pertains to reverse supply chain. Lastly, to the best of the researcher knowledge, this is the first of a kind where collaboration dimensions will be empirically analysed in the reverse supply chain domain.

1.1 Benefits of collaboration in reverse supply chain

1.1.1 Improving performance operationally

SCC improves the performance of the companies [28, 29]. Through working closely, organisations boost the outcomes of working with supply chain partners [30].

Antecedents	Collaboration dimension
Managerial support	Information sharing
Internal alignment	Resource sharing
Resource investment and development	Decision synchronisation
Relationship building and management	Goal congruence
Free information flow & system integration	Incentive alignment
Formalisation	Collaborative communication
Rationalisation	Joint knowledge creation

Table 1.
Antecedents and collaboration dimensions.

The benefits of working closely with supply chain stakeholders consist of a higher responsiveness level and changes in service levels from their joint supply chain projects [22, 28, 29]. The reverse supply chain's performance came into question because of the uncertainty in the return of goods, the volume of return goods, and other instances.

1.1.2 Increasing service quality

Organisational performance is dependent on exact and timely SC information [8, 23]. Organisations expect a better degree of service level development from the supply chain partnership initiatives. An added benefit of supply chain cooperation is that it contributes to supply chain cost reductions often associated with intercompany transactions, output and inventory [29].

1.1.3 Improving logistical performance

Several investigations have shown that an advanced level of collaboration in the supply chain can improve a firm's performance [31], in logistics activities [32], which could lead to the future to more collaborative actions because of the success of collaboration [33].

1.1.4 Mitigating risks

In addition, collaboration in the supply chains could also reduce gaming and rationing. It is one of the bullwhip effect's primary triggers [34]. Additional benefits may well be a higher level of cooperation, which can be attributed to removing the bullwhip effect, reducing inventory levels, the efficient use of transport capacity, and risk mitigation [22].

Collaborative strategies such as information sharing, reward coordination, and decision synchronisation explored in several dimensions [35–37]. Simatupang and Sridharan [17] reported that partnership dimensions could be prioritised using several measures, such as mutual objectives, information sharing and reward alignment, among many. When organisations work together, they manage their inventory and ordering policies effectively [22]. In a bid to achieve a solution where everyone benefits, collaboration can be applied to pricing strategies [38]. Reducing supply chain costs through a decentralised supply chain approach is one of many reasons why businesses are partnering with their suppliers.

1.1.5 Information sharing

The sharing of information is one of the leading collaborative activities; it is described as necessary [39], foundation [40] and a fundamental prerequisite [23]. Any relationship must have a flow of information [39]. Information sharing decides the direction and extent of the flow of material from product returns and end-user repairs. Therefore, the flow of information involves transactional data exchange and customer feedback on product research and development problems, which is critical for the movement of information flowing from customers to manufacturers through the suppliers.

The sharing of information, as said by Hudnurkar et al. [40], is the glue that binds the relationships between partners, allowing the RSC to be more reactive in addressing competitive advantage issues. Moreover, as supply chain partners are increasingly growing and practising across various parts of the globe, organisations must share accurate and factual time information among partner organisations to achieve common goals. Additionally, this will lead to proper handling among supply chain partners of returned goods.

Crook et al. [41] advocated that small firms must collaborate and share information. They can attain advantages from what accomplished in exchanging information when adopting an adversarial relationship. Nonetheless, companies are worried about sharing information that is so important within their supply chain, as there is worry about the leakage of important information such as demand projections, adoption of emerging technology and new product innovations, and returning to their rivals for goods. Consequently, the interaction between the collaborative RSC should help lessen the likelihood of information leakage. Trust between SC associates plays a vital role in sharing information via the supply chain [42].

1.1.6 Goal congruence

The combining of supply chains with individual preferences has recently been considered [43, 44]. Goal congruence is the level at which SC participants are satisfied with supply chain targets [40, 45]. The goal congruence consists of the definition of the roles and responsibilities of an individual partner, the establishment of goals, specific targets, performance measures, IT standardisation, mutual knowledge formalisation and the joint implementation of the strategy [26].

True partnership is defined as one of the target congruence responsibilities [46] and requires an understanding of the needs and competencies of each member of the RSC to ensure to focus the efforts of individuals working in the supply chain. As a result of specific strategic goals. Moreover, the value of the organisation's strategic direction and vision raised concerns, as reported by Lambert et al. [47]. Lambert et al. [47] suggested that RSC members buy into the RSC vision and critical business processes. Collaborative relationships should be a key focus for achieving changes and incentives through industries [48]. Lastly, the desires and requirements of the RSC members must consider the RSC strategies and operations to best benefit each member, cash flow and return on investment [49].

1.1.7 Decision synchronisation

Decision synchronisation was conceptualised by Simatupang and Sridharan [12], where the researchers described it as "the extent to which the chain members are able to coordinate critical decisions at planning and execution levels for optimising supply chain profitability. It occurs when the forward and reverse chain partners orchestrate supply chain decisions and combine RSC with operations that create better RSC benefits [12]. Planning decisions are crucial to deciding the most efficient and effective ways of managing the enterprise's resources to achieve the targets set. These decisions are as follows: strategy planning, demand management, production planning and scheduling, procurement promise delivery and distribution management [50]. Joint planning aligns collaborative partners and makes organisational decisions.

1.1.8 Incentive alignment

This collaborative concept refers to cost-sharing, risk, and the benefits of the supply chain among partners [51]. The alignment of incentives involves assessing the cost, risks, benefits, and designing incentives. Successful SCC requires each member's ability to share profits and losses equally, and the collaborative result must be quantifiable [52]. Incentive alignment entails a careful interpretation of gain-sharing arrangements, ensuring profits are proportionate to risk and investment [53].

1.1.9 Resource sharing

Resource planning is the utilisation of the supply chain members' abilities and assets and the supply chain members investing in the capabilities and assets. Physical resources such as manufacturing equipment, return facilities, and technologies are those assets in which participants of the reverse supply chain need to invest [54]. Activities such as vendor-managed inventory (VMI) allow suppliers to take stock-level data through electronic data interchange (EDI) and take the required top-up action in sectors such as retail [55].

1.1.10 Collaborative communication

This process is the process of communication and transmission of messages between SC partners about the duration, direction, mode, and strategy of control. Two-way communication is usually an indicator of close inter-organisational relationships [56, 57]. This research examines communication patterns from the mechanistic perspective of communication theory. They fit with the term 'collaborative communication strategy,' which refers to the main communication attributes, including frequency extent of bidirectional flows, informal modes and indirect content.

1.1.11 Joint knowledge creation

SC partners can develop improved market knowledge and response and the competitive environment through working together [58]. The two kinds of knowledge creation activities are knowledge exploration (searching and acquiring new and essential knowledge) and knowledge exploitation (integrating and applying relevant knowledge) [59]. Capturing, exchanging, and assimilating knowledge (e.g., process, technology, or market knowledge) among supply chain partners enables innovation and the supply chain to be competitive in the long term [54].

1.2 Reverse supply chain and sustainable development goals

The review of literature reveals that there has been a lack of study that comprehensively discussed issues in relation to reverse supply chain and cover the market, workplace, environment, and society. These areas are found within the 17 UN Sustainable Development Goal (SDG). The integration of RSCM practices into the SDG will enable entrepreneurs to develop an advanced and complex reverse supply chain management which could lead to a more efficient and ethical reverse supply chain. It is well to note that the goals of sustainable development (SD) have been designed to interact with organisations and stimulate economic effects [60]. While the SDGs do represent a different approach, their potential for transforming the dominant governance approaches to sustainability remains an open question. Thus, global collective action does not end when decisions are reached, but these decisions introduce new practices in a complex political process that can bring in new actors, new ideas, and new action for sustainability in rural developments [61]. It seems, therefore, that the role of RSCM may be decisive in the successful implementation of SDGs in rural development, if development goals are understood as a process in which all components interact with each other [62]. In a similar vein, [63] emphasise the key role of the links that co- create supply chains are embedded within the SDGs. In turn, Russell et al. [64] suggests that SDGs have undoubtedly been successful in broadening the awareness of entities co- creating supply chains, yet their implementation in the reverse supply chain may be problematic due to their very wide scope, hence the need for collaboration among all stakeholders to try to narrow the scope. It should therefore be emphasised that due to the complexity of SDGs, management decision makers may encounter many barriers and limitations at the stage of implementation of objectives in supply chains, this can be eliminated if there is adequate collaboration among the stakeholders in the reverse supply chain, one of the challenges faced is that everyone in the supply chain want to work independently forgetting that they are all interlinked. This situation is influenced by the fact that the development of an integrated supply chain management system aligned with the SDGs is a highly complicated undertaking and requires significant involvement, thus bringing about collaboration in this process of these to achieve the goals to attain the SDG goals.

2. Research design and methodology

According to Denzin and Lincoln [65], paradigms are a broad framework of perceptions, beliefs, and feelings with which theories and practice operate. For Guba and Lincoln [66] research philosophies are the set of feelings about how the world works (ontology) and how it should be understood (epistemology) and studied (methodology). Whereas ontology raises questions about the nature and form of reality to be known, epistemology raises questions about the nature of the Knower (researcher) relationship and what can be known (the problem under investigation). Finally, methodology refers to general principles that underline how we investigate the social world and demonstrate that the knowledge generated is valid [67–69].

According to Guba and Lincoln [66], positivism, post positivism, critical theory and constructivism or interpretivism are the four schools of thoughts that underline the significant paradigms that structure social science research. For this paper, a positivism paradigm was selected, this approach was selected based on the following points. Quantitative research was adopted for this study, the criterion for selecting is as follow, to test the following collaboration strategies, more extensive and randomly selected respondents, numbers and statistics, Single reality; Objectivity is critical (precise measurements using validated data-collection instruments), the researcher cannot influence the participants. The characteristics of the participants are hidden intentionally from the researcher. The scientific method is confirmatory. A survey research approach was selected to use for this study; this was selected for the following reason: The survey approach is associated with the research using positivist quantitative methodologies [70]. Since a large amount of data is being collected using the survey approach, the findings are generalised to the entire population. The study examines collaboration strategies and confirms them to be used for a confirmatory analysis within the South African manufacturing industry, which involves collecting data from many participants, especially when using structural equation modelling (SEM) technique in data analysis, employing another research approach will be costly and time-consuming [71]. The data was collected using the google forms survey; the following reasons were behind the use such as data can be collected from many participants simultaneously in a quick, easy, efficient and economical way compared with other methods such as interviews [72–74], It is designed and administrated.

For example, interviews usually require much administrative skills [69], Higher privacy of respondents because issues such as anonymity and confidentiality were dealt with in the cover letter, collecting the questionnaires at once after being completed will assure a higher response rate [75].

2.1 Research design

According to Nachmias and Nachmias [76], research design supplies overall guidance and framework for the data collection and analysis of the study. It is critical to link the theory and the empirical data collected to answer the research questions [77]. A choice of a suitable research design will influence the use and type of data collection, sampling techniques, and the budget [71]. Additionally, when designing a study, the researcher should make a sequence of rational decisions about the purpose of the study, location of the study, the investigation type, role of the researcher, time horizon and the level of data analysis [75]. Before going ahead to the data collection process, the sampling technique is considered a critical concern to the research to be the targeted population and eliminate the bias in the data collection methods and thus generalise the results [76, 77]. According to Tharinger et al. [78], there are four critical issues to be considered when designing the sample as follows; (1) the choice of probability or non-probability sample technique; (2) the sample frame; (3) the size of the sample; (4) the response rate. The choice of probability or non-probability sample technique selected for this study was the non-probability sampling technique, the convenience sampling method was selected for this study, as the convenience sampling methods allow the researcher to select the sample subjects from the targeted population based on who is willing and easily accessible to be recruited and included in the research. This method is the least expensive, least time-consuming among all other techniques. Convenience sampling method is the most used in behavioural and social science studies. The justification for the selection of convenience sampling is as follows; it was not feasible to access data to allow random sampling to take place, as well as time and budget constraints led to a decision to employ the non-random approach with the potential to significantly collect the sample sizes needed for the analysis. This research employed a convenience sampling technique in collecting data that assumes a homogeneous population, and thus, generalisation of results to the entire population should be made with caution. The research must specify the sample size within the targeted population. According to Zikmund [72], using a large sample within the study cannot guarantee precision and thus will waste time and money.

On the contrary, significantly, when statistical data analysis such as SEM is required, using a small size will lower the results accuracy [71]. Therefore, the sample size was determined based on the rules of thumb for using structural equation modelling within AMOS. According to Roscoe [79], the following rules of thumb should be considered when considering the sample size: Sample size >30 and < 500 are appropriate for most research; in multivariate research (e.g., SEM), the required sample size should exceed by several times (preferably ten times) the number of variables within the proposed framework or study. Similarly, Kline [80] suggested that a sample of 200 or larger are appropriate for a complicated path model. In contrast, a sample size varies between 50 and 1000, of which 50 as very poor and 1000 as excellent.

3. Findings and discussion

Table 2 reveals the descriptive analysis. From the analysis, it was revealed what the respondents thought of enablers of collaboration. The respondents agreed that relationship building and management for implementing collaboration with a mean score of 4.05, resource investment and development in the reverse supply chain with a mean of 3.98, which reveals that the respondents agreed with the extent to which this construct enable collaboration. The respondents revealed with a mean of 3.97 that free information flow in the organisation enables collaboration, this means that the respondents agreed with the variables. The respondents revealed with a mean of 3.92 that Internal alignment of the organisation and partners to the implementation of collaboration, which means that the respondents agreed with the variable. Lastly, top management support for collaboration revealed that the respondents agreed with the variable with a mean of 3.79, this means that the respondents agreed with it.

Furthermore, a descriptive analysis for collaboration practices revealed that quick response on returned goods revealed that the statement was agreed with a mean of 4.06, rapid processing of order returns revealed that a mean of 4.01 which means that the respondents agreed with the statement. Information sharing with suppliers on the returned products revealed that that the statement was agreed by the respondents with a mean of 3.99. Joint knowledge creation among the stakeholders on reverse supply chain revealed that the respondents agreed with the statement with a mean of 3.98. lastly, the respondents agreed with the statement close relationship with customers who purchase the products revealed that the respondent agrees with the statement with a mean of 3.71.

Table 3 reveals the convergent validity and internal consistency of the collaboration construct. The threshold of average variance extracted (AVE) is above 0.5, as recommended by Fornell and Larcker, Hair et al. [81, 82]. The composite reliability (CR) threshold as recommended by Litwin [83] is 0.5, but Fornell and Larcker [81]

Collaboration enablers	Mean	Std Dev
Relationship building and management for implementing collaboration	4.05	1.088
Resource investment and development in the reverse supply chain	3.98	1.139
Free information flow in the organisation	3.97	1.142
Internal alignment of the organisation and partners to the implementation of collaboration	3.92	1.04
Top managerial support for collaboration	3.79	1.099
Collaboration practices		
Quick response on returned goods	4.06	1.072
Rapid processing of order returns	4.01	1.037
Information sharing with suppliers on the returned products	3.99	1.089
Joint knowledge creation among the stakeholders on reverse supply chain	3.94	1.132
Participation of suppliers in product return inventory control	3.94	1.091
Information sharing with customers on returned product	3.93	1.116
Close relationship with customers who purchase the products	3.71	1.131

Table 2.
Descriptive statistics for collaboration.

			Estimate	AVE	CR
COL1	←	COL	0.927	0.81	0.98
COL2	←	COL	0.913		
COL3	←	COL	0.886		
COL4	←	COL	0.908		
COL5	←	COL	0.908		
COL6	←	COL	0.907		
COL7	←	COL	0.867		
COL8	←	COL	0.901		
COL9	←	COL	0.931		
COL10	←	COL	0.903		
COL11	←	COL	0.916		
COL12	←	COL	0.845		
COL13	←	COL	0.894		

Table 3.
Convergent validity and internal consistency.

recommended 0.7. The AVE finding of this study is above 0.5, meeting the cut-off criteria. The CR for this construct is 0.96 because they show that all the indicator variables measure the same phenomenon [77].

Furthermore, the regression weights of the variables measuring collaboration showed that they were all significant, which means that they accurately measured collaboration.

Although several measures for deciding the fitness of a model exit. Hu and Bentler [84] suggested that use of the ML-based standardised root mean squared (SRMR)

Measure	Threshold	COL
X2	—	266.082
Df	—	65
p-value	—	0.000
X2/Df	< 3 is good; < 5 is acceptable	4.094
GFI	0 to 1 (0 = no fit; 1 – perfect fit)	0.966
CFI	> 0.95 to >0.80	0.965
NFI	0.60 to 1.00	0.955
RFI	0.90 to 1.00	0.937
TLI	> 0.95	0.952
RMSEA	0.05 to 0.10 acceptable; < 0.05 is	0.099
	good	
SRMR	< 0.08	0.07

Table 4.
Model of fit.

along with any supplemental fit index such as Tucker-Lewis's index (TLI), comparative fit index (CFI), Gamma Hat, McDonald's centrality index (Mc), or root mean squared error of approximation (RMSEA).

This fit index will help reduce the possibility of committing a Type I error (the probability of rejecting a null hypothesis when it should be accepted) or a Type II error (the probability of accepting a null hypothesis when it should be rejected [84]. The X^2 divided by the degree of freedom (Df) revealed a good fit of 3 to 5 as noted by Tharinger et al. [78], while the goodness-of-fit (GFI) must be from 0 to 1, as saw by Doloi et al. [85]. Similarly, the CFI met the minimum threshold of at least 0.80 set by Hu and Bentler [84] and a cut-off of 1 as set by Singh [86], while the normed fit index (NFI) can fall within the 0.6 to 1.0 threshold observed in Van Dijk, and Akkermans et al. [87, 88]. The chi-square was 4.0, which makes it a good fit. The GFI, CFI, NFI, RFI and TLI all met the cut-off criteria of >0.90 and > 0.95, respectively. The RSMEA gave a fit of 0.99, which is acceptable, and the SRMR is 0.07. The fit index shows that the following cut-off criteria for all collaboration variables are fit for acceptance into the final structural equation model (**Table 4**).

3.1 Discussion

This study develops a measurement model which was adapted from earlier studies [87]. Collaboration is a strategy for effective supply chain management [26]. But this has not be the case for reverse supply chain, the findings from this study are consistent with findings from the forward supply chain. Enablers of collaboration such as relationship building and management for implementing collaboration is consistent with findings whereby [88] stated that in order to enable collaboration, it requires high level of trust and information transparency which can only be achieved by building relationship of trust to enable trust and information strategy. Free information flow in the organisation also is one of the several enablers of collaboration, this finding is consistent with [14] who said that information exchange in addition to trust is very key to enabling collaboration. Further the findings of Top management play a significant role in enabling collaboration stated by Fawcett et al. [27], who stated that without top management support it is almost impossible to enable collaboration.

One of the challenges seen from literature, is the ability for returned goods to be processed quickly and with the help of efficient collaboration enablers there will be quick response on returned goods, as information sharing with the suppliers and customers will lead to quick response on returned goods. This is pertinent as information sharing will lead to a joint knowledge creation among the stakeholders in the reverse supply chain but also enhance a rapid processing of order returns. Furthermore, information sharing with suppliers and customers will bring about an enhancement in the decision making across the reverse supply chain, as well as establishing stronger partnerships and closer integration among the customer, supplier and manufacturer [89, 90]. Information sharing with suppliers and customers will lead to increased visibility, velocity and flexibility within the reverse supply chain. In this regard, the type of information being shared, the frequency, direction and mode of information sharing are particularly important for the growth of collaboration among reverse supply chain partners. Prajogo and Olhager [44] showed in their research of suppliers and manufacturers that information sharing improves logistics integration in inventory management, as these findings is somewhat related to the variable as there is a need for suppliers to participate in the inventory control of product return.

3.2 Implication of findings

From the practical point of view, several valuable managerial implications could provide valuable insights for organisations seeking to get involved with reverse supply chain not only in the distribution networks of reverse supply chain, but in the other types of the reverse supply chain networks. The research results proved that reverse supply chain partners practising collaboration should ensure to improve information sharing, decision synchronisation, incentive alignment, resource sharing, collaborative communication, joint knowledge creation and goal congruence to ensure the capability to achieve and support a prominent level of collaborative advantage for their supply chain. The following capabilities of collaboration is key to the development of sustainable rural development as information sharing is key as there has to be a.

Furthermore, using a structural modelling approach, the issue of collaboration in reverse supply chain was examined to gain an understanding of collaboration in reverse supply chain. Information sharing, decision synchronisation, incentive alignment, resource sharing, collaborative communication, joint knowledge creation and goal congruence. The study also suggests that reverse supply chain collaboration in the driving of rural development increases firm performance by enhancing inter-firm trust and commitment, which then reduce transaction costs in the reverse supply chains. This research offers a managerial insight for the reverse supply chain managers in terms of the various aspects of reverse supply chain partners toward the relationships of the partners in the rural areas of South Africa, with emphasis on the trust building mechanism, making long-term commitment more important for reverse supply chain partners. Another important finding of this study is the effect of collaboration to improve the operational performance of the reverse supply chain in rural areas, it is important that manufacturing organisations willing to improve operational performance of the reverse supply chain, should ensure that there is an achievement of the collaboration dimensions. There are different definitions and measures of collaborative advantages, which can help managers to improve shared reverse supply chain processes and achieve benefits for all members.

This study is consistent with the research by Van Dijk, and Cao and Zhang [21, 87], confirms that the use of such collaborative dimension offers flexibility, process efficiency, innovation and business synergy is the most efficient. Ignoring collaborative dimension may be one of the reasons why so many firms failed to develop effective collaboration in their supply chains. Obtaining collaborative advantages may help overcome the challenges and complexities in inter-firm collaboration that a variety of companies have faced. Collaboration is referred to as inter-organisational competitive advantage, which seeks to maximise a common profit for all reverse supply chain members. This synergetic effect of collaboration is what drives the organisational performance improvement. It arises due to collaboration efforts of the reverse supply chain partners, and it is obtained only through joint action and close collaboration. Thus, suggestions that, for a reverse supply chain to perform well, firms should try to create a win–win situation that all participants collaborate to achieve business cooperation and compete with other chain.

According to Cao and Zhang [21], competitive intentions make individual firms promote their own interests at the expenses of others, which is very insidious for collaboration and can worsen or destroy the relationships. Long-term relationships such as reverse supply chain collaboration have to be motivated by the mutuality of intent, goal congruence, and benefit sharing. Thus, managers need to align goals and benefits with reverse supply chain partners for creating collaborative advantage. Such collaborative advantage indeed directly increases the performance for each partner in the chain. As the empirical results of this study show, the main instrument of obtaining collaborative advantages is the dimensions of supply chain collaboration. Under the conditions of the growing uncertainty of business environment and increasing competition, decision synchronisation, incentive alignment and information sharing come at the forefront. Practicing these collaborative dimensions allow firms to improve process visibility and reduce the uncertainty level in decision-making. Furthermore, the benefits of collaboration practices could bring about a smooth implementation of reverse supply chain, thus making available the cores required for remanufacturing, recycling practices. It is noted that this practices are sustainable practices and will help in the attainment of sustainable development in the rural areas, which are often neglected. The aforementioned practices will bring help key into the triple bottom dimension of sustainability which are economic, social and environmental. It will boost rural development economically because it will bring about more exposure and in terms of revenue in the form of tax in the rural areas, socially, it would bring about job creation to the rural area, environmentally it will bring about lower landfill as an there will be increase in awareness of the population not just disposing of their used goods. In addition, SDGs have a strong link to practices implemented under RSCM and their integration into the forward supply chain management process can stimulate synergistic effects in the attainment of sustainable rural development. Managers need guidance on implementing SDGs in the supply chain. The integration of the SDGs and RSCM provides new areas of research and reflection. The implementation of SDGs in a supply chain require a new level of commitment from all the links that co-create value, inform strategic choices, and provide actionable options for daily tasks that align supply chains, firms, and society with goals of sustainable development.

4. Conclusion and limitations

In summary, this study contributes to the knowledge base of collaboration. Firstly, collaboration dimensions are likely to translate into greater management of returns. Furthermore, there has not been a lot of studies on collaboration in manufacturing reverse supply chain and in addition in the South African manufacturing context in relation to rural development, hence it is with believe that this study will contribute to the reverse supply chain industry in South Africa in driving the development of rural development. Managers need guidance on implementing SDGs in the supply chain. The integration of the SDGs and RSCM provides new areas of research and reflection.

The limitation of this study is that the respondents were from the Gauteng province, which means that a generalised statement cannot be made of the findings. It is for this that the author recommends that this finding should be carried out in other provinces. Furthermore, recommendations that research must be carried out on collaboration practices and its impact on the performance of reverse supply chain in the manufacturing industry and furthermore maybe the study should be streamlined to industries within the manufacturing industry.

Conflict of interest

There is no conflict of interest.

Author details

Ifije Ohiomah*, Clinton Aigbavboa and Nita Sukdeo
University of Johannesburg, Johannesburg, South Africa

*Address all correspondence to: Ifije93@gmail.com

References

[1] Kocabasoglu C, Klassen RD, Prahinski C. Linking forward and reverse supply chain investments: The role of business uncertainty. Quality Control and Applied Statistics. 2008;**53**(4):469-471

[2] Kannan D, Diabat A, Shankar KM. Analyzing the drivers of end-of-life tire management using interpretive structural modeling (ISM). The International Journal of Advanced Manufacturing Technology. 2014; **72**(9-12):1603-1614

[3] Shaik M. Comprehensive performance measurement methodology for reverse logistics enterprise. 2015

[4] Govindan K, Azevedo SG, Carvalho H, Cruz-Machado V. Lean, green and resilient practices influence on supply chain performance: Interpretive structural modeling approach. International Journal of Environmental Science and Technology. 2015;**12**(1):15-34

[5] Bouzon M, Govindan K, Rodriguez CM, Campos LM. Identification and analysis of reverse logistics barriers using fuzzy Delphi method and AHP. Resources, Conservation and Recycling. 2016; **108**:182-197

[6] Liao SH, Hu DC, Ding LW. Assessing the influence of supply chain collaboration value innovation, supply chain capability and competitive advantage in Taiwan's networking communication industry. International Journal of Production Economics. 2017;**191**:143-153

[7] Fu Y, Piplani R. Supply-side collaboration and its value in supply chains. European Journal of Operational Research. 2004;**152**(1):281-288

[8] Mentzer JT, DeWitt W, Keebler JS, Min S, Nix NW, Smith CD, et al. Defining supply chain management. Journal of Business Logistics. 2001;**22**(2):1-25

[9] Whipple JM, Lynch DF, Nyaga GN. A buyer's perspective on collaborative versus transactional relationships. Industrial Marketing Management. 2010;**39**(3):507-518

[10] Lumsden K, Stefansson G, Tilanus B. Collaboration in logistics. European Journal of Operational Research. 2003;**144**(2):235-236

[11] Sriram V, Krapfel R, Spekman R. Antecedents to buyer-seller collaboration: An analysis from the buyer's perspective. Journal of Business Research. 1992;**25**(4):303-320

[12] Simatupang TM, Sridharan R. The collaborative supply chain. The International Journal of Logistics Management. 2002;**13**(1):15-30

[13] Raweewan M, Ferrell WG Jr. Information sharing in supply chain collaboration. Computers and Industrial Engineering. 2018;**126**:269-281

[14] Barratt M. Understanding the meaning of collaboration in the supply chain. Supply Chain Management. 2004;**9**(1):30-42

[15] Lafferty G, van Fossen A. Integrating the tourism industry: Problems and strategies. Tourism Management. 2001;**22**(1):11-19

[16] Agrawal S, Singh RK, Murtaza Q. Prioritizing critical success factors for

reverse logistics implementation using fuzzy-TOPSIS methodology. Journal of Industrial Engineering International. 2016;**12**(1):15-27

[17] Simatupang TM, Sridharan R. An integrative framework for supply chain collaboration. The international Journal of Logistics management. 2005;**16**(2):257-274

[18] Nokkala T, Heller-Schuh B, Paier M, Wagner-Luptacik P. Internal integration and collaboration in European R&D projects. In: First ICC Conference on Network Modelling and Economic Systems. Lisbon, Portugal: ISEG; 2008. pp. 9-11

[19] Kampstra RP, Ashayeri J, Gattorna JL. Realities of supply chain collaboration. The International Journal of Logistics Management. 2006;**17**(3):312-330

[20] Cao M, Vonderembse MA, Zhang Q, Ragu-Nathan TS. Supply chain collaboration: Conceptualisation and instrument development. International Journal of Production Research. 2010;**48**(22):6613-6635

[21] Cao M, Zhang Q. Supply chain collaboration: Impact on collaborative advantage and firm performance. Journal of Operations Management. 2011;**29**(3):163-180

[22] Holweg M, Disney S, Holmström J, Småros J. Supply chain collaboration: Making sense of the strategy continuum. European Management Journal. 2005;**23**(2):170-181

[23] Sheu JB, Chou YH, Hu CC. An integrated logistics operational model for green-supply chain management. Transportation Research Part E: Logistics and Transportation Review. 2005;**41**(4):287-313

[24] Ho GTS, Choy KL, Lam CHY, Wong DWC. Factors influencing implementation of reverse logistics: A survey among Hong Kong businesses. Measuring Business Excellence. 2012;**16**(3):29-46

[25] Kim D, Lee RP. Systems collaboration and strategic collaboration: Their impacts on supply chain responsiveness and market performance. Decision Sciences. 2010;**41**(4):955-981

[26] Min S, Roath AS, Daugherty PJ, Genchev SE, Chen H, Arndt AD, et al. Supply chain collaboration: What's happening? The International Journal of Logistics Management. 2005;**16**(2):237-256

[27] Fawcett SE, Magnan GM, McCarter MW. A three-stage implementation model for supply chain collaboration. Journal of Business Logistics. 2008;**29**(1):93-112

[28] Squire B, Cousins PD, Lawson B, Brown S. The effect of supplier manufacturing capabilities on buyer responsiveness: The role of collaboration. International Journal of Operations & Production Management. 2009;**29**(8):766-788

[29] McLaren T, Head M, Yuan Y. Supply chain collaboration alternatives: Understanding the expected costs and benefits. Internet Research. 2002;**12**(4):348-364

[30] Wilding RD. Understanding collaboration: Generating the multiplication effect. Introduction. 2006

[31] Robson MJ, Katsikeas CS, Bello DC. Drivers and performance outcomes of trust in international strategic alliances: The role of organizational complexity. Organization Science. 2008;**19**(4):647-665

[32] Ha B, Park Y, Cho S. Suppliers' affective trust and trust in competency in buyers: Its effect on collaboration and logistics efficiency. International Journal of Operations & Production Management. 2011;**31**(1):56-77

[33] Ramanathan U. Supply chain collaboration for improved forecast accuracy of promotional sales. International Journal of Operations & Production Management. 2012;**32**(6):676-695. DOI: 10.1108/01443571211230925

[34] Lee HL, So KC, Tang CS. The value of information sharing in a two-level supply chain. Management Science. 2000;**46**(5):626-643

[35] Holweg M, Pil FK. Theoretical perspectives on the coordination of supply chains. Journal of Operations Management. 2008;**26**(3):389-406

[36] Akintoye A, McIntosh G, Fitzgerald E. A survey of supply chain collaboration and management in the UK construction industry. European Journal of Purchasing & Supply management. 2000;**6**(3-4):159-168

[37] Spekman RE, Kamauff JW, Myhr N. An empirical investigation into supply chain management: A perspective on partnerships. Supply Chain Management. 1998;**3**(2):53-67. DOI: 10.1108/13598549810215379

[38] Chen J, Zhang H, Sun Y. Implementing coordination contracts in a manufacturer Stackelberg dual-channel supply chain. Omega. 2012;**40**(5):571-583

[39] Chopra S, Meindl P. Supply chain management. Strategy, planning & operation. In: Das Summa Summarum des Management. Gabler; 2007. pp. 265-275

[40] Hudnurkar M, Jakhar S, Rathod U. Factors affecting collaboration in supply chain: A literature review. Procedia—Social and Behavioral Sciences. 2014;**133**:189-202

[41] Crook TR, Giunipero L, Reus TH, Handfield R, Williams SK. Antecedents and outcomes of supply chain effectiveness: An exploratory investigation. Journal of Managerial Issues. 2008;**1**:161-177

[42] Fawcett SE, Magnan GM, Ogden JA. Achieving world-class supply chain collaboration: Managing the transformation. Center for Advanced Purchasing Studies. 2007

[43] Coccia M, Wang L. Evolution and convergence of the patterns of international scientific collaboration. Proceedings of the National Academy of Sciences. 2016;**113**(8):2057-2061

[44] Prajogo D, Olhager J. Supply chain integration and performance: The effects of long-term relationships, information technology and sharing, and logistics integration. International Journal of Production Economics. 2012;**135**(1):514-522

[45] Kohli AS, Jensen JB. Assessing effectiveness of supply chain collaboration: An empirical study. Supply Chain Forum: An International Journal. 2010;**11**(2):2-16

[46] Poirier CC, Houser WF. Business Partnering for Continuous Improvement: How to Forge Enduring Alliances Among Employees, Suppliers & Customers. Berrett-Koehler Publishers; 1993

[47] Lambert DM, Cooper MC, Pagh JD. Supply chain management: Implementation issues and research opportunities. The International Journal of Logistics Management. 1998;**9**(2):1-20

[48] Krishnapriya V, Baral R. Supply chain integration-a competency based perspective. International Journal of Managing Value and Supply Chains. 2014;**5**(3):45

[49] Ramanathan U, Gunasekaran A. Supply chain collaboration: Impact of success in long-term partnerships. International Journal of Production Economics. 2014;**147**:252-259

[50] Lockamy A, McCormack K. Linking SCOR planning practices to supply chain performance: An exploratory study. International Journal of Operations & Production Management. 2004;**24**(12):1192-1218. DOI: 10.1108/01443570410569010

[51] Simatupang TM, Sridharan R. The collaboration index: A measure for supply chain collaboration. International Journal of Physical Distribution & Logistics Management. 2005;**35**(1):44-62. DOI: 10.1108/09600030510577421

[52] Manthou V, Vlachopoulou M, Folinas D. Virtual e-chain (VeC) model for supply chain collaboration. International Journal of Production Economics. 2004;**87**(3): 241-250

[53] Lee HL, Whang S. E-business and supply chain integration. In: Standford Global Supply Chain Management Forum. Vol. 2. 2001

[54] Harland C, Zheng J, Johnsen T, Lamming R. A conceptual model for researching the creation and operation of supply networks 1. British Journal of Management. 2004;**15**(1):1-21

[55] Lamming R, Hampson J. The environment as a supply chain management issue. British Journal of Management. 1996;7(1):S45-S62

[56] Goffin K, Lemke F, Szwejczewski M. An exploratory study of 'close'supplier–manufacturer relationships. Journal of Operations Management. 2006;**24**(2):189-209

[57] Tuten TL, Urban DJ. An expanded model of business-to-business partnership formation and success. Industrial Marketing Management. 2001;**30**(2):149-164

[58] Majchrzak A, Malhotra A, John R. Perceived individual collaboration know-how development through information technology–enabled contextualization: Evidence from distributed teams. Information Systems Research. 2005;**16**(1):9-27

[59] Bhatt GD, Grover V. Types of information technology capabilities and their role in competitive advantage: An empirical study. Journal of Management Information Systems. 2005;**22**(2):253-277

[60] Willis K. International development planning and the sustainable development goals (SDGs). International Development Planning Review. 2016;**38**(2):105

[61] Stevens C, Kanie N. The transformative potential of the sustainable development goals (SDGs). International Environmental Agreements: Politics, Law and Economics. 2016;**16**(3):393-396

[62] Campagnolo L, Eboli F, Farnia L, Carraro C. Supporting the UN SDGs transition: Methodology for sustainability assessment and current worldwide ranking. Economics. 2018;**12**(1):1-13

[63] Sudusinghe JI, Jayaratne RP, Kumarage AS. UN SDGs shaping sustainable supply chains: The case of apparel manufacturers in developing countries. In: 2018 IEEE International

Conference on Service Operations and Logistics, and Informatics (SOLI). IEEE; 2018. pp. 102-107

[64] Russell E, Lee J, Clift R. Can the SDGs provide a basis for supply chain decisions in the construction sector? Sustainability. 2018;**10**(3):629

[65] Denzin NK, Lincoln YS, editors. The Sage Handbook of Qualitative Research. Sage; 2011

[66] Guba EG, Lincoln YS. Competing paradigms in qualitative research. Handbook of Qualitative Research. 1994;**2**(163-194):105

[67] Simon L. Approaches to social enquiry: Advancing knowledge. Sociological Research Online. 2008;**12**(6):186-187

[68] Mingers J. A classification of the philosophical assumptions of management science methods. The Journal of the Operational Research Society. 2003;**54**(6):559-570

[69] Orlikowski WJ, Baroudi JJ. Studying information technology in organizations: Research approaches and assumptions. Information Systems Research. 1991;**2**(1):1-28

[70] Saunders M, Lewis P, Thornhill A. Research Methods for Business Students. Pearson Education; 2009

[71] Hair JF Jr, Black WC, Babin BJ, Anderson RE. Multivariate Data Analysis. New Jersey: PrenticeHall; 2012

[72] Zikmund WG. Business Research Methods. 7th ed. Mason: Thomson, South- Western; 2003

[73] Bryman A. The end of the paradigm wars. In: The SAGE Handbook of Social Research Methods. Sage; 2008. pp. 13-25

[74] Sekaran U, Bougie R. Business Research Methods: A Skill-Building Approach. Chichester: John Wiley& Sons Ltd.; 2011

[75] Bell E, Bryman A, Harley B. Business Research Methods. Oxford University Press; 2018

[76] Nachmias D, Nachmias C. Content Analysis. Research Methods in the Social Sciences. 1976. pp. 132-139

[77] Preskill H, Russ-Eft DF. Building Evaluation Capacity: Activities for Teaching and Training. Sage Publications; 2015

[78] Tharinger DJ, Finn SE, Gentry L, Hamilton A, Fowler J, Matson M, et al. Therapeutic assessment with children: A pilot study of treatment acceptability and outcome. Journal of Personality Assessment. 2009;**91**(3):238-244

[79] Robin H. What sample size is "enough" in internet survey research. Interpersonal Computing and Technology: An Electronic Journal for the 21st Century. 1998;**6**(3-4):1-12

[80] Kline RB. Promise and pitfalls of structural equation modeling in gifted research. 2010

[81] Fornell C, Larcker DF. Structural equation models with unobservable variables and measurement error: Algebra and statistics. 1981:382-388

[82] Hair JF, Page M, Brunsveld N. Essentials of Business Research Methods. Routledge; 2019

[83] Litwin MS. How to Measure Survey Reliability and Validity. London: Sage Publication; 1995

[84] Hu LT, Bentler PM. Cutoff criteria for fit indexes in covariance structure

analysis: Conventional criteria versus new alternatives. Structural Equation Modeling: A Multidisciplinary Journal. 1999;**6**(1):1-55

[85] Doloi H, Iyer KC, Sawhney A. Structural equation model for assessing impacts of contractor's performance on project success. International Journal of Project Management. 2011;**29**(6):687-695

[86] Singh A. A confirmatory factor analysis of student engagement at the University of Hail, Saudi Arabia. Amazonia Investiga. 2020;**9**(26):181-190

[87] Van Dijk M. Cross-border collaboration in European–Russian supply chains: Integrative approach of provision on design, performance and impediments. Contributions to Game Theory and Management. 2016;**9**(0):118-169

[88] Akkermans H, Bogerd P, Van Doremalen J. Travail, transparency and trust: A case study of computer-supported collaborative supply chain planning in high-tech electronics. European Journal of Operational Research. 2004;**153**(2):445-456

[89] Du L. Acquiring competitive advantage in industry through supply chain integration: A case study of Yue Yuen Industrial Holdings Ltd. Journal of Enterprise Information Management. 2007;**20**(5):527-543

[90] Khan M, Hussain M, Saber HM. Information sharing in a sustainable supply chain. International Journal of Production Economics. 2016;**181**:208-214

Section 3

Strategies, Cases and Recent Advances

Chapter 7

The Hambach Forest in the German Debate on Climate Protection: Is There a Symbolic Value beyond the Environmental Value?

Meike Henseleit, Sandra Venghaus and Wilhelm Kuckshinrichs

Abstract

In the late summer of 2018, the Hambach Forest (North Rhine Westphalia/ Germany) appeared prevalently in the media due to massive protests against its clearance for lignite mining with for the power generation. Because coal power as a form of energy supply is extremely climate intensive, the Hambach Forest rapidly became a symbol of the fight against climate change and the ongoing destruction of nature and its resources for economic reasons. Due to the extra-ordinarily prominent role of the Hambach Forest in the public opinion across Germany, this research addresses values of the forest to the population in monetary terms as well as the underlying factors that determine those values. For the analysis, a contingent valuation survey was conducted in December 2019 in Germany. The proposed amounts for the preservation of the Hambach Forest are mostly in accordance with previous evaluation studies of woods and forests, although this time almost only passive-use values are decisive. Further, a conversion of the WTP values to the area of the Hambach Forest results in an extra-ordinarily high per-hectare value of about 3.6 million. Thus, the symbolic value of the forest is remarkable and should be considered in future political decisions.

Keywords: symbolic value, environmental valuation, climate movement

1. Introduction

In the late summer of 2018, the Hambacher Forst (Hambach Forest) in Germany appeared prevalently in the media covering the strong protests against the intention of the utility company Rheinisch-Westfälisches Elektrizitätswerk AG (RWE) to grub large parts of the remaining forest in order to mine the lignite underneath. Up to 50,000 people from across Germany as well as neighboring countries gathered for protest marches in order to save the forest and express their position against lignite mining and to demand more political action regarding climate protection. The grubbing was suspended when in October 2018 the Higher Regional Court of Munster issued a provisional stop until there was going to be a decision as to whether the Hambach Forest falls into a category of the Habitats Directive.[1] This was an option since the forest is habitat to 13 species considered in the Directive, among them the Bechstein's bat (*Myotis bechsteinii*) and eight other bat species, two species of toad, the agile frog (*Rana dalmatina*), and the endangered common dormouse (*Muscardinus avellanarius*). In January 2020, following the recommendations of a caucus, the German government decided to phase out coal power by 2038 at the latest and to preserve the Hambach Forest [2]. Despite this decision, however, scientists expect the forest to be endangered if RWE upholds its plans to continue grubbing in the coming years, thus requiring the further lowering of groundwater (lignite mining requires the lowering of the groundwater levels) and increasing temperatures caused by the declining surrounding vegetation [3]. So far, nearly 60 villages have been relocated or are in the process of relocation at present within the Rheinische Revier due to the exploitation of the lignite underneath. Considering this, the magnitude of the protests against the clearance of the Hambach forest was surprising.

Several factors seemed to be of relevance for the massive protests and the unexpected, high commitment of the population, either as participants in rallies or as debaters in social networks or the like. First, the forest on its own as an old forest and as a habitat for endangered species might be seen as valuable and irreplaceable [4]. Second, worries about climate change have gained in importance and were fostered not least by the extremely hot summer of 2018, and the increasing awareness that coal-fired power generation is among the largest sources of carbon emissions in Germany. Third, for many people, RWE turned into an enemy image, as the company has been made responsible not only for massive carbon emissions, but also for the destruction of nature and villages, despite the efforts of RWE to compensate relocated people and to re-cultivate large areas. This position, however, neglects proprietary rights, which RWE holds based on the acquisition of the Hambach Forest in 1978, operating permits, and mining rights, all negotiated with the provincial government in Dusseldorf, represented by different parties (Social Democrats, Christian Democratic Union, and Green Party). Since 1978, the size of the Hambach Forest has shrunk from 4100 ha to only about 500 ha in 2020, and since 2012, the Hambach Forest has been occupied by about 20–100 people permanently living there in self-constructed tree houses. However, the ownership based on the legal definition may be in contrast to what people perceive as legacy or moral ownership here [5]. The Hambach Forest eventually became a symbolic battleground for climate activists from Germany and other countries. Ten thousands of people have visited the Forest since, and its publicity goes well beyond the borders of Germany [4, 6, 7].

Accordingly, the Hambach Forest has become a location of "meaning," and this meaning can be attributed on vastly different scales. Such "meaning" does not manifest itself in particular physical characteristics, but is instead attributed by humans and may be closely linked to notions of identity and sense of "belonging" [5]. Only places identified as symbolic by a certain number of individuals are socially recognized as such, and a group can form and give itself an identity within this movement of recognition [8]. Most generally, a place can be considered "symbolic" whenever it contributes significantly to giving a group an identity—for example, the stadium of "their" club is something meaningful for football fans. Members of a particular scene are aware of this, and the symbolic meaning of a place is common understanding

[1] This is a European Directive as an EU response to the Berne convention, which intends to protect nature and wildlife. It requires national governments to specify areas that are expected to ensure the conservation of flora and fauna species [1].

among them. Accordingly, the symbolic character of a location is both, a powerful matter and a power instrument: the person who manipulates symbols can also manipulate processes of identification and thus take an influence on the constitution of the group [8]. Further, a symbolic place does not have the same meaning seen from nearby or from a distance, by a small group or by a large community, from inside or from outside, by "us" or by "others," and through time [8]. This symbolic aspect may partly explain the fierce fight over the Hambach Forest, since the topic activates the identification with either one side: following traditional rules or claiming change in order to protect nature and climate.

Against this background and given the described unusually high empathy for the forest, the question about its value for the German population arose. We therefore intended to find out whether it was possible to measure the meaning of the forest in the view of the population and to translate it into quantifiable values in order to make it comparable. These values are to represent its role in the controversial political debate on climate protection, transformation strategies, and coal phase-out in Germany. It thus may indicate the non-use and probably symbolic value of the Hambach Forest. Furthermore, stated values can also be considered as an indicator of how important the protection of the forest for single individuals is. Since no similar case is known to us so far, this study has a rather explorative character.

The structure of the paper is as follows: In section 2, we provide a literature review about valuation studies of forests. Methods and procedure are described in detail in section 3, followed by the results in section 4. This paper ends with a discussion and conclusion section.

2. Literature review

In order to elucidate the values of environmental goods, several methods have been developed depending on the values to be considered. Although the importance of environmental goods to humankind has many dimensions (e.g., ecological, sociocultural, or economic), the values are usually expressed in monetary units as an important tool to raise awareness and convey the (relative) importance of ecosystems and biodiversity to policymakers [9]. Economists have recognized the possibility that individuals who make no active use of a particular forest, river, certain species, or other such natural resources may, nevertheless, derive satisfaction from their mere existence, even if they never intend to make active use of them [10–16]. This concept has come to be known as "existence value," and it is the major element of what is now referred to as "non-use" or "passive-use" values [17]. The most common methods for the evaluation of environmental goods, which comprise also non-use values, are direct methods such as the contingent valuation method (CVM) or Choice Experiments (CE). As part of these methods, individuals are presented a hypothetical scenario for which they are asked to state their willingness to pay (WTP) and/or their preferences for a change in the provision of a specific environmental good [18].

Many surveys have been carried out during the last four decades about values of sylvan ecosystems and their ecological services. However, the WTP values are hardly transferable due to several reasons: First, they are scenario-dependent. Second, WTP values normally depend on individual characteristics, such as attitudes and sociodemographic variables. Third, the values may change quite a bit over time according to circumstances [19]: for example, about 40 years ago, there were hardly any protests against the cutting of the Hambach Forest, because, on the one hand, there was still a large part of the forest remaining, and on the other hand, climate change and the impact of coal-fired power generation were almost unknown, at least for large parts of the population. Fourth, by conducting a survey, previously unknown and/or unexpected correlations may be revealed. Nevertheless, procedures and results of comparable studies are useful for the design and the interpretation of new surveys.

In a first step, we analyzed a database of more than 80 surveys about preferences for wooden areas applying CVM, CE, travel-cost method (TCM), or benefit transfer method (BTM) within German-speaking countries during the last three decades [20]. Most of the studies measure recreational values by directly asking for the WTP for entrance fees [21–25], or they evaluate minor changes in attributes such as the introduction of environmental protection programs by asking for additional taxes or the like [26–28]. None of the studies deal with pure existence values only, and thus, no directly comparable values could be extracted from the studies listed in the database.

In a second step, we searched the Environmental Valuation Reference Inventory (EVRI)[2] database, which compiles environmental valuation studies from all over the world, for studies about values of woods and forests in order to find comparable studies to the case of the Hambach Forest. As search criteria, we chose "plants," since this comprised both forests and woodlands, "willingness to pay," "passive uses," and "stated preference" or "simulated market price" in order to identify comparable studies. Altogether, 182 studies were found (January 2020), of which 94 indicated "forest" as environmental asset, 88 "trees," 50 "woodland," and 16 "rainforest." Since some studies consider more than one environmental asset, overlaps occurred. After deleting those, 171 studies remained. Most of them address preferences for specific aspects such as species diversity, infrastructure, preferences for leisure activities, or forest protection schemes. In most of the studies on rain forests, the researchers surveyed the willingness to pay for the preservation of a certain minimum area. Only five studies dealt with the total value of a forest rather than values for single characteristics.

The first study in chronological order was a cost-benefit analysis about the option to log the Aorangi-Awarua-Forest in New Zealand (**Table 1**). A CV among 500 New Zealanders was conducted via mail in 1991 by Beanland [30] in order to find out whether the total economic value of the forest was higher than the revenues from logging it. The mean WTP to preserve the forest was 13.12 New Zealand $ as a yearly payment, with 41% of the respondents willing to pay at all. This amount is comparable to roughly 10€ currently when accounting for exchange rates and inflation. However, since in this New Zealand mail survey, the return rate of questionnaires was just around 50%, and normally those who are less interested in the topic are more likely not to send back the questionnaire [35], an interpolation of the mean WTP to the total population did not appear advisable.

The second study by Kniivilä et al. [31] assessed the regional and local user and non-user benefits of the current conservation of old forests in the region of Ilomantsi/Finland in 1999 by surveying 800 people in North Karelia. The response rate was 59.2%, the median WTP was 19€, and the mean WTP 48.6€ per person/year, which corresponds to 25€ and 65€, respectively, in 2019. The WTP values were taken by the dichotomous choice (DC) question format, which normally leads to higher WTP values [33]. However, 18.5% of the respondents chose an "I don't know" option when they were asked whether they would be willing to pay a certain amount for the preservation of the forest and were excluded from further analysis. About 45% of the

[2] https://www.evri.ca/, last time accessed in September [29].

	1	2	3	4	5
	Beanland [30]	**Kniivilä et al. [31]**	**Amirnejad et al. [32]**	**Veisten and Navrud [33]**	**Broberg [34]**
Survey year	1991	1999	2004	1995	2005
Country	New Zealand	Finland	Iran	Norway	Sweden
Method	Mail survey	Mail survey	Personal interview	Mail survey	Mail survey
Selected sample	500	800	n.a.	2,498	2,000
Participation rate	50%	59%	n.a.	71%	49%
Sample size	225	472	950	1776	930
WTP method	Open ended	DC	DC	DC/open ended	Open ended
% pos. WTP	41%	36.5%	65%	25–39%	45%
WTP value	10€	65€	40€	27–61€	35€
Forest size	5142 ha	20,000 ha	1,900,000 ha	n.a.	126,000 ha

Source: own compilation.

Table 1.
WTP surveys about forest areas with mainly non-use values.

remainder had a true zero WTP. Indeed, recreational values of the evaluated forests there are non-negligible, since the forests are popular tourist destinations [31].

Amirnejad et al. [32] conducted a CV in order to estimate the existence value of north forests in Iran. By analyzing the answers from personal interviews of about 950 residents of Iran, the mean WTP for the protection of the forests was 30.12 US$ annually (corresponding to roughly 40€ in 2020), which is quite high considering that the GDP per capita in Iran in 2004, when the interviews for this survey were conducted, was only 2500 US$. However, the WTP values appear more valid when considering that the sample is highly biased in terms of education and income. The rate of respondents with a positive WTP is indicated with 65%, of whom 80% have already visited the north forests of Iran. The rather high mean amount can at least partly be reasoned by the survey design (talking about and showing pictures of the beauty of the forest and of future scenarios of damage) and the double-bounded DC questionnaire, and the occurrence of direct use values due to the high rate of visitors cannot be excluded. Also an interviewer effect cannot be denied, since the WTP in personal interviews usually is higher than in e-mail or mail surveys [36]. Furthermore, cultural conditions in this country may have influenced the stated WTP positively [37].

Veisten and Navrud [33] analyzed the WTP for the protection of old forests in Norway, a good nearly exclusively linked to passive-use values, using a mail survey among 2498 people in Norway conducted in 1995. The efficient total sample and overall response rate were 1792 and 71.7% respectively. According to the payment question format (open ended or dichotomous choice), estimated WTP values ranged from a mean of 20.5 US$ to 41.6 US$ in form of a one-time payment to the WWF's Forest Fund. This corresponds to roughly 27€ to 61€ in 2020 with a rate of positive WTP of 29–46%. With an additional invoice for the stated WTP value, this value changed to 24–37% and a mean WTP value of 5$ to 11$, corresponding roughly to 7€ and 14€ in 2020. These values indicate that CV values are not only sensitive to the question format, but also to the scenario setting and the payment vehicle.

Broberg [34] used contingent valuation to estimate the public benefit derived from preserving 126,000 ha of state-owned old-growth forest in the sub-mountainous region of Sweden. In this mail survey, the response rate was 49%. About 45% of the 905 respondents had a positive WTP with an average of approximately SEK 300 (35€ in 2020) for the preservation program as an annual tax increase over the next 5 years. Males were significantly less likely to hold a positive WTP, and the likelihood decreased with age for both, males and females. Education, income, and membership in any environmental NGO were correlated positively with the likelihood of observing a positive WTP.

Table 1 provides an overview of the studies.

Although we searched broadly for comparable studies about environmental objectives with a symbolic character, besides few direct use values, we did not find any. Laplante et al. [38] surveyed the value of the Armenian lake Sevan for US American Residents with Armenian origin. They asked 6000 people about their willingness to participate in a mail survey, of which 1325 agreed to participate, but only 389 returned a completed questionnaire. The WTP was surveyed in form of DC as a one-time donation and led to a mean value between 80 US$ and 280 US$, which corresponds to 47€–118€ nowadays. Most significant variables for the WTP were past visits and the option of future visits of the lake. Thus, although the lake has a highly symbolic character, use values seemed to be most important for the WTP.

Even though the population of Germany is known for its love of wild forests [39, 40], there is currently no German study that explicitly addresses non-use values of forests. Compared with the studies above, the Hambach Forest is rather small given its size of only about 500 ha, and to almost all Germans, its value comprises passive-use values only. Therefore, even though it may be ecologically valuable, it is hardly comparable to the other evaluated woods and forests from an ecological point of view. Instead, it is especially its symbolic character that makes it a highly interesting research subject that justified its evaluation. Since no comparable situation analyzed by an evaluation survey so far was found, our study is of highly explorative character.

3. Methods and procedure

According to the literature review, CVM proposed itself to be the method of choice, since a high rate of existence value of the Hambach Forest was presumed for the German population. Ideally, the surveyed sample corresponds to the distribution of these variables across the basic population. In mail surveys, the sample selection bias is usually stronger than in telephone or personal inquiries [41, 42], making the latter more advisable as survey methods, although they are normally more expensive. We therefore decided to conduct the survey via telephone with strong instructions regarding the representativeness concerning age, gender, education, and place of residence (federal state). Furthermore, questions regarding attitudes toward environmental values and behavior, renewable and nonrenewable energy systems as well as political issues were included in our questionnaire.

Moreover, protest responses occur regularly in environmental valuation surveys [43]. They can be reduced by the survey design, however, since they are usually lower when voluntary payment schemes are provided [44]. As a procedure, it appeared useful in our case to contact participants personally following a random selection scheme in order to guarantee the representativeness due to the self-selection bias of online and mail surveys. Voluntary contributions to a fund seemed to be most adequate as a payment vehicle for several reasons: First, because of an expectedly high rate of passive-use values, payment vehicles linked to a certain kind of use dropped out. Second, due to a high level of politicization of the issue, a rather "neutral" instrument for the protection of the forest seemed to be preferable to a tax, for example. Third, the voluntary character of the payment scenario fit better to the climate protection movement, since the latter is strongly characterized by the perception that "policy isn't doing enough to fight climate change" and that it is instead the people who need to take action now.

Therefore, the developed CV scenario was the following: "A bit over a year ago, the Hambach Forest was prevalently in the media, because it was uncertain whether vast parts of it should be cleared in favor of lignite mining and its electricity generation. Assume that a private forest conservation initiative would be founded, which relies on private donations to buy and maintain the forest, thereby preventing the lignite below it from being mined. Would you be willing to donate to such an initiative?"[3] If respondents answered with "yes," they were asked to indicate their hypothetical donation in Euros. Furthermore, based on the observed factors of relevance for the WTP from the literature survey, we also included questions about attitudes and habits regarding climate change, energy, and environmental issues, as well as the usual sociodemographic queries in our questionnaire.

The initial idea of our study was to conduct a cost-benefit analysis (CBA) under consideration of the use- and non-use values evaluated by the CV as well as of opportunity costs arising from the preservation of the forest for RWE and the region. The intention was to provide a rather neutral perspective on the highly politicized issue and support decision-makers in finding solutions by considering all aspects adequately. However, only four weeks after the survey was conducted in December 2019, the German government decided the early coal phase-out, also proposing that the preservation of the Hambach Forest should be guaranteed. Therefore, the focus of our analysis was broadened from the evaluation of the forest as such to the assessment of factors for a positive WTP for the Hambach Forest and the role of attitudes toward different forms of energy generation. Since the CV was part of a bigger survey about the German energy transition and the bioeconomy, questions regarding preferences for power generation technologies, methods from the field of bioeconomy as well as general attitudes regarding the environment were also included.

4. Results

The survey was conducted by a professional agency in December 2019. Altogether, 1,002 people participated in the telephone survey. Data obtained from the national survey were analyzed using the statistical package IBM SPSS 19. The sample is almost representative of the German population in terms of age, gender, education, and city size and residence in the 12 German federal states. Regarding the number of people per household, single households were underrepresented in our sample (27.9% as compared with 41.9%), whereas households with two persons were overrepresented (40% as compared with 33.8%) [45]. The household income was approximately representative of the German population, with the restriction that 10% of the respondents refused to answer this question. Accordingly, the lowest and the highest

[3] Since the survey was conducted among the German population, the original language of the survey was German.

income classes are underrepresented in our survey, which is in line with the common observation that people in extreme classes of income more often refuse to reveal their household income [46].

4.1 Willingness to pay—yes or no

Altogether, 47.2% of the surveyed people stated to be willing to pay an amount between 3€ and 1000€ for the preservation of the Hambach Forest. Furthermore, 49.7% answered the question about their WTP with "no," and 3.1% refused to provide an answer. In this case, a clear distinction between true zeros and protest zeros was not possible, since we did not ask for the reasons based on which they would refuse to pay. However, an indication for the occurrence of protest answer can be seen in the fact that 20% of the survey participants who stated a WTP of zero refused the further run of coal-fired power plants completely. For those with a positive WTP, the rate was 38%. According to a meta-analysis of Meyerhoff and Liebe [44], the mean share of protest responses in CV surveys is about 18% with a median value of 16%. The share is higher in scenarios with taxes or entrance fees as payment vehicle, and also the survey method may have an impact with onsite-, web-, and phone surveys leading to a lower share of protest zeros [44].

In order to observe differences between those with and those without a positive WTP for the preservation of the forest, we excluded all participants who did not answer this question with either "yes" or "no," which resulted in a total of 971 cases. We observed no differences regarding the WTP per se in terms of sociodemographic aspects except that females were more likely to state a positive WTP than males ($p = .05$). Highly significant differences between the two groups were observed for revealed activities listed in question 5.2 (**Table 2**): "Which of the following did you do within the last 12 months?" People with a positive WTP were much more likely to confirm those. Correlation coefficients are calculated as Pearson's r, since the variables were dichotomous (yes-no).

Participants with a positive WTP also favored nearly all surveyed aspects of a renewable energies, including the application of biogas, biofuels, and renewables in the industry, more than those who were not willing to pay. Those who stated they were willing to pay rated themselves as more informed regarding the energy

Measurement	Pearson's r	Significance
(a) Selective buying of regional food	.185	.000
(b) Selective buying of packages made of renewable materials	.247	.000
(c) Purchase of green electricity	.164	.000
(d) Avoidance of packaging	.185	.000
(e) Adaptation of the mobility behavior (e.g., to abandon the car or to use the bicycle more often)	.234	.000
(f) Purchase of bio-products	.271	.000
(g) Carbon compensation (payment to a specific organization for carbon offsetting projects)	.079	.014

Source: own.

Table 2.
Correlation between reported pro-environmental behavior (Q5.2) and a positive WTP.

transition. Furthermore, they had a more positive attitude toward solar energy, wind turbines, water turbines, and energy from biomass. Accordingly, the same was observed with a negative correlation for coal-fired and nuclear power generation. Furthermore, they rated themselves as more informed regarding the coal phase-out. All of those correlations were significant at the 0.01%level. No significant differences between the two groups were found for attitudes toward conventional and nonconventional natural gasoline.

Those who refused to answer the question whether they would be willing to pay at all apparently have strong pro-environmental attitudes, even compared with those who stated a positive WTP. For example, all of the 31 respondents from this category answered Q5.2e whether they changed their mobility behavior, e.g., by taking more often the bicycle instead the car, with "yes," while around 65% of those with a positive and 42% with a negative WTP affirmed this question. A similar distribution was observed for Q5.2d, whether respondents had consciously forgone packaging while doing groceries during the last 12 months. Those who refused and those who stated a positive WTP answered most other questions regarding attitudes and behavior toward environmental issues similarly.

4.2 Willingness to pay: amount

For the analysis of the amount of the WTP, we omitted those 31 cases with no answer as to whether they would be willing to pay. Of the 971 considered cases, 498 (51.3%) declined a willingness to pay. For further analyses, we treated these values as true zeros, although it cannot be ruled out that, by doing so, protest zeros are neglected. Therefore, the results should be considered a conservative estimate, and real preferences might be higher. In a first step, we checked the theoretical validity of the stated amounts by analyzing some of the variables, such as the income and attitudinal variables [47]. The mean WTP for all participants, including the zero values, was 26.83€, and the most frequently stated positive value was 50€, provided by 124 participants (12.8%). The mean WTP considering only positive values was 55.08€, whereas about 10% of the sample had a WTP higher than 50€. The highest stated amount was 1000€, expressed by three survey participants. Since none of the three profiles provided an indication for unreliable values, due to high income, high education, and a strong attitude toward environmental issues, we did not exclude them from further analyses. The same correlations were also checked for the whole sample. As expected, significant positive correlations were observed between stated pro-environmental behavior of Q5.2 and the level of income. Furthermore, also significant positive correlations were observed for preferences for renewable energy technologies such as solar, wind, biomass, and hydro, and negative correlations for coal power and nuclear engineering. Here, again, no significant differences in attitudes regarding conventional and nonconventional natural gas were found. People who rated themselves as more informed regarding the coal phase-out also had a significantly higher WTP. When considering only cases with a positive WTP, no significant differences regarding preferences for energy technologies were observed. Regarding stated pro-environmental behavior, only minor significant correlations were observed for the acquisition of green energy Q5.2c (positively) and the adaptation of the mobility behavior Q5.2e (negatively). Furthermore, people living in an owned house or flat, bigger household size, and higher income were linked to a significantly higher WTP. Although females had a significantly higher WTP overall, males with a positive WTP stated significantly higher amounts.

4.3 Willingness to pay: extrapolation

Even though the Hambach Forest seems to no longer be immediately threatened in the near future, an extrapolation of the stated WTP values is of high interest, for example, in order to be able to compare the stated value to those of other natural goods. As discussed, the estimation is rather conservative, since we did not exclude potential protest zeros, which generally account for between 5% and 50% [43, 48]. Given the representative character of the survey for the German population, an extrapolation of the mean value of 26.83€ for the full-age population of Germany, which is about 67 million people [49], would result in roughly 1.8 billion € without considering benefits for future generations and people outside of Germany. This correlates to roughly 3.6 million € per ha. However, the following points may have led to a higher estimated value:

- Protest zeros are not considered.

- Non-respondents hold above-average pro-environmental values in this case, which could be an indication for a high estimation for the preservation of the Hambach Forest, although they apparently would not or could not monetarize their preferences.

- People with preferences for the scenario outside of Germany are not considered.

On the other hand, we did not provide the opportunity to state a quasi-negative WTP for our scenario to account for the fact that people might also be willing to pay in order to avoid the preservation of the Hambach Forest and favor instead the continuation of the original lignite mining plan. Just to compare these values, in North Rhine-Westphalia, the prices for forest areas—mainly working forest—in 2020 range from 10,000 to 30,000€ per ha according to a sales platform.

4.4 Willingness to pay: yes or no—a binary logistic regression

In order to generate an understanding of the importance of different factors influencing the willingness or non-willingness to pay, we conducted a binary logistic regression. Our hypotheses were that people with more pro-environmental behavior, pro-environmental attitudes, preferences for fossil-free power generation, and younger people are more willing to pay. Accordingly, attitudes toward the environment and energy technologies appeared to be important as impact factors, as well as gender and age. We also tested both, schooling and vocational education, but neither had significant impact on the dependent variable within the binary logistic regression model and were thus left out (**Table 3**). Our final model consists of six variables and considers 971 cases, representing roughly 97% of the sample. The remaining cases were excluded due to missing values regarding the dependent variable. A check for multicollinearity of the explaining variables showed no critical values. Using our binary logistic regression model, the rate of correctly predicted values rose from 51.3% to 66.9%. The pseudo-R squared (Nagelkerke) is 0.207, which is "acceptable" according to Backhaus et al. [50].

	B	S.E.	Wald	df	Sig.	Exp(B)
Attitudes toward renewables Q2.1 a-d	.300	.112	7.189	1	.007	1.350
Age (years) Q8.1.2	−.009	.004	4.078	1	.043	.991
Gender (male) Q8.1.1	−.245	.144	2.899	1	.089	.783
Pro-environmental behavior Q5.2	.371	.045	68.672	1	.000	1.449
Eco-centered conviction Q7.8.3	.128	.051	6.375	1	.012	1.136
Attitude toward lignite Q2.1e	−.156	.043	12.953	1	.000	.856
Constant	−1.397	.398	12.350	1	.000	.247

B: coefficient for the standard; S.E.: standard error for the coefficient around the constant; Wald: Wald Chi Square statistics; df: degree of freedom for the Wald Chi Square test; Exp (B): exponentiation of B coefficient, which is an odds ratio. Source: own.

Table 3.
Parameter estimate of the binary logistic regression.

Table 3 illustrates the influence of nearly each predictor variable, except gender, to the logistic model and the statistical significance ($p < .05$) of the Wald Chi Square test, which is obtained by squaring the ratio of the regression coefficient (B) to its standard error (S.E.). According to our analysis, the stated pro-environmental behavior, which normally correlates with strong pro-environmental attitudes, has the strongest impact on the WTP. This observation is not surprising, and this strong relationship has been shown by various studies (e.g., [51, 52]). The Odds Ratio (Exp(B)) indicates that, if the stated pro-environmental behavior increases by one unit, the probability to state a positive WTP increases by roughly 45%. The second largest impact factor is the attitude toward renewable energies (solar, wind, hydro, and biomass), which is also correlated significantly with stated pro-environmental behavior; an increase of one unit here implicates an increase of probability by 35%. This observation supports the assumption that not only values of the forest, but also a favor for the energy transition may have played a role in the decision to state a positive WTP. Further, an eco-centered conviction, measured by a statement regarding the perception of the vulnerability of the earth (Q7.8.3), has a clear positive impact on the WTP, although it is comparatively small. The acceptance of the use of lignite as an energy source (Q2.1e), instead, has a negative impact on the WTP; an increase by one unit of acceptance implies a decrease of roughly 14% in the probability to state a positive WTP. A negative impact on the probability to state a positive WTP can be observed for age and being male, although on a rather low level of significance. According to the literature, no general impact of gender on the WTP for environmental goods can be observed, since other factors such as attitudes, education, or income are generally more important [53]. However, a recent survey found that females probably are more pro-environmental in both Germany and the Netherlands [54]. By separating age from other factors, sometimes a negative impact can be observed, which means that older people are less willing to pay for environmental issues [53]. In these regards, our results are thus in line with previous findings.

4.5 Willingness to pay—ordinary least squares (OLS) regression analysis

In order to detect significant impact factors on the stated monetary values for the protection of the Hambach Forest, we conducted an OLS regression analysis. The dependent variable, the stated amount of those willing to pay, was rather log-normal than normally distributed. Therefore, we logarithmized the dependent variable and applied a semi-log model. The following explanatory variables were considered:

- stated behavior regarding the environment (Q5.2),
- household income (Q8.6),
- age (Q8.1.2),
- gender (Q8.1.1),
- stated preferences for the renewables (Q2.1),
- attitudes toward lignite (Q2.1e),
- trust in the national and federal government (Q6.1a and Q6.1b),
- political party, which was chosen during the national election in 2017 (Q7.3),
- satisfaction with the way political decisions are conducted in Germany (Q7.4),
- perception of the vulnerability of the earth (Q7.8c).

However, only a very low rate of explained variation with an R^2 of .081 and a corrected R^2 of .057 could be reached using an OLS regression approach. The results are displayed in **Table 4**. No more than three variables with significant impact on the stated amount of the WTP were observed: gender ($p = .01$), household income ($p = .001$), and the degree of agreement with statement Q7.8c about the vulnerability of the earth ($p = .05$). As generally the case in CVs, household income has a positive effect on the WTP: the higher the income, the more easily people can afford to pay for environmental goods and services. Compared with the other variables, income had the strongest impact on the WTP, but only when considering the average household income. In other models, when per-capita income was considered the explaining variable, no significant impact was observable. Furthermore, in our case also being male had a positive impact on the WTP. It is important to note that males had a significantly higher household income in our survey. However, testing for collinearity did not reveal problematic values. Finally, the impact of perceived higher vulnerability of the earth on the WTP can be considered as an indicator for a higher estimation of existence values as well as fears of losses of environmental goods and services. Stronger agreement with this statement normally goes along with a more nature-centered point of view, which also could explain a higher WTP. The low rate of explained variation leads to the assumption that other factors, which we did not consider in our survey, may be of strong relevance for the stated amount. Imaginable are issues regarding personal budget constraints, a general estimation of woods and forests, or dissatisfaction with climate policies, as well as embedding effects resulting from our payment scenario. Alternatively, the group of respondents with a positive WTP is more homogeneous regarding attitudinal values compared with the same values over all respondents, as an analysis of variance showed.

Modell	Non-standardized coefficients		Standardized coefficients	T	Sig.
	Regression coefficient B	Standard error	Beta		
Constant	2.401	.434		5.531	.000
Attitude toward lignite Q2.1e	.031	.030	.051	1.021	.308
Trust in the national government Q6.1a	−.027	.047	−.049	−.570	.569
Trust in the federal government Q6.1b	.069	.049	.120	1.415	.158
Perception of the vulnerability of the earth Q7.8c	.088	.035	.124	2.512	.012
Age (years) Q8.1.2	.003	.003	.054	1.085	.279
Summary of pro-environmental behavior Q5.2	.001	.032	.002	.045	.964
Summary of attitudes toward solar power, wind power, water power, and energy from biomass Q2.1	−.001	.056	−.001	−.010	.992
Voted the Green party at the last national election Q7.3	.087	.119	.036	.726	.468
Gender: Male Q8.1.1	.272	.097	.137	2.793	.005
Household income Q8.6	.000	.000	.184	3.799	.000
Attitude toward the end of lignite mining Q4.3	−.143	.097	−.077	-1.470	.142

[a]*Dependent Variable: LogNormal_0 WTP.*
Source: own.

Table 4.
Parameter estimate of the OLS regression.

5. Discussion

Striking outcome of our survey results is that, against our expectations, no clear differences in the results were found compared with earlier surveys about stated WTP values of forests. All of the results are more or less in line with previous findings about values for environmental goods [20, 55]. This is especially interesting, since in this case nearly exclusively existence values are of relevance, whereas the other cited surveys mostly consider also direct use values. Kriström [56] found that respondents who expressed only a use motive stated a higher WTP on average than respondents stating only a non-use motive. Those who expressed both, use and non-use motives, stated the highest WTP. Since the stated values here are quite similar to those of the other cases described in the literature review section, this may be seen as an indicator for the existence of a premium for a symbolic value on top of already known non-use values. The variable with the highest impact on the WTP was income, which can be interpreted as an indicator for reliability of the stated amounts [57]. However, in contrast to most other surveys about environmental goods, this time only passive-use values were of relevance. Further, the area of the Hambach Forest is rather small compared with other woods and forests evaluated so far. Thus, WTP values referring to the area lead to an exceptionally high value compared with previous surveys [20] based on passive-use values only. There is a long-lasting discussion about the sensitivity of scope in contingent valuation surveys [55, 58–65]. Inconsistencies are quite often a result of a lack of spatial sense: only a minority has a clear image of, e.g., 10,000 ha or 50,000 cormorants. However, in cases where existence values dominate over use values, scope is usually of minor relevance for the stated WTP [61, 63]. Lindhjem [35] found in a meta-analysis of Scandinavian WTP surveys for woods and forests only minor scope effects, whereas he argues that woods and forests are complex environmental goods, and simplified indicators such as area size or percentage may not easily capture their scope. Hjerpe et al. [63], in contrast, found that individuals are typically sensitive to the scope of ecosystem service provision, in both quality and quantity. Also Ojea and Loureiro [65] found that CV results are sensitive to the scope of the good being valued, but the results depend on how the environmental change is measured: absolute sizes are preferable over relative ones. Further, Barrio and Loureiro [55] found out that, among others, recreational aspects play an important role for the WTP of people for the preservation of forests. It is therefore possible that the following two effects might have led to "normal" WTP values in our case: the nearly total absence of direct use values might have lowered the WTP for the Hambach Forest, whereas the threat of a drastic change in form of a complete annihilation might have had an opposite effect. Further, the already mentioned symbolic value is reflected in the stated WTP values.

Regarding the results from the regression analyses, the remarkable difference in the rate of explained variance is astonishing: while the decision to be willing to pay or not could reach an acceptable level of explained variance by a regression model, the explained variance for the stated amount remained on a low level. This means that some underlying factors seem to exist, which are not covered by the survey. Imaginable are attitudes toward the procedure of the policy administration in this case, which initiated a strong and disproportionate police operation, in which the Hambach Forest should be freed from occupants and thus also be prepared for a quick clearance, in case of need. Another possible explanation may be a considerable rate of protest responses in our survey. Furthermore, in the view of parts of the population, RWE is often portrayed as voracious based on the fact that it is responsible for having already effaced the largest part of the Hambach Forest and for planning to annihilate it completely. Therefore, the will to contribute to a fund, which prevents the complete annihilation of the forest, may also result from a desire to stop RWE in following its operating plan. The concern of climate change and its mitigation may have played an additional role, since 2018 and 2019 were the first and the third hottest year in Germany since the beginning of the weather recordkeeping. Furthermore, this development was accompanied by the Fridays for Future movement initiated by Greta Thunberg and may thus also have contributed to the wish to combat climate change. Altogether, an interplay of different influencing factors on an individual basis seems to have led to the rather low level of explained variance of the stated WTP values.

6. Conclusions

We conducted a representative CV survey within the German population in order to find out whether there is a symbolic value of the Hambach Forest. It is a remarkable statement for the preservation of the Hambach Forest that at least 47% of the respondents stated a positive WTP, considering that the forest does not provide any direct use values to almost all of the surveyed people. The mean value of the positive WTP was rather high with 55.08€, whereas the standard deviation of the stated values with about 76€ indicates considerable differences regarding attitude and estimation toward the Hambach Forest. Due to the almost complete absence of use values, the stated WTP can be seen as a premium for the existence of the Hambach Forest and as a vote against political decisions regarding lignite mining operation, climate protection, and the acceptance of coal fired power plants, as our regression analyses showed. Further, since the Hambach Forest provides nearly exclusively passive use values, and the WTP values are at least at the same level as in previous surveys with larger forests comprising also direct use values, this can be considered an indication for a premium for its symbolic value. Not least the strong media coverage with very different views reflects the broad spectrum of attitudes toward the Hambach Forest. People who are trying to act more environmentally benign also were significantly more likely to be willing to pay for the preservation of the Hambach Forest. Furthermore, their attitudes toward renewable energies were significantly more positive compared with those without a positive WTP. The symbolic status of the Hambach Forest can be characterized by opposites. For those who rather support the protests, the contrasts might be nature—destruction, climate protection—climate catastrophe, small (population)—big (RWE and provincial government), commons/common welfare—greed/profit. For those who are rather critical toward the protests, the case may represent the defense of jobs, welfare, law, and order against chaos, cadgers, and violent anarchists. Thus, a stated positive WTP can be an indicator for the self-identification of the respondents, whereas due to the unknown rate of protest zeros, the opposite is not so easy to state. A more in-depth analysis would be needed in order to answer the question about the motives of the respondents more profoundly.

Through our survey, we were able to show that the Hambach Forest holds a high valuation among the German population, which is nearly exclusively based on non-use values only, in which a premium for its symbolic value might be included. Considering the results of the CV question and bearing in mind that there is probably a non-negligible proportion of protest zeros, the valuation of the forest as a symbol against climate change is remarkable and should be considered in future political decisions. Especially for the background of the European Green Deal, which means that the European Union aims to become the world's first "climate-neutral bloc" by 2050, these values might be understood as a hint for people's support of an ambitious environmental and climate policy.

Author details

Meike Henseleit[1]*, Sandra Venghaus[2] and Wilhelm Kuckshinrichs[1]

1 Forschungszentrum Jülich, Germany

2 Forschungszentrum Jülich and RWTH Aachen University, Germany

*Address all correspondence to: m.henseleit@fz-juelich.de

References

[1] Council Directive 92/43/EEC of 21 May 1992 on the conservation of natural habitats and of wild flora and fauna. 1992

[2] Presse- und Informationsamt der Bundesregierung. Einigung zum Kohleausstieg [Press release]. 2020

[3] Ibisch P, Blumenröder J, Kriewald S. Hambacher Forst in der Krise. Hamburg, Germany: Greenpeace; 2019

[4] Donahue MZ. Ancient Forest Home of Squatter Communities Is Doomed by Coal. Washington D.C, USA: National Geographic; 2018 Available from: https://web.archive.org/web/20190914181247/https://www.nationalgeographic.com/news/2018/04/hambach-forest-germany-logging-coal-conservation-science/

[5] McLachlan C. Technologies in Place: Symbolic interpretations of renewable energy. The Sociological Review. 2009;**57**(2):181-199

[6] Graham-Harrison E. Greta Thunberg takes climate fight to Germany's threatened Hambach Forest. London, UK: The Guardian; 2019. Available from: https://web.archive.org/web/20190912045758/https://www.theguardian.com/environment/2019/aug/10/greta-thunberg-climate-change-fight-germany-hambach-forest

[7] Smith-Spark L. Hambach Forest Clearance Halted by German Court. Atlanta, USA: CNN International; 2018. Available from: https://edition.cnn.com/2018/10/05/europe/germany-hambach-forest-court-intl/index.html

[8] Monnet J. La symbolique des lieux: pour une géographie des relations entre espace, pouvoir et identité. Cybergeo: European Journal of Geography [Online], Politique, Culture, Représentations, document 56; 1998. DOI: 10.4000/cybergeo.5316. Available from: http://journals.openedition.org/cybergeo/5316

[9] de Groot R, Brander L, van der Ploeg S, Costanza R, Bernard F, Braat L, et al. Global estimates of the value of ecosystems and their services in monetary units. Ecosystem Services. 2012;**1**(1):50-61. DOI: 10.1016/j.ecoser.2012.07.005

[10] Hampicke U. Kosten und Wertschätzung des Artenschutzes und Biotopschutzes. Berlin: Schmidt; 1991

[11] Henseleit M, Holm-Müller K. Vergleich nachfrageorientierter Methoden zur Ausgestaltung der Honorierung ökologischer Leistungen der Landwirtschaft im Rahmen einer ergebnisabhängigen Honorierung im Vertragsnaturschutz, Bonn, Germany. Biological Conservation. 2006

[12] Jax K, Heink U. Searching for the place of biodiversity in the ecosystem services discourse. Biological Conservation. 2015;**191**:198-205. DOI: 10.1016/j.biocon.2015.06.032

[13] Kotchen MJ, Reiling SD. Environmental attitudes, motivations, and contingent valuation of nonuse values: A case study involving endangered species. Ecological Economics. 2000;**32**(1):93-107. DOI: 10.1016/s0921-8009(99)00069-5

[14] Krutilla JV. Conservations reconsidered. The American Economic Review. 1967;**57**(4):9

[15] Marggraf R, Streb S. Ökonomische Bewertung der natürlichen Umwelt:

Theorie, Politische Bedeutung, Ethische Diskussion/Economic Evaluation of the Natural Environment: Theory, Political Significance, Ethical Discussion. Heidelberg: Spektrum, Akad. Verl.; 1997

[16] Pearson RG. Reasons to conserve nature. Trends in Ecology & Evolution. 2016;**31**(5):366-371. DOI: 10.1016/j.tree.2016.02.005

[17] Arrow K, Solow R, Portney PR, Leamer EE, Radner R, Schuman H. Report of the NOAA Panel on Contingent Valuation. Washington D.C., USA: National Oceanic and Atmospheric Administration; 1993

[18] Carson RT. Valuation of tropical rainforests: philosophical and practical issues in the use of contingent valuation. Ecological Economics. 1998;**24**(1):15-29

[19] OECD. Cost-Benefit Analysis and the Environment. Paris: OECD Publishing; 2018

[20] Elsasser P, Meyerhoff J, Weller P. An Updated Bibliography and Database on Forest Ecosystem Service Valuation Studies in Austria, Germany and Switzerland. Braunschweig, Germany: Thuenen Institute; 2016

[21] Bernath K, Roschewitz A, Studhalter S. Die Wälder der Stadt Zürich als Erholungsraum. Birmersdorf, Switzerland: Eidg. Forschungsanstalt für Wald, Schnee und Landschaft WSL; 2006

[22] Elsasser P, Englert H, Hamilton J. Landscape benefits of a forest conversion pro-gramme in North East Germany: Results of a choice experiment. Annals of Forest Research. 2010;**53**(1):37-50

[23] Elsasser P, Weller P. Current and potential recreation value of forests in Germany: Monetary benefits of forest recreation from the population's perspective. Allgemeine Forst Und Jagdzeitung. 2013;**184**(3-4):84-96. Retrieved from: <Go to ISI>://WOS:000321028300006

[24] Kleiber O. Monetäre Bewertung von Erholungsnutzen und Nutzerkonflikten in stadtnahen Wäldern: Konzeption und empirische Prüfung am Beispiel des Allschwiler Waldes. Marburg, Germany: Tectum Verlag; 2006

[25] Weller P. Preferences for forest structural attributes in Germany—Evidence from a choice experiment. Forest Policy and Economics. 2018;**93**:1-9

[26] Bastian OSC, Lupp G, Behrens J, Renner C, Grunewald K. The appreciation of nature and landscape by tourism service providers and visitors in the Ore Mountains (Germany). Landscape Online. 2015;**41**:1-23

[27] Getzner M. Willingness to pay for nature conservation policies in state-owned forests: An Austrian case study. Forests. 2018;**9**(9):537

[28] Schönbäck W. Nationalpark Donauauen: Kosten-Nutzen-Analyse. Austria: Wien; 1997

[29] Civita PD, Filion F, Frehs J. Environmental Valuation Reference Inventory EVRI. 2020. Available from: https://www.evri.ca/en

[30] Beanland RA. Implementation of Sustainable Resource Management: A Process for Environmental Evaluation, Aorangi Awarua Case Study (Master of Resource and Environmental Planning). Massey, New Zealand: Massey University; 1992

[31] Kniivilä M, Oskainen V, Saastamoinen O. Costs and benefits of forest conservation: Regional and local

comparisons in Eastern Finland. Journal of Forest Economics. 2002;**8**(2):131-150

[32] Amirnejad H, Khalilian S, Assareh MA, Ahmadian M. Estimating the existence value of north forests of Iran by using a contingent valuation method. Ecological Economics. 2006;**58**(4):665-675

[33] Veisten K, Navrud S. Contingent valuation and actual payment for voluntarily provided passive-use values: Assessing the effect of an induced truth-telling mechanism and elicitation formats. Applied Economics. 2006;**38**(7):735-756

[34] Broberg T. Assessing the Non-timber Value of Old-growth Forests in Sweden. Umea, Sweden: Umeå Economic Studies, Umeå University, Sweden; 2007;**12**(4):251-277

[35] Lindhjem H. 20 years of stated preference valuation ofnon-timber benefits from Fennoscandianforests: A meta-analysis. Journal of Forest Economics. 2006;**12**(4):251-277

[36] Noonan DS. Contingent valuation and cultural resources: A meta-analytic review of the literature. Journal of Cultural Economics. 2003;**27**(3):159-176. DOI: 10.1023/A:1026371110799

[37] Jacobsen JB, Hanley N. Are there income effects on global willingness to pay for biodiversity conservation? Environmental and Resource Economics. 2009;**43**(2):137-160. DOI: 10.1007/s10640-008-9226-8

[38] Laplante B, Meisner C, Wang H. Environment as Cultural Heritage: The Armenian Diaspora's Willingness to Pay to Protect Armenia's Lake Sevan. Washington D.C., USA: World Bank; 2005

[39] Borchmeyer D. A Very Special Relationship, Germans and their Forest. Berlin, Germany: German Times; 2019

[40] Bundesministerium für Umwelt Naturschutz Bau und Reaktorsicherheit, & Bundesamt für Naturschutz. Naturbewusstsein 2014—Bevölkerungsumfrage zu Natur und Biologischer Vielfalt. Bonn, Germany: 2014

[41] Whitehead JC. Environmental interest group behaviour as self-selection bias in contingent valuation surveys. Growth and Change. 1991;**22**(1):10-21

[42] Whitehead JC, Groothuis PA, Blomquist GC. Testing for non-response and sample selection bias in contingent valuation: Analysis of a combination phone/mail survey. Economics Letters. 1993;**41**(2):215-220. DOI: 10.1016/0165-1765(93)90200-V

[43] Jorgensen BS, Syme GJ, Bishop BJ, Nancarrow BE. Protest responses in contingent valuation. Environmental and Resource Economics. 1999;**14**(1):131-150. DOI: 10.1023/a:1008372522243

[44] Meyerhoff J, Liebe U. Determinants of protest responses in environmental valuation: A meta-study. Ecological Economics. 2010;**70**:366-374

[45] Statistisches Bundesamt. Bevölkerung und Erwerbstätigkeit. Wiesbaden, Deutschland: Statistisches Bundesamt, Germany; 2019

[46] Yan T, Courtin R, Jans M. Trends in income nonresponse over two decades. Journal of Official Statistics Stockholm. 2010;**26**(1):145-164

[47] Riera P, Signorello G, Thiene M, Mattieu P-A, Navrud S, Kaval P, et al. Non-market valuation of forest goods and services: Good practice guidelines. Journal of Forest Economics. 2012;**18**:4

[48] Halstead JM, Luloff AE, Stevens TH. Protest bidders in contingent valuation.

Northeastern Journal of Agricultural and Resource Economics. 1992;**21**(1):160-169

[49] Statistisches Bundesamt. 14. Koordinierte Bevölkerungsvorausberechnung für Deutschland. Germany: Wiesbaden; 2019

[50] Backhaus, Erichson, Plinke, Weiber. Multivariate Analysemethoden. Vol. 10. Heidelberg, Germany: Springer; 2003

[51] Choi AS, Fielding KS. Environmental attitudes as WTP predictors: A case study involving endangered species. Ecological Economics. 2013;**89**:24-32

[52] Filippini M, Martinez-Cruz A. Impact of environmental and social attitudes, and family concerns on willingness to pay for improved air quality: A contingent valuation application in Mexico City. Latin American Economic Review. 2016;**25**(1):1-18

[53] Farreras V, Riera P, Mogas J. Does gender matter in valuation studies? Evidence from three forestry applications. Forestry: An International Journal of Forest Research. 2005;**78**(3):238-248

[54] Economou A, Halkos G. The Gender Environmentalism Gap in Germany and the Netherlands. Social Science Quarterly. 2020;**101**(3):1038-1055. DOI: 10.1111/ssqu.12785

[55] Barrio M, Loureiro ML. A meta-analysis of contingent valuation forest studies. Ecological Economics. 2010;**69**(5):1023-1030. DOI: 10.1016/j.ecolecon.2009.11.016

[56] Kriström B. Spike models in contingent valuation. American Journal of Agricultural Economics. 1997;**79**(3):1013-1023

[57] Schläpfer F. Survey protocol and income effects in the contingent valuation of public goods: A meta-analysis. Ecological Economics. 2006;**57**(3):415-429

[58] Borzykowski N, Baranzini A, Maradan D. Scope effects in contingent valuation: Does the assumed statistical distribution of WTP matter? Ecological Economics. 2018;**144**:319-329

[59] Desvousges W, Johnson F, Hudson S, Wilson K, Boyle K. Measuring natural resource damages with contingent valuation: Tests of validity and reliability. In Hausman JA editors. Contingent Valuation: A Critical Assessment. Contributions to Economic Analysis. 1993;**220**:91-164

[60] Hampicke U. Die monetäre Bewertung ökologischer Güterzwischen ökonomischer Theorie und politischer Umsetzung. Agrarwirtschaft: Zeitschrift für Betriebswirtschaft, Marktforschung und Agrarpolitik. 2003;**52**(8):408-417

[61] Henseleit M. Möglichkeiten der Berücksichtigung der Nachfrage der Bevölkerung nach Biodiversität am Beispiel von Grünland bei der Ausgestaltung eines ergebnisorientierten Honorierungskonzepts im Rahmen des Vertragsnaturschutzes. Göttingen, Germany: Cuvillier Verlag; 2006

[62] Henze A, Kämmerer S, Schmitz P. Die monetäre Bewertung positiver und negativer externer Effekte der Landwirtschaft-Erfahrungen und Perspektiven. In: Linckh G, Sprich H, Flaig H, Mohr H, editors. Nachhaltige Land- und Forstwirtschaft. Exoertisen: Springer Verlag; 1996

[63] Hjerpe E, Hussain A, Spencer P. Valuing type and scope of ecosystem conservation: A meta-analysis. Journal of Forest Economics. 2015;**21**(1):32-50

[64] Meyerhoff J. Die Bedeutung nutzungsunabhängiger Werte

für die ökonomische Bewertung von Umweltgütern. In: Elsasser P, Meyerhoff J, editors. Die ökonomische Bewertung von Umweltgütern. Weimar/Marburg, Germany: Metropolis Verlag; 2001

[65] Ojea E, Loureiro ML. Identifying the scope effect on a meta-analysis of biodiversity valuation studies. Resource and Energy Economics. 2011;**33**(3):706-724

Chapter 8

The Dynamics of Taro (*Colocasia esculenta*) through Value Chain Analysis and Crop Accounting in Partido District, Camarines Sur, the Philippines

Emmanuel A. Onsay, Kevin C. Baltar, Eleanor R. Galicia and Ivan Ruzzel C. Pesino

Abstract

This paper scrutinizes and evaluates the value chain of taro in Partido district, Camarines Sur, the Philippines. Taro (Colocasia Esculenta) is rich in carbohydrates and also a good replacement for wheat flour. However, it is well-known to farmers for its ability to produce a reasonable yield in poor soil conditions with less or no farm inputs. It is this reason why Taro is often grown by resource-limited farmers and regarded as a good source of food security. To tap its full potential, this study was conducted to understand and analyze the flow of Taro. This study employed participatory techniques, crop accounting, and financial analysis. The Taro value chain has varied gender roles, according to this study. It also highlighted why, despite having a choice, farmers frequently chose the less profitable transaction path. It also solved the mystery of low productivity in this locale as compared with the national production. The results also showed the cost build-up of Taro and the profitability of each player in the chain. Farmers usually end up at a 22% profit-to-cost ratio by selling the harvest to the middleman, while sellers usually end up at a 47% profit-to-cost ratio. The existing entry barriers in each chain were also identified with the researchers' recommendations on how to possibly eliminate or mitigate them. This extensive analysis can be valuable to stakeholders in the Taro value chain in the area, as well as government entity and non-governmental organizations in developing initiatives or projects on behalf of the players.

Keywords: Taro, value chain analysis, crop accounting, productivity, the Philippines

1. Introduction

Taro (Colocasia Esculenta), locally known as Natong or Linsa, is the most widely cultivated species of several plants in the Araceae family. It is thought to be native to Southern India and Southeast Asia but is widely naturalized [1–4]. The majority of taro research has focused on its anatomy, biology, and physiology. As a result, a wide spectrum of taro value chain evaluations, including agricultural accounting and financial analysis, is required. Other value chain players, such as intermediaries, processors, and sellers were included in this analysis to account for the dynamics of the value chain beyond the hands of the farmers. Furthermore, no research on the taro value chain has ever been done in this exact location, making this study even more unique and significant.

The term "value chain" refers to a series of activities through which items move in order and gain value at each stage [5–7]. It encompasses the entire range of operations from production to consumption [8]. Taro is a tropical perennial plant that is primarily farmed for its edible starchy corm and as a leaf vegetable. It's a common food in African, Oceanic, and Indian cultures, and it's said to be one of the first cultivated plants [9]. The plant can be used as a sweetener in beverages, candies, and pastries, and can also be used as a substitute for wheat flour. It's also a resilient crop that provides farmers with a steady income. It's one of the few staple crops that can be farmed on a small budget and in difficult conditions where other crops would fail. In the tropical zone, it is also high in carbohydrates and a good source of calories.

The Philippines' taro plantation area is shrinking, with a total size of 14,992.84 hectares as of 2020. However, in Camarines Sur, where the Partido district is located, the area of plantation remains nearly constant at 281 hectares. In terms of taro production in the country, it is normally declining, although it grew in 2020 to 107,422.18 tons. The production of taro in Camarines Sur is dropping, with a total of 2,433.53 tons, and the province ranks second in the Bicol area after Albay. Taro output in the Bicol region is relatively low and limited in comparison to other regions and provinces. From 2015 to 2020, the farmgate price of taro in the Philippines is PhP15-25.00, whereas it is PhP16-26 in Camarines Sur. From 2014 to 2020, the retail price of taro in the Philippines is PhP29-49.00, while it is PhP42.00-48.00 in the province of Camarines Sur [10, 11]. Currently, there is no specific policy in the country in general or in the district in particular regarding the potential expansion of taro production. As a result, the goal of this study was to produce data that could be used as input for efficient and productive taro cultivation and distribution throughout the district. Achieve long-term productivity and sustainability, it is also in consonance with the UN Sustainable Development Goals (SDF), specifically Goal No. 1 No Poverty and Goal No. 2 Zero Hunger. It may also be indirectly related to Goal No. 12 Responsible Consumption and Production [9]. In Asia, taro cultivation is important because the crop serves as a staple food that ends the hunger of many households. In Oceania, taro cultivation is important because it plays a significant role in the national food security that alleviates poverty. In the Pacific islands, taro cultivation draws substantial attention on cultural aspects and socio-economic dominance [12–16]. Taro cultivation is important in Solomon Island because it is part of daily diet and custom [17]. In the Philippines, the plant is vital as a source of food and is also being cultivated for domestic purposes [18–20]. According to the UN Food and Agriculture Organization 2018, taro cultivation increases food security and enhances the livelihood of low-income households amidst climate change when sustainable utilization and conservation of the crop are made. Taro is an agricultural crop that can greatly contribute to gross value added in agriculture of a country when utilized properly [21]. The output of the study aims to eliminate poverty and reduce hunger in the district through the efficient utilization and production of taro. Considering the foregoing, value chain analysis is needed to understand how these potentials and contributions to the economy travel from their origins to the market.

2. Research methods

2.1 Materials

The analysis was carried out in 2020, during the emergence of CoVid-19 in the region but before the impact of three major typhoons that hit the region in the same year, destroying taro output severely.

Key informant interviews (KII) and focus group discussions (FGD) were undertaken in order to acquire primary data. Furthermore, document review was oriented to collect extensive data that might be used to compare study findings to current assertions on a certain issue. Because there are no existing records of taro players in the Partido district, a purposive technique using snowball sampling was utilized to find the responders.

The investigation includes every recognized, operational, and identifiable entity of farmers, merchants, and middlemen. Ten producers (planters/farmers), seven middlemen and processors, and twelve sellers of the commodity responded. Other major informants who contributed significantly to the study's completion were Municipal Agriculturists, Local Government Unit (LGU) officials, Department of Trade and Industry (DTI) employees, and Department of Agriculture (DA) officers. In order to collect data, the researcher employed a structured questionnaire as the major tool.

The preliminary inquiry was first carried out in conjunction with the Municipal Mayor's Office in several municipalities within the Partido area. The researcher worked with several officials to find the respondents after determining possible barangays. Crop accounting spreadsheets were used to construct summaries of pertinent accounting information using Stata and Excel.

2.2 Methods

To evaluate the roles and relationship dynamics of participants in the network, a value chain analysis and value chain mapping were conducted. Crop accounting techniques were utilized to assess, account for, and examine productivity, logistics, and marketing costs, value-added, and returns.

The profitability of total activity in each chain was determined using the Return on Revenue (ROR) method. It calculated the association between net income and revenue generated by the activity. The Gross Profit Rate (GPR) was computed to examine the link between cost of sales and gross revenue in determining the degree of company risk. The Cost to Revenue Ratio (EC/R) was calculated to examine each player's expected income in a certain transaction chain.

Furthermore, the profit-to-cost ratio was utilized to analyze the overall attractiveness of the activity by measuring the interaction between the profit generated and the costs expended. The value-added in each chain was computed to determine the activity's liquidity.

The equations and formulae below were used [22]:

$$\text{Gross Sales} = \text{Quantity Sold x Unit price}; \quad (1)$$

$$\begin{aligned}\text{Cost of Goods Sold} = {} & \text{Direct Materials} + \text{Direct Labor} \\ & + \text{Variable Overhead Costs} + \text{Fixed Overhead Costs}; \quad (2)\end{aligned}$$

$$\text{Gross Margin}(\text{in pesos}) = \text{Sales} - \text{Cost of Goods Sold;and} \quad (3)$$

$$\text{Gross Margin Percentage} = \text{Gross Margin} / \text{Sales x } 100; \quad (4)$$

$$\text{Value Added}(VA) = \text{Costs incurred} - \text{selling price}; \quad (5)$$

$$ROR = \frac{netincome}{grossincome}; \quad (6)$$

$$GPR = \frac{grossincome}{grossrevenue}; \quad (7)$$

$$\frac{EC}{R} = \frac{TotalExpenses}{grossrevenue}; \quad (8)$$

$$ProfittoCostRatio = \frac{Profit}{Costs}. \quad (9)$$

3. Results and discussion

3.1 Socio-demographic profile of each player and level of involvement of farmers in the taro value chain in Partido district, the Philippines

The table shows that the average age of the farmers is 55.3 years old and that the majority are in the age bracket of 41–50. As compared with the result of the PSA study which revealed an average age of 48, it can be interpreted that this study does not present any notable difference against the PSA's as far as age is concerned. The average age for middlemen/processors and sellers is 42 and 52.70, respectively and the majority of the respondents are also in the bracket of 41–50. **Table 1** also shows that 60% percent of the farmers are male, 85% of middlemen are male and 58% of the sellers

Profiles	Taro farmers	Taro middlemen & processors	Taro sellers
Count of entity	10	7	12
Gender	Male	Male	Female
Marital status	Married	Married	Married
Average age	55.3	42	52.75
Average monthly family income (PhP)	9,800.00	13,400.00	15,625.00
Family size	7	8	6
Highest educational background	High School	High School	High School

Table 1.
Players in the taro value chain's socio-demographic characteristics (Partido district, the Philippines).

are female which reflects the different gender roles in production and marketing. The findings of the study are congruent with the results as corroborated in Cassava post-harvest systems. Women are responsible for marketing while men are responsible for farming [23]. However, this study does not deal with the issues about gender involvement in technology along with the VC, specifically the level of gender involvement in Taro processing when commercialization and mechanization increase.

The level of farmers' income was found below the country's poverty threshold (5.41 × Php 1,813 = PhP9,800) while sellers' income hovers above the threshold (8.25 × Php 1,813 = Php 14, 957 & 8.618 × 1,813 = 15,625). It indicates that taro farmers are resource-limited which a typical scenario is because taro can be grown with less or no farm inputs [14]. On the other hand, taro sellers live slightly above the poverty threshold. It's also worth noting that both actors have significantly more family members than the country's average household size of 5.2 individuals. Both players have the same civil status and educational attainment model classes.

3.2 Identification of value chain characteristics pertaining to the relationships of participants from farmers to purchasers using farm to market analysis

Farmers, middlemen/processors, and sellers make up the Taro VC in Partido, Camarines Sur. The sellers' prices are frequently cheaper than the middlemen's. The majority of farmers sold their produce to the intermediary since it was profitable and the crop is perishable. Fresh taro tubers decay in 2 to 3 weeks, therefore selling it at a lesser price but in bulk is preferable to selling it (to sellers) at a much higher price but in pieces. Unlike the farmer-to-intermediary value chain, where only the middleman determines the price, the seller-to-market value chain contains a number of variables that might influence Taro's price.

3.3 Profitability and liquidity of the players through crop accounting and financial analysis

3.3.1 Costs and returns of taro farmers

Based on transaction path 3 and 4 (please refer to **Figure 1**), switching buyer from middleman to seller will yield a substantially greater profit; that is, from 3.395 to 9.651 profit-to-cost ratio. However, other factors are present such as bulkiness of harvest, which it is likely that great portion of the harvest will not be sold and set aside to rot. The resulting loss to farmers can likely outweigh the benefit of a higher price. This provides little or no options to farmers but to sell it to middlemen even though it will

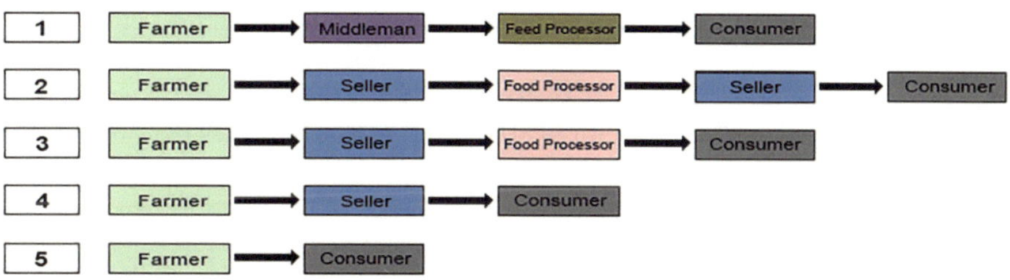

Figure 1.
Distinct taro value chain transaction paths in partido district, the Philippines.

Items	Per hectare			Per farm	Per Kg.	Cost/ kg.	Percentage
	Quantity	Unit	Value (₱)				
Production	727.732	kg.	25,526.655	67,262.737	35.077		
Area harvested 2.64 hac.							
EXPENSES							
Materials (Input Stage)							
Planting Materials			314.253	828.056	0.432		
Fertilizers			15.275	40.250	0.021	0.453	6%
Labor (Production Stage)					–		
Hired Labor			3,276.630	8,633.919	4.503		
Rentals - Machinery			610.859	1,609.614	0.839		
Rentals - Animals			283.882	748.028	0.390	5.732	74%
Others (Production Stage)					–		
Land Rentals			217.263	572.488	0.299	0.080	1%
Transportation (Logistics Stage)					–		
Transportation of Materials (to Farm)			25.021	65.929	0.034	0.034	0%
Transportation of Harvest (Farm to Coop)			1,064.948	2,806.138	1.463	1.463	19%
Total Expenses			5,808.130	15,304.422	7.981	7.763	100%
Gross Returns			25,526.655	67,262.737	35.077	35.077	439%
Net Returns			19,718.525	51,958.314	27.096	27.314	
Net Profit to Cost Ratio			3.395	3.395	3.395	3.519	

Table 2.
Average production costs and returns of taro farmers (Partido district, the Philippines).

result to, if not loss, lesser profit. It implies that in **Figure 1**, transaction paths 2, 3, and 4 are more profitable to farmers, yet they often choose transaction path 1.

The study also revealed that farm production averages 2,727.730 kilograms per hectare, which is too far below the quantity a result of the study conducted by PSA. This could be due to the non-application of organic fertilizer [14]. In fact, the respondent with the highest production per hectare was the only one using fertilizer. It is followed by the two respondents having the highest time spent per week in Taro farming (**Table 2**).

Taro sellers, on the other hand, exhibit a reasonable profit of Php39.218 per kilogram of Taro tubers or 9.56 margin-to-cost ratio. It can also be noted in the figure that

Items		Sales/quarter (₱)	Per kg.	Cost/kg.	Percentage
Production		**17,106.030**	**45.000**		
Average sale per quarter: kg.	280.134				
EXPENSES					
Direct Materials (Input Stage)					
Taro Tubers		1,103.728	3.940	3.940	68%
Marketing Costs (Marketing Stage)		–	–		
Rentals - Place of Operation		27.033	0.097		
Pasada		20.465	0.073		
Payment to Municipal Office		71.006	0.253	0.423	7%
Transportation (Logistics Stage)		–	–		
Transportation of Harvest (Farm to Seller)		397.579	1.419	1.419	25%
Total Expenses		1,619.812	5.782	5.782	100%
Gross Returns		17,106.030	45.000		778%
Net Returns		15,486.218	39.218		
Net Profit to Cost Ratio		9.561	6.782		

Table 3.
Average production costs and returns of sellers (Partido district, the Philippines).

logistics, a non-value-adding activity, takes up much of the portion of the total cost for farmers and sellers. This is due to the bulkiness and low value of Taro tubers which cannot take substantial expenses without suffering a net loss.

3.3.2 Costs and returns of taro sellers

Taro merchants, on the other hand, appear to maximize profit, earning PhP24.38 per kilogram of taro produce, or 303% gross returns. Farmers have the largest gross returns, followed by sellers, processors, and intermediaries, according to the prior and subsequent financial statements. Furthermore, it appears that taro intermediaries do not necessarily advance in the district, thus the actors are mostly farmers and merchants. Middlemen must outsource and transact other root crops or agricultural items other than taro since their revenues are lesser (**Table 3**).

3.4 Taro value chain diagram

According to **Figure 2**, farmers typically engage with two types of traders: sellers, processors, and intermediaries. There have been instances where they have sold it straight to customers; however, this occurs seldom and in extremely little quantities. Sellers, on the other hand, can sell it directly to customers or to food processors, depending on the price agreed upon. This is most common in the markets of central Goa, San Jose, Tigaon, and Lagonoy, where marketing activities take place twice

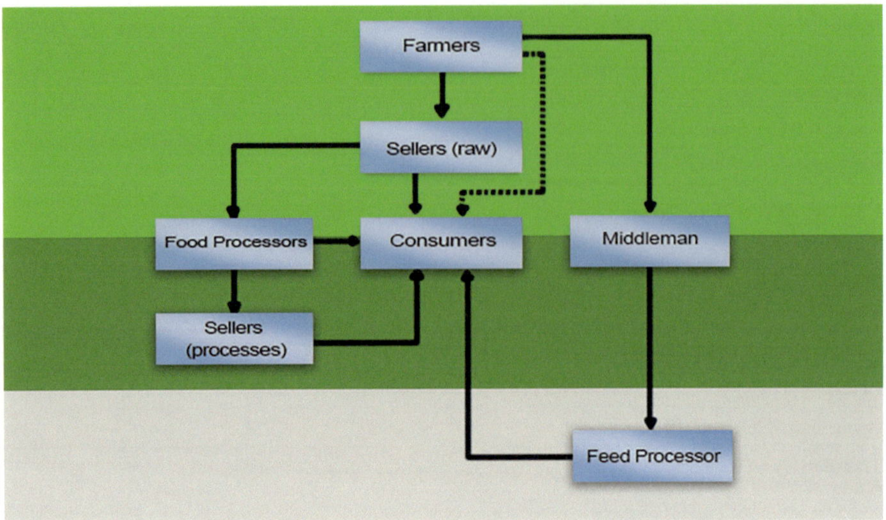

Figure 2.
Diagram of the Taro value chain in the Partido District, the Philippines.

a week. Taro processing begins when taro is sold to a customer or a food processor, as indicated by the dark green part of the map.

During the data collection, it was discovered that the intermediary was acting in two roles: as an assembler and as a processor. Farmers sell raw Taros, which are then processed into food and/or culinary additives. Taro is transported from the farm to the storage facility using the middleman's vehicle. Granulation is the process of cutting tubers into little pieces using a granulating machine. Then it's dried for 2 to 3 days on a drying pavement or in a drying facility. It is sold to a feed processor outside of the Municipality of Goa after processing, as indicated by the gray part of the map.

The distribution of Taro in Partido, Camarines Sur is also depicted in **Figure 1** as transaction pathways. The main line of the local Taro value chain, Transaction Path 1, provides minimal returns to growers. The most profitable transaction paths for farmers are 2, 3, and 4, with path 2 being the longest and least sorted. Sellers, on the other hand, have three options: transaction 2, 3, or 4, all of which give an acceptable and equal return.

3.5 The taro value chain's assets, liabilities, interest, revenue, and costs

In a structured interview by the researchers with the respondents of this study, the only accounts they recognized for the asset section are cash, receivables, machinery and equipment, loan receivables, supplies, and furniture and fixture. For the liability section, accounts payable only. For the equity section, they have an unrestricted fund which they used in farming operations or in selling activities, they also recognized expenses (outflows), such as transportation expense/logistics, expenses, miscellaneous expenses, rentals, registration expenses, fertilizers expenses, and tubers expense. They only have sales income for income/revenues (inflow). They do not, however, account for non-cash expenses, hence depreciation was calculated using the straight-line approach rather than the diminishing balance method. Crop accounting analysis was used to assess all of the accounts concerned.

3.6 Existing taro value chain entry constraints

Since the location is located in the Philippines' super typhoon capital zone, typhoons are the natural entry barriers. Most of the farmers claimed that they have lower production than the previous harvest. They attributed it to the poor soil condition which is typical to the Taro farmlands when the soils are not properly managed or used. This study revealed that all of the respondents employ conventional tillage which is not a sustainable farming practice and can adversely affect Taro production. This practice can increase the initial year of farm production; however, it destroys the soil composition and kills the microorganisms, which nourishes Taro in times of nutrient depletion, by exposing it to the scorching sun [14].

Most of the farmers also identified the lack of capital as their major difficulty in farming Taro. This also resulted in a lower farm yield because of the inability to buy the necessary farm inputs. As stated earlier, only one respondent used fertilizer and produced the highest yield. It is due to the fact that, in addition to its ability to produce a reasonable yield on low-fertility soils, Taro responds very well to fertilization. Being aware of their capital deficiency, most of the respondents recommended financial support as a key to improving their Taro production.

Farmers also complained that the low price of the product is their major marketing difficulty. This finding is in consonance with the study conducted and data provided by authoritative agencies [24, 25]. On the other hand, sellers complained that instability of price is their major difficulty in marketing Taro. It is primarily due to several factors that can lead to price determination. However, their recommendation was on the provision of a proper place of operation. Sellers in the market have no permanent location and have to find their own place every time they want to sell goods. They usually conduct their marketing activities under their makeshift tents which poses a big problem when bad weather comes.

3.7 Assess the accounting practices applied by each player in the value chain.

In this part, the researcher determined whether the players maintained financial records of their transactions, the books being maintained, and the time basis for keeping a record. It also includes the accounting method used in recording, the accounting system, the bookkeeping system, the financial statements being prepared, and the time interval in preparing the financial report. Apparently, most of the players are not keeping accounts or financial transactions. They simply record their income and expenses but they did not keep this record in the long run.

3.8 Propose accounting enhancement program for different players of the value chain for University Extension

Necessary mechanisms should be proposed to improve the recordkeeping and financial reporting practices of the players. Based on the documentary analysis, most respondents are not maintaining financial records. Thus, accounting enhancement program for different players in the value chain is proposed. It is proposed by the researcher to maintain book of accounts because chronological records of transactions that explain economic event happening in the organization will be monitored, the accounts being affected and the amount of it. They also need to maintain general ledger because it is the master set of accounts that summarized all transactions occurring within the organization. However, the balance between cost and benefit should be considered.

4. Conclusions

Those who live below the poverty line usually engage in taro farming, while those who live just over the poverty line mainly engage in marketing. Men are associated with manufacturing, whereas women are associated with marketing. Farmers frequently choose the less profitable transaction path due to the perishability of the crop. Agriculture and processing technology are both deficient and unavailable. It is consequently advised that the governments and NGOs' efforts and activities for farmers to achieve higher production levels be shifted. Low production indicates that the taro industry in the Partido district is not being fully utilized. As a result, government funding should be made available to help farmers obtain the farm supplies they need to increase productivity. Apart from that, it may be deduced that logistics, as a non-value-adding component, has reduced returns while providing no advantages. As a result, a technical mechanism can lower or eliminate non-value-added costs.

Crop accounting is based on past expenses, but it does not account for the implications of imputed costs. As a result, it is suggested that a study be conducted that includes family labor. Another suggestion is to provide transportation infrastructure and amenities, as inefficiency in logistics is difficult for the chain's players to pay. The farmers advocated for financial assistance and the provision of processing machinery, while the sellers advocated for an appropriate operating location and training workshops. In the creation and implementation of government policies and NGOs' initiatives, entry barriers should be considered. Players' financial accounts are not kept up to date. As a result, it is strongly advised that participants keep financial records using simple bookkeeping. The institution may offer an accounting upgrading program to aid the players with their recordkeeping procedures.

Acknowledgements

This project was funded by Partido State University (ParSU) – College of Business and Management (CBM), and the Research & Development Office (R&D). The authors express their heartfelt thanks to *Pres. Raul G. Bradecina, Prof. Patricia Candelaria, Dean Rina A. Abner, Dir. Luisa Lanciso, and Prof. Rolan Jon G. Bulao* for holistic support. Likewise, to *Dean Arlene Inocencio, Dr. Dickson Lim, and Dr. Alellie Sobreviñas* of De La Salle University (DLSU) – School of Economics (SOE), for statistical and econometric methods. The researchers also convey their great appreciation to *IntechOpen* for the opportunity. Furthermore, *Mark Rey*, the research enumerator, deserves special commendation for his generous support. *To God be the glory!*

Author details

Emmanuel A. Onsay[1,2*], Kevin C. Baltar[3], Eleanor R. Galicia[4] and Ivan Ruzzel C. Pesino[5]

1 Faculty of Accountancy and Economics, Partido Institute of Economics, College of Business and Management, Partido State University, The Philippines

2 Applied Economics, School of Economics, De La Salle University, Manila, The Philippines

3 Faculty of Economics, Partido Institute of Economics, College of Business and Management, Partido State University, Camarines Sur, The Philippines

4 Faculty of Accountancy, Business, and Management, San Rafael National High School, Camarines Sur, The Philippines

5 Faculty of Accountancy, College of Business and Management, Partido State University, Camarines Sur, The Philippines

*Address all correspondence to: emmanuel.onsay@parsu.edu.ph

References

[1] Rashmi DR, Raghu N, Gopenath TS, Palanisamy P, Bakthavatchalam P, Karthikeyan M, et al. Taro (Colocasia esculenta): an overview. Journal of Medicinal Plants Studies. 2018;**6**(4):156-161

[2] Srinivas T, Nedunchezhiyan M, Misra RS. Marketing of taro in India. NSCFT Proceedings. 2011;**2011**:609-612

[3] Kolchaar K. Economic Botany in the Tropics. Bangalore, India: Macmillan; 2006

[4] Suminarti NE, Ariffin GB, Rayes ML. Effect of fertilizer application and plant density on physiological aspect and yield of taro (Colocasia esculenta (L.) Schott var. Antiquorum). International Journal of Agricultural Research. 2016;**11**(1):32-39

[5] Hempel E. Value Chain Analysis in the Fisheries Sector in Africa. 2010. [Accessed: July 25, 2017]

[6] Trienekens JH. Agricultural Value Chains in Developing Countries A Framework for Analysis. 2011. [Accessed: July 10, 2017]

[7] Webber M, Labaste P. Using Value Chain Approaches In Agribusiness and Agriculture in Sub-Saharan Africa. 2008. [Accessed: July 28, 2017]

[8] Hellin J, Madelon M. Guidelines for Value Chain Analysis. 2006. [Accessed: July 20, 2017]

[9] United Nations Sustainable Development Goals (UNSDGs). 2020. Available from: https://www.un.org/sustainabledevelopment/poverty/

[10] Philippine Statistics Authority. Area, Volume, Cost and Returns of the Taro Production in the Philippines. 2018.

[11] Philippine Statistics Authority Open Statistics. Cost and Returns of the Taro Production in the Philippines. 2020

[12] Akwee PE, Netondo G, Kataka JA, Palapala VA A Critical Review of the Role of Taro Colocasia esculenta L. (Schott) to Food Security: A Comparative Analysis of Kenya and Pacific Island taro germplasm. Scientia Agriculturae. Pakistan. Available from: www.pscipub.com/SA. E-ISSN: 2310-953X, P-ISSN: 2311-0228 2015.

[13] FAOSTAT. Chicago, USA: FAO; 2008. Available from: www.fao.org/ag/agpc/gcds/ [Accessed: July 20, 2017]

[14] FAO. Save and Grow: Taro. Food and Agriculture Organization of the United Nations. 2013. Available from: http://www.fao.org/3/a-i3278e.pdf [Accessed: July 5, 2017]

[15] FAOSTAT. 2017. Food and Agriculture Organization of the United Nations. Available from: http://www.fao.org/faostat/en/#data/QC/visualize [Accessed: July 22, 2017]

[16] Onwueme I. Taro cultivation in Asia and the Pacific. RAP Publication. 1999;**16**:1-9

[17] Liloqula R., Saelea J., Levela H. Traditional taro cultivation in the Solomon Islands. 1993

[18] Kreike CM, Van Eck HJ, Lebot V. Genetic diversity of taro, Colocasia esculenta (L.) Schott, in Southeast Asia and the Pacific. Theoretical and Applied Genetics. 2004;**109**(4):761-768

[19] Matthews PJ, Agoo EMG, Tandang DN, Madulid DA. Ethnobotany and ecology of wild taro (Colocasia esculenta) in the Philippines:

Implications for domestication and dispersal. Senri Ethnological Studies. 2012;**78**:307-340

[20] Villanuev MR, Tupas, PR. Taro Production in the Philippines – Its Prospects and Problems 1. 2021

[21] Food and Agriculture Organization of the United Nation. Conservation and Sustainable Utilization of Underutilized Taro to Increase Food Security and Improve Livelihoods of Marginalized Communities Faced with Climate Change.pdf. 2018

[22] Onsay EA. Productivity value chain analysis of cassava in the Philippines. In: IOP Conference Series: Earth and Environmental Science. Bristol, England and Philadelphia, United States: IOP Publishing; 2021

[23] Martin A, Forsythes L, Butterworth R. Gender implications of developing cassava postharvest systems. Expert Consultation on Cassava Processing, Utilization, and Marketing. 2008

[24] Meridian Institute. Innovation for Agricultural Value Chain in Africa: Applying Science and Technology to Enhance Cassava, Diary, Maize Value Chain. 2009. [Accessed: July 20, 2017]

[25] Philippine Statistics Authority Open Statistics. 2004. Available from: https://openstat.psa.gov.ph/

Chapter 9

Building New Rural Areas in Vietnam

Nguyen Hay, Le Quang Huy and Pham Van Kien

Abstract

The study focused on the orientations of building sustainable new rural areas associated with urbanization. This was very necessary, in order to develop a concentrated, commodity agriculture and to form residential areas and infrastructure, which were suitable to the process of industrialization and modernization of agriculture and rural areas. The agricultural sector was restructured, and the living conditions of rural people were improved close to those in urban areas. The project of the National Target Program on building new rural areas in Vietnam was analyzed. The program has achieved many expected results. In which, the construction of new rural areas has achieved great achievements. The basic socioeconomic infrastructure in the rural areas was strengthened in a synchronous direction. The sustainable development of the rural economy was in the direction of increasing added values and raised people's incomes and gradually narrowed the gap between rural and urban areas. The work of landscape construction, implementation of environmental sanitation had a remarkable change.

Keywords: new rural areas, infrastructure, industrialization, national target program, socioeconomic, sustainable development

1. Introduction

1.1 Some concepts of new rural area construction

1.1.1 The concept of new rural area

A new rural area is a modern civilized rural area but still retains the traditional beauty. New rural area is the result of a new strategic-oriented rural development process, meeting new economic, social, environmental, and institutional development requirements in rural area under specific conditions of each rural area [1]. The new rural area is expressed through the following basic contents:

- The people's material, cultural, and spiritual life of people in rural area is constantly improved, which gradually reduces the gap between rural and urban areas.

- A rural area has spacious and civilized villages with modern and synchronous infrastructure, developed according to the planning to ensure favorable conditions for production and daily life of rural residents, combined between agriculture and industry, services and cities.

- A rural area has reasonable economic structure, comprehensive development. The economic fields are developed toward efficient and sustainable production of goods. Farmers are trained, absorbed advanced technical engineering.

- A stable rural area is clean, beautiful, and the ecological environment is protected. The political security and social order remain.

1.1.2 New rural area construction with the goal of sustainable development

The criteria reflecting the sustainable development goals in Vietnam rural area cover all economic, social, and environmental aspects [2] as below:

- Economical sustainable development is fast, safe, and qualitative development. One sustainable economy needs to meet the following requirements: (i) having high GDP growth and GDP per capita; (ii) GDP structure is a criterion of evaluating economic sustainable development; (iii) economic growth must be growth with high efficiency and the growth with safe environment.

- Social sustainable development is assessed by criteria such as: HDI (Human Development Index), income equality coefficient, indicators on education, health, social welfare, cultural enjoyment. In addition, social sustainability is the guarantee of a harmonious social life; there is equality between social classes, gender equality; the gap between richness and poorness is not too high and tends to be small; the difference of life between regions is not large.

- Sustainable development on the environment includes: (i) effectively using resources, especially nonrenewable resources; (ii) development does not exceed the load-bearing threshold of the ecosystem; (iii) protecting biodiversity and ozone layer; (iv) controlling and reducing greenhouse gas emissions; (v) protecting sensitive ecosystems; (vi) minimizing discharge, overcoming pollution (water, gas, soil, food), improving and restoring the environment of polluted areas.

1.1.3 Sustainable new rural construction associated with urbanization

The overall objective of the whole program is to build a new rural areas to improve the material and spiritual life of the people; Socioeconomic infrastructure is appropriate; economic structure and forms of production organization are reasonable, agricultural development and industry and service are linked; rural development combines with urban areas; a rural society is democratic, equal, and stable; ecological environment is protected; the political security and social order remain.

Sustainable new rural construction associated with urbanization is an integrated process in which the construction is associated with sustainable development goals and the urbanization process [3].

1.2 The experiences in rural development and new rural construction in the world

1.2.1 China

China government demanded that it was necessary to gradually change "the disparity between industry and agriculture, between urban and rural areas, between regions," "unifying urban socio-economic development planning, building modern agriculture, developing rural economy, increasing income for farmers" [4].

Up to now, China has achieved many great achievements such as: agriculture fields and rural areas have developed strongly. China's agriculture has formed many high-value agricultural products such as: food, livestock, natural rubber, and fruits. Many agricultural products are cultivated on a large area with high output and high economic efficiency. The number of large enterprises operating in the agricultural sector as well as the number of cooperatives and associations constantly increased. As a result, the rural agricultural economy has developed rapidly, the average income of farmers has increased significantly. The infrastructure of rural areas has been significantly improved. Building a cultural life in rural areas has achieved many important achievements.

Useful experiment:

- Attracting businesses to invest in agriculture and rural areas
- Forming linkages along the value chain and implementing commitments between farmers and enterprises.
- Encouraging many large economic groups to invest in agriculture

1.2.2 Japan

The agricultural economy is prioritized for development by the Japanese Government. Therefore, at each stage, Japan applied a different plan and "agricultural extension policy" such as: "Socio-economic development plan," "Building rural area as an attracting and comfortable living space," "one village one product" in order to carry out the construction of new rural areas, create a foundation for agricultural economic development with the solidarity of people [5].

Japan developed agricultural economy from the first small-scale villages from 900 to 1000 farming households, then replicated the model to 4548 villages. Japan government has applied a financial support policy to build new rural areas in addition to local revenue and the loans from agricultural credit funds.

The Japanese government has launched the movement as "One Village One Product (OVOP)." OVOP is a movement with three main principles: (i) localization and then globalization; (ii) autonomy, independence, and creation; (iii) human resource development. The government is in charge to do the basic construction items of the rural areas, improve the environment, bring water, electricity, road construction, information (telephone) to the people. The primary education was completely free, rural and urban areas supported each other for sustainable development.

Useful experiment:

- Three main principles: (i) localization and then globalization; (ii) autonomy, independence, and creation; (iii) human resource development.
- Applying to craft villages, regions with specialty products and implementing the branding, traceability production, and product consumption.
- Preserving and developing traditional craft villages under the motto "One village, one product," developing crafts according to local strengths was one of the core contents to realize the goal of "transforming structure, economic development and increasing income of people"

1.2.3 South Korea

The SU (Saemaul Undong) movement was established with three criteria: (i) diligence (hard work); (ii) self-reliance to overcome difficulties; (iii) cooperation (community synergies). As a result, many projects have been implemented and completed with the aim of developing infrastructure in rural areas. South Korea has applied high technology in agricultural production in order to increase productivity and value of agricultural products. The number of economical autonomous villages in South Korea reached 98%. Thus, the effectiveness of the SU movement made an important contribution to the sustainable development of the South Korean rural economy [6, 7].

Useful experiment:

- The leader's determination
- The role of training and building capacity of managers.
- Promoting the role of the people at the village.
- Creating motivation instead of pressure that associated with the spirit of "industrious, self-reliant, reunited."

1.2.4 Taiwan

The important role of farmers' organizations was one of the factors that made up the success of agricultural development. Taiwan has four farmers' organizations such as: the Agricultural Association, the Fruit Cooperative, the Irrigation Association, and the aquaculture Association. Basically, they were economic organizations that cooperate each other to provide nonagricultural services, including supplying materials and consuming agricultural products. The main function of these organizations was to help farmers in trading activities. All four organizations were registered to operate under the state management of the government [8].

The government focused on supporting many aspects of the Agricultural Association. First of all, 50% of the farm's capital was provided by the government. Besides, the government implemented many direct investments in rural areas such as building infrastructure, transferring new varieties and technology through development programs. Basing on the activities of the Agricultural Association, Taiwanese farmers have mastered the entire supply chain of input materials and output products of agricultural production.

Useful experiment:

- The model of farmer association, forming farmers' organizations, not only organized production but also represented the voice of farmers in policy criticism and policy implementation.

1.2.5 Other ASEAN countries

Policy directions for rural development in ASEAN countries focused on many aspects such as: from the program of promoting the rural economy to the program of infrastructure development; from improving the life quality of rural residents to developing high-quality human resources and protecting natural resources and the environment in rural areas. Mainly, the Rural Development Program of ASEAN countries focused on three main points: (i) strengthening the legal authority for rural people; (ii) training and developing human resources in rural areas; and (iii) agricultural and rural development associated with environmental protection and sustainable development [9, 10].

In generally, the useful experiment, which was applied in building new rural area in Vietnam, could be considered as: i) encouraging many large economic groups to invest in agriculture to form the linkages between farmers and enterprises. ii) preserving and developing the traditional rural villages' values including both cultural life and production. iii) Implementing the human resource development, in which, the role of the rural people was promoted and the capacity of managers in building new rural area should be considered. iv) Implementing the process of building new rural area step by step in accordance with the Government's strategy.

2. The strategy of building rural areas in Vietnam

2.1 The process of forming the national target program on building rural areas in Vietnam

The process of forming the National Target Program on building new rural areas in Vietnam through the periods was shown in **Figure 1** [11, 12].

In which, in 1988, the Vietnamese government implemented the renovation policy, which created significant developments for Vietnam's agriculture in the renovation period. Orientations for building a new rural areas have been set out in terms of rural planning, infrastructure development, production organization, building a prosperous rural life, solidarity, democratization and publicization, and promoting the mastery of the working people.

In the period of 2001–2009, there were two pilot programs of building new rural areas such as: the program on building a model of rural development in the direction of industrialization, modernization, cooperation, and democratization in the period of 2001–2005 and the pilot program on building new rural areas at village levels in the period of 2006–2009. The practical experience from the pilot programs showed that in order to achieve the goals of building a new rural area, it was necessary to

Figure 1.
The formation of the National Target Program on building new rural areas in Vietnam.

develop in a harmonious manner in different fields, not just to focus on building infrastructure.

In the period of 2009–2011, the Vietnamese government directed the implementation of the pilot program on building a new rural area in the period of accelerating industrialization and modernization in the period of 2009–2011. In particular, the new rural area criterion was promulgated that has created a comprehensive rural development approach and become a prominent feature of rural development in Vietnam.

The lessons learned from the pilot programs were the basis for officially implementing the National Target Program on building new rural areas for the period of 2010–2020. Thus, after more than 20 years of innovation, basing on lessons learned and practical needs as well as the suitability of the general context, the National Target Program on building new rural areas was born to implement the building new rural areas through all rural areas of the Vietnam and to contribute the foundations for sustainable rural development.

2.2 The objectives, principles, and scope of the National Target Program on building new rural areas for the period of 2010: 2020

The National Target Program on building new rural areas took the commune level as the implementing unit with the goal that the whole country would have 50% of communes meeting the new rural area standard by 2020. Besides, there were other objectives such as: encouraging each province and city to have at least one district meeting the new rural area standard; basically completing essential works to meet the requirements of production development and life of rural residents as: traffic, electricity, water, daily life, schools, commune health stations; improving the life quality of rural residents, creating many production models associated with stable jobs for people, increasing income at least 1.8 times than that of 2015 [11, 12].

The basic principle of building a new rural area was to promote the role of the local population community, the government played the role of orientation and promulgation of criteria, standards, policies, support mechanisms, staff training, and implementation guidance. The community of people in villages and communes democratically discussed and implemented the specific activities. In addition, building a new rural area must be carried out on the basis of inheriting and integrating national target programs, targeted support programs, and other programs and projects being implemented in the rural area that associated with the local socioeconomic development plan. The ownership role of the people and the community was promoted and the process of planning, organizing, implementing, monitoring, and evaluating must be implemented democratically [11, 12].

It could be said that the National Target Program on building new rural areas was a comprehensive rural development program that paid attention to most of the different fields in rural areas. The scope of implementation was all communes across the country and the beneficiaries were residential communities in rural areas; The subjects of implementation were the communities in rural areas, the government, enterprises, and socioeconomic organizations.

2.3 Criteria and contents of new rural area construction

The new rural area criteria set of the program has promulgated 11 group of contents, each group of content included specific goals and contents such as: [13].

- Content 1: New rural area construction planning;
- Content 2: Development of socioeconomic infrastructure;
- Content 3: Developing production in association with the agricultural sector, restructuring rural economy, raising income for people;
- Content 4: Poverty reduction and social security;
- Content 5: Development of education in rural areas;
- Content 6: Developing basic health care, improving the quality of health care for rural people;
- Content 7: Improving the quality of cultural life of rural people;
- Content 8: Implementing rural environmental sanitation, pollution solution, and environmental improvement in craft villages;
- Content 9: Improving the quality and promoting the role of authorities and social organizations in building a new rural area;
- Content 10: Maintaining national defense, security, and rural social order;
- Content 11: Improving the capacity of building new rural area and carrying out the communication on building new rural area

3. Results and discussions

3.1 Summarization of the results of construction of new rural area

In the period of 2011–2020, the achievement of new rural area standard increased sharply in the second phase, many important targets of new rural area construction completed in 2019. By July 2021, 195/664 districts (29.6%) belonging to 52 provinces and cities were recognized as meeting new rural area standard. The number of communes meeting the new rural area standard was 5331/8267 communes (64.8%) [11, 12]. The result of new rural area construction by 2020 is given in **Table 1**.

In general, the construction of new rural areas has achieved great achievements over the past 10 years. That significantly changed the face of the rural areas, especially in rural infrastructure. In the construction of new rural areas, the promotion of urbanization in rural areas has not been clearly shown.

3.2 The result of the basic socioeconomic infrastructure in the rural areas

The basic socioeconomic infrastructure in the rural areas was strengthened in a synchronous direction, which significantly changed the face of the rural areas and connected with the urban areas step by step. The basic socioeconomic infrastructure gradually met the living and production needs of the rural population and gradually caught up with the needs of socioeconomic development in rural areas. That was shown below [11, 12]:

- The rural transport infrastructure: by 2020, over 206,743 km of roads has been built and upgraded, which increased the rate of hardened rural roads to 68.7%; Over 97% of communes had plasticized and hardened roads; The quality of roads has been improved and quite synchronous, which has contributed to the formation of large production areas and created favorable conditions to attract investors to the rural areas including large enterprises. Up to now, there have been 6460 communes (78.2%) meeting the criteria of rural transport

Province/ City	Total number of communes of the province	Number of districts meeting new rural area standard	Number of communes meeting new rural area standard	Percentage of communes meeting the new rural area standard (%)	Average of criteria per commune
+ Total	8,902	173	5,555	62.4	16.4
+ Northern Midlands and Mountains	2,280	18	828	36.3	13.5
+ Red river delta	1,882	69	1,805	95.9	18.9
+ North Central	1,585	18	1,019	64.3	16.8
+ South Central Coast	825	12	476	57.7	16.4
+ Highlands	599	3	270	45.1	15.2
+ South East	445	22	354	79.5	18.0
+ Mekong Delta	1,286	31	782	60.8	16.9

Table 1.
The result of new rural area construction by 2020.

(increasing 41.8% compared with 2015 and exceeding 23.2% compared with the 5-year target for the period of 2016–2020);

- The irrigation system has been gradually completed. The thousands of small and large irrigation projects have been built, repaired, and upgraded. More than 80% of the productive area was actively irrigated, which contributed to the transformation of crop structure and improved production value and promoted the restructuring of the agricultural sector. Up to now, the area that applied the advanced and water-saving irrigation was 288,620 hectares (for upland crops, reaching 17.5%), 1,320,118 hectares (for rice, reaching 18%). Thus, the crop productivity and irrigated areas were increased and adapted to water-scarce areas and effectively that formed the large-scale production areas with high economic value. Some regions have developed advanced water-saving irrigation system such as the Southeast (40%), the Central Highlands (27%), the Mekong River Delta (18%). There were 7934 communes (96%) reaching the irrigation criteria (increasing 34.6% compared with 2015 and exceeding 19% compared with the 5-year target for the period of 2016–2020);

- The rural electricity system, up to now, 100% of communes and 99.25% of rural households have had electricity, in which, 7729 communes (93.5%) have met the electricity criteria (increasing 11.1% compared with 2015). The quality of electricity in rural areas was increasingly improved and stable and basically met people's daily-life needs and developing production. The quality of electricity has created favorable conditions for the application of hi-tech agricultural production, including in highland districts. In particular, remote, extremely difficult, and border areas and island communes were also focused on investing in electricity supply.

- The education system at all levels in rural areas has received special attention of government. There were 31,016 schools at all levels in rural areas in the whole country. Many localities have prioritized resources to invest in building new synchronous schools including: supplementing classrooms and function rooms, enhancing teaching equipment, libraries, physical or skill training areas, and toilets, improving the green-clean-beautiful landscape. In which, many localities have promoted the socialization of education and invested in school systems in the direction of modernity, meeting the new requirements of education. There were 6375 communes that met the school criteria (77.2%, and increasing 35.1% compared with 2015).

- The system of rural health facilities in recent years has continued to be invested, upgraded, and improved and highly appreciated by international organizations. 100% of communes had health stations, of which, about 76% of commune health stations met the National Health Standards (an increase of 8.9% compared with 2015); about 87.5% of commune health stations had doctors working; 95.0% of villages, hamlets had medical staff working.

- The system of rural commercial infrastructure has developed in both quantity and scale. The types and levels of markets were diversified. Many forms of modern retail infrastructure, different types of convenience shops and mini markets have also been formed and developed to meet the consumption needs of rural people. There were 7763 communes that met the criteria for rural commercial infrastructure (accounting for 93.9%, increasing 36% compared with the end of 2015 and 23.9% higher than the 5-year target for the period of 2016–2020).

3.3 The sustainable development of the rural economy

The sustainable development of the rural economy was in the direction of increasing added values and raised people's incomes and gradually narrowed the gap between rural and urban areas [11, 12].

In the period of 2010–2020, basing on the construction of new rural areas and the process of urbanization, the rural economy had many drastic changes, active transfer structure in the right direction. The industries, construction, trade, and services in rural areas developed rapidly and accounted for an increasing proportion; agricultural, forestry, and aquicultural production has restructured in production type and scale.

Industry and services in the rural areas have had positive changes, the value of rural industrial production has grown well. Services in rural areas developed diversely with the participation of all economic sectors. Industrial production value in rural areas tended to grow significantly and reached 12.2% in the period of 2010–2018. That contributed to job creation, restructuring of rural labor (the proportion of agricultural laborers decreased from 49.5% in 2010 to 32.8% in 2020) and raised incomes of rural households.

Restructuring the agricultural sector has achieved many important achievements in terms of both scale and production level. Agriculture has shifted strongly to international competitive commodity production. The productivity and product quality increased and firmly ensured national food security, and exports have increased rapidly. The agricultural sector has a sustainable growth, the average GDP growth rate of the industry is 2.85%/year in the period 2011–2019, the agricultural production value will reach VND 99.5 million/ha in 2020 (an increase of 82% compared with 2010). Vietnam's agricultural, forestry, and aquicultural exports in 2020 reached over $41.25 billion, ranking in the top 15 in the world and second in ASEAN. Agriculture continued to be Vietnam's strength. Many specific agricultural regions were formed according to functions as: peri-urban agriculture, large-scale commodity agriculture, agriculture adapting to climate change, and agriculture with application of high technology.

The forestry industry has had a remarkable development in the past period with a stable growth rate. The national forest coverage rate in 2020 reached 42%, which formed the second largest forestry product processing industry in Asia and the fifth largest in the world. Over the past 10 years, aquiculture had the highest growth with an average growth of 5.2% per year in production value. In 2020, the total output was 8.4 million tons, the production value obtained on 1 hectare of aquaculture land reached 10,462 USD (2.3 times higher than 2010). The level of processing technology and hygiene safety of aquacultural product were invested to meet the requirements of the market in the world.

Types of production organizations are renewed in a more appropriate and effective manner; household economy continues to be supported and organized in the direction of increasing production scale, gradually adapting to the market mechanism. Strongly develop agricultural product value chains, form more and more product consumption linkage chains between producers, cooperatives, and enterprises; some large corporations have joined the linkage chain in agriculture such as Dabaco, Ba Huan, Saigon Coop, Masan Group...). Local authorities are increasingly interested in agricultural development, well performing the role of "State" in linking "4 houses" to organize production and consumption of agricultural products.

By the end of June 2021, 57/63 provinces have issued the policies to encourage linkages in production and consumption of local agricultural products, in which,

44/63 provinces and cities approved projects and plans on linkage under the guidance of the Ministry of Agriculture and Rural Development. As a result, up to now, more than 27,000 value chain-linked production models have been built; 1644 safe agricultural product chains were developed with 3267 places of sale of controlled products according to the chain and 2038 places of sale of controlled products according to the value chain. By 2020, there were about 49,600 enterprises investing in agriculture (of which, more than 11,800 enterprises invested directly), accounting for about 8% of the total number of enterprises operating in the country.

The speed of cooperative development has increased sharply year by year. By the end of 2020, there were 17,642 agricultural cooperatives (beyond the target to 2020 assigned by the government). The cooperatives gradually promoted their effective support role for farmers by actively linking with enterprises under the form of that the enterprises provided inputs, production techniques, and product consumption that created stable output for agricultural products.

3.4 The income and living standards of rural people

The income and living standards of rural people were continuously improved and enhanced and the gap between rural and urban areas was narrowed [11, 12].

Average income per person per year in rural areas increased faster than the growth rate of urban people's income. The income increased from 676 USD/person in 2010 to about 1807 USD/person in 2020. Basically, the target by 2020 was achieved. The income gap between rural and urban areas tended to decrease from 1.99 times in 2010 to 1.61 times in 2020. People in rural areas are less and less dependent on agriculture. The structure of income from agricultural, forestry, and aquicutural activities decreased from 33.5% in 2010 to 18.5% in 2020. The rate of poor households in rural areas decreased by 1% per year on average. By the end of 2020, the rate reached 7.1%.

Along with the increase in income, the spiritual life in rural areas has also been significantly improved. The people had easier access to basic social services, especially rural people in remote and isolated areas and ethnic minority areas. The countryside was the place where they maintained and developed the vibrant culture, artist, physical training, and sports. The civilized lifestyle was implemented and the unsound customs and superstition were eliminated. The maintenance of security and order was ensured. The evils of drugs, theft, gambling, addiction were controlled and managed. Many activities of planting trees, lighting up rural roads and protecting the environment have been actively organized by the community in order to create many fresh, bright, green, clean, and beautiful rural areas.

3.5 Environmental protection

The work of landscape construction, implementation of environmental sanitation had a remarkable change, representing the achievements of building a new rural areas [11, 12] . Many models of landscape improvement in villages and hamlets have been creatively applied according to actual conditions; Thousands of kilometers of flower routes have been formed, and many districts have had the percentage of rural roads planted with trees and flowers that reached over 50%.

By the end of 2020, 51% of rural households used standard clean water; There were 6222 communes (75.3%) meeting the criteria on Environment and Food Safety (increasing by 32.9% compared with 2015, completing 5.3% more than the 5-year target for the period of 2016–2020).

The protection of the rural environment, especially in industrial production, services and craft villages has always been concerned, and environmental pollution has been gradually overcome. Up to now, 59/63 provinces and cities have approved the solid waste management planning in the areas; 42/63 provinces and cities have plans for concentrated waste treatment in rural areas, of which, a number of localities have implemented that throughout the province; 16/63 provinces and cities have approved the investment policy of rural solid waste treatment plants.

Solid waste collection has been promoted, and most villages and communes have formed a domestic waste collection team. The proportion of daily-life solid waste collected has increased significantly year by year, from 44.1% in 2011 to 66.0% in 2020. The scale and methods of solid waste treatment have also changed significantly. Common treatment methods included landfilling, incineration, fertilizer, and fuel pellet production. Although the current method of landfilling still accounted for a relatively high rate (about 70%), the trend of incineration became more common in many localities. There were about 425 domestic solid waste incinerators, of which, there were more than 100 incinerators with a capacity of over 300 kg/h, meeting the requirements of the National Technical Regulation.

The environment in craft villages has been significantly improved. There have been 33 provinces and cities that have issued the policies on environmental protection of craft villages. Many craft village waste treatment models have been implemented. Many craft villages applied advanced technology and production processes to limit waste emissions into the environment. The percentage of craft villages with centralized wastewater collection accounted for 27.6% of the total number of craft villages with industrial wastewater. The percentage of craft villages with concentrated wastewater treatment meeting environmental standards accounted for 16.1%; the percentage of craft villages with industrial solid waste collection places accounted for 20.9% of the total number of craft villages with industrial solid waste.

3.6 The cultural life of the people and social security and defense

A healthy and colorful cultural environment was created in the cultural and spiritual life of rural people. Cultural activities, arts, and sports in residential areas were promoted. The preservation and promotion of cultural values have made a practical contribution to the construction of cultural life of the new rural areas.

Social security was guaranteed. The combat and timely prevention of crimes and social evils were effectively implemented in accordance with the Government's strategy. The security and defense continued to be maintained. Thus, the people felt secure and they worked, created, and dedicated themselves to the task of new rural area construction and development.

4. Conclusions

Based on the objectives, principles, and scope of the National Target Program on building new rural areas for the period of 2010–2020, Vietnam has implemented sustainable new rural construction associated with urbanization. The results showed that Vietnam rural areas have changed both in terms of the basic socioeconomic infrastructure, the income, and living standards of rural people. Vietnam rural areas have had economic restructuring. In which, the occupation was shifted from agriculture to nonagricultural occupation, and the population, labor, and employment were shifted from rural to urban areas. Thus, millions of new jobs were created. Besides, the rural infrastructure system was invested and met the needs of production, travel, education, health care, goods circulation, service development. People's income and living standards were improved and enhanced. The security and defense continued to be maintained.

Author details

Nguyen Hay[1], Le Quang Huy[2] and Pham Van Kien[3*]

1 Nong Lam University, Ho Chi Minh City, Vietnam

2 Cao Thang Technical College, Ho Chi Minh City, Vietnam

3 Faculty of Automobile Technology, Van Lang University, Ho Chi Minh City, Vietnam

*Address all correspondence to: kien.pv@vlu.edu.vn

References

[1] Zhang X, Zhang B, Jiang L, Xie M. A study on planning and construction of new rural communities. Advanced Materials Research. 2013;**838**:2938-2941

[2] Palliyaguru R, Karunasen G, Ang S. Review on sustainable building design and construction in the rural context: the case of building ampara, Sri Lanka. New York: Springer International Publishing AG; 2018. p. 978. DOI: 10.1007/978-3-319-73293-0_29

[3] Fangyu F, Cao Y. Study on the "3F-in-1" Sustainable Reconstruction of Rural Architecture from Placeality Perspective--A Case Study of Caiyuan Village in Jingmen City, Hubei Province, 2017. In: International Conference on Environmental and Energy Engineering (IC3E 2017); 19-22 October 2017. Shanghai, China: IOP Publishing; 2017. pp. 1-7

[4] Na Z. Study on the construction of new rural communities in China. International Journal of Humanities and Social Sciences. 2016;**6**(2):173-179

[5] Xu F, Kai W, Keni M. Research on Japanese rural planning, construction, and governance. Strategic Study of Chinese Academy of Engineering. 2019;**21**(2):34-39

[6] Mo W, Ren N. The study of Korean new rural residential construction in "Changchun-Jilin-Tumen" district. In: International Conference on Multimedia Technology; 26-28 July 2011. Hangzhou; China: IEEE Publisher; 2017. pp. 735-746

[7] Brandt VSR. Rural development in South Korea. Asian Affairs An American Review. 2012;**6**(3):148-163

[8] Zhang Y, Jingjing WANG. Research on the construction strategy of rural community space in China. In: 2nd International Symposium on Architecture Research Frontiers and Ecological Environment (ARFEE 2019); December 20-22. Guilin, China: E3S Web of Conferences Publishing; 2020. pp. 1-5

[9] Thanh VT, Duong NA. Promoting rural development, employment, and inclusive growth in ASEAN [project]. Viet Nam: Central Institute for Economic Management; 2016

[10] Rigg J. Rural Development in Southeast Asia. London, England: Cambridge University Press, Cambridge University; 2020. 400 p. DOI: 10.1017/9781108750622

[11] Luan NN. Report of summarizing 10 years of implementation of the National Target Program on building new rural areas in the period of 2010-2020, The Central Steering Committee for National Target Programs in the period of 2016-2020. Ha Noi, Vietnam: Institute of Policy and Strategy for Rural Development, Vietnam Government, Vietnam Academy of Agriculture Publisher; 2019

[12] Hung TA. Research and evaluate the results of new rural construction in the period 2010-2020 from a scientific perspective and identify basic problems in sustainable rural development in the period after 2020. Ha Noi, Vietnam: Institute of Policy and Strategy for Rural Development, Vietnam Government, Vietnam Academy of Agriculture Publisher; 2020

[13] Linh PNM, Long PT, Khanh TLQ, Thu NT, Nang DT, Sang NQ. State management in building a new rural area in Vietnam: A Research in Muong Tra District, Dien Bien Province. Environmental Management and Sustainable Development. 2021;**10**(2):58-78

Chapter 10

Biogas Generation from Co-Digestion Waste Systems: The Role of Water Hyacinth

Adedeji A. Adelodun, Temitope M. Olajire and Ochuko Mary Ojo

Abstract

Using biomass as a renewable energy source has earned tremendous interest from researchers in recent decades, especially because the technology is environmentally benign. This article reviews the recent methods for generating biogas from water hyacinth (WH, *Eichornia crassipes*), arguably the world's most evasive aquatic macrophyte. Therefore, various economic, environmentally benign, and renewable procedures that enhance biogas production from WH biomass are reviewed. WH has been co-digested with numerous waste types, including poultry droppings, municipal wastes, animal tissue wastes, pig wastes, cow dungs, etc., recording varying success degrees. Other studies focused on optimizing the operation parameters, such as mixing ratio, contact time, pH, temperature, organic loading rate, etc. We observed that most attempts to generate biogas from WH alone were not promising. However, when co-digested with other biomasses or wastes, WH either increases the process rate or improves the methane yield content. Also, the potential of WH as a phytoremdiator-cum-biogas source was investigated. This chapter provides mathematical models, scale-up installation models, and specific experimental results from various studies to guide future study plans toward optimizing CH_4 generation from WH co-digestion.

Keywords: *Eichornia crassipes*, biomethanation, methanogens, biogas yield, biogas purity

1. Introduction

Biogas, an energy source comprising CH_4, CO_2, and traces of some gaseous impurities, is generated via biomethanation, i.e., anaerobic digestion of substrates. Irrespective of the substrate, typical biogas is composed of 50–80% CH_4, 20–50% CO_2, 5–10% of H_2, 1–2% of N_2, ≈0.3% water vapor, and traces of H_2S and $H_2O_{(g)}$ [1, 2]. Regardless of their proportions, CO_2 and H_2S are the major impurities in biogas. Therefore, post-production cleanup processes are required to remove them for optimum performance of the final product. Usually, CO_2 is absorbed into hydroxides of

Ca, K, or Ba (Eq. (1)), while $CuSO_4$ removes H_2S, $FeSO_4$, $Pb(NO_3)_2$, or $FeCl_3$ (Eq. (2)). For CO_2 removal, NaOH is an efficient absorbent, although KOH is 27% more effective, using only 125 kWh/Tor CO_2 energy [3]. Otherwise, to minimize the cost and avoid additional waste generation, the pristine gas stream could be bubbled through water to remove both gases, albeit with less efficiency [4].

$$Ca(OH)_{2(aq)} + CO_{2(g)} \longrightarrow CaCO_3 + H_2O \tag{1}$$

$$(CH_3COO)_2Pb_{(aq)} + H_2S_{(g)} \longrightarrow 2CH_3COOH_{(aq)} + PbS_{(s)} \tag{2}$$

2. Biodigestion process

A typical biodigester is made of concrete, metal, or other material that permits anaerobic biomass fermentation [5]. For optimum performance, the operational and ambient conditions must be diligently considered. Several factors that affect biogas production efficiencies include pH, temperature, type and quality of the substrate, mixing speed and consistency organic loading, formation of highly volatile fatty acids, and inadequate alkalinity [6]. The retention (turn-over) time is the period required for organic materials to be decomposed entirely toward achieving maximum biogas yield. Fertilizers and mineralized water are the usual valuable by-products of this process [5].

Research into biogas technology in Africa gained momentum in the last decade. For instance, in Nigeria, biogas production from Bambara nut chaff [6], agricultural waste [7], and abattoir waste [8], and the performance evaluation of a biogas stove for cooking [9] have been reported. Furthermore, biogas generation from co-digested substrates, such as spent grains and rice husk [10], banana and plantain peels [11], pig waste and cassava peels [12], sewage and brewery sludge [13], have also been experimented. In most cases, co-digestion enhances methane yield by ≈60%. Similar studies were carried out in other African countries such as Uganda [14], South Africa [15], Sudan [16], etc.

Generally, plant-based biofuels are environmentally clean energy, with a high potential of lowering fossil fuel consumption to the barest minimum in the near future [17]. Over the past decade, several studies have focused on producing biomethane using lignocellulosic residues of high abundance and low cost [18, 19]. According to Bekkering et al. [20] and Holm-Nielsen et al. [21], biogas can be used as fuel and fuel cells to generate heat, steam, electricity, produce chemicals, upgrade natural gas grids via injection, etc. Elsewhere, Jantsch and Mattiasson [22] discussed how anaerobic digestion could treat wastewater and organic wastes, yielding biogas as a valuable by-product. The four major sources of biogas production are livestock waste, landfill gas (LFG), activated sludge from wastewater treatment plants, and IIC (industrial, institutional, and commercial waste) [22–24].

Biomethanation occurs in four main steps (**Figure 1**) viz. hydrolysis [23] , acidogenesis [24], acetogenesis [26], and methanogenesis [27]. Methane is the main component of biogas (50–70%). Other components include CO_2 (30–40%) and traces of H_2S and $H_2O_{(g)}$ [28]. The respective equations for the four steps are provided as Eqs. (3)–(6):

$$\text{Hydrolysis}: (CsH_{10}O_5) + nH_2O \longrightarrow n(C_6H_{12}O_6) \tag{3}$$

$$\text{Acidogenesis}: n(C_6H_{12}O_6) \longrightarrow 3nCH_3COOH \tag{4}$$

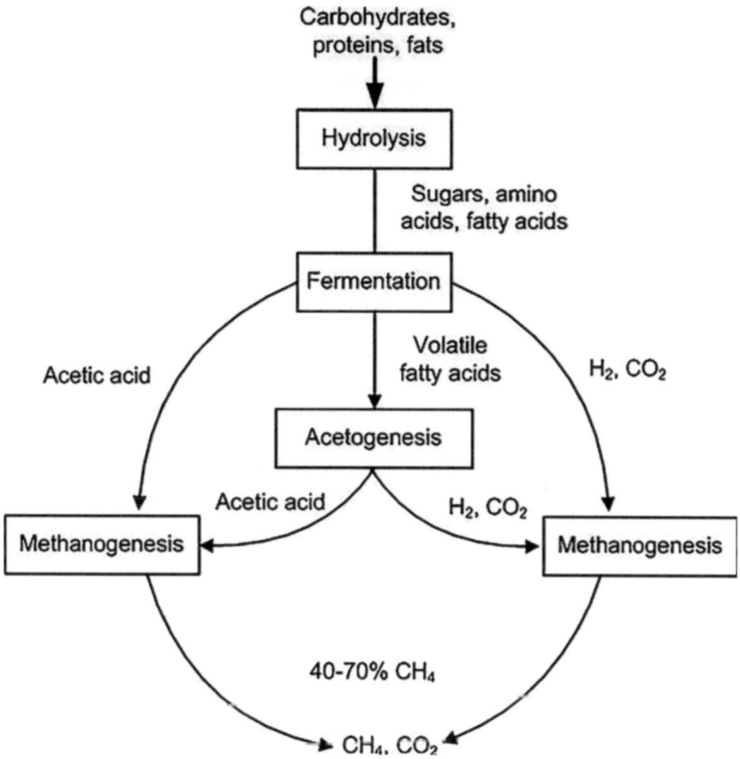

Figure 1.
Process flow of the degradation of organic material through anaerobic digestion [24, 25].

$$\text{Acetogenesis}: CH_3COOH \longrightarrow CH_4 + CO_2 \tag{5}$$

$$\text{Methanogenesis}: CO_2 + 4H_2 \longrightarrow CH_4 + 2H_2O \tag{6}$$

Four major microbial groups are respectively involved: the hydrolytic-fermentative bacteria (hydrolyze complex organic compounds into simple ones), fermentative bacteria (convert the simple organic compounds into volatile fatty acids, yielding H_2 and CO_2), acetogenic bacteria (convert the fatty acids into acetic acid), and methanogenic archaea (produce CH_4 either from acetate or from H_2 and CO_2) [24, 25].

3. Factors affecting biogas yield and production

The quality of manure influences the methanogenic diversity in a reactor, and the overall conversion efficiency of manure to CH_4 is influenced by the retention time [29], pH, oxygen level, NH_3-N, and volatile fatty acids (VFA) contents, and temperature [30, 31]. Biogas can be produced under psychrophilic (10–30°C), mesophilic (20–50°C), and thermophilic (50–75°C) conditions [2]. Mesophilic and thermophilic conditions present different reactor designs, operational advantages, and drawbacks. Most anaerobic digesters are designed to operate at mesophilic (40°C) or thermophilic (55°C) temperature to maximize biogas yield [32], whereas, between 40 and 50°C, methanogens are inhibited. Elsewhere, anaerobic digestion temperature was optimized at 25–38°C (mesophilic conditions), with temperatures near 38°C showing more excellent (≥95%) digestion stability. Likewise, a mesophilic treatment at 38°C reportedly destroys 99.9% of pathogens [33].

Similarly, the C/N [34, 35], slurry concentration, mixing rate, and bacteria type (starter) are other crucial parameters that influence biogas quality and yield [36]. Typically, the organic loading rate (OLR) ranges between 0.5 and 3 kg VS (volatile solids) per m^3 per day [29]. **Table 1** lists the average C/N of various substrates used for biogas production [70–72]. Typical C/N values/ranges for biogas production are as follows: liquid cattle manure (6–20), chicken manure (3–10), liquid swine manure (5), straw (50–150), grass (12–26), potatoes (35–60), sugar beet/beet foliage (35–46), cereals (16–40), fruits and vegetables (7–35), mixed food waste (15–32), slaughterhouse waste—soft tissue (4), slaughterhouse waste—guts (22–37), food waste (3–17), distillery waste (8), etc. An increasing C/N (10–30) increases the formation of fatty acids in the process [34, 35]. If the fatty acid concentrations are not sufficiently high, methanogenesis could result.

Methanogens are sensitive to rapid temperature change, while thermophilic methanogens are more temperature-sensitive counterparts. Therefore, temperature should be kept exactly at ±2°C [2].

Some researchers have investigated the optimal pH for microbial performance during anaerobic digestion. According to Yadvika et al. [73], the pH within the digester should be kept within 6.8–8.0, whereas Thy et al. [74] concluded that 6–8 pH range is the optimal pH. At the onset of the acid-forming stage of the digestion, the pH may be <6.0. However, it could be >7.0 during methane formation and maintained because it is sensitive to acidity. In a properly operating anaerobic digester, a pH of 6.8–7.2 converts volatile acids to CH_4 and CO_2 [75]. The pH is the most suitable indicator for plausible digester instability after gas production [29]. Initially, the pH would decrease as the organic matter undergoes acetogenesis. However, as the methanogens rapidly consume the acids, the pH rises, stabilizing the digester performance. Fermentative bacteria require a pH > 5.0 to become enzymatic, while methanogenic activity takes place at a pH range of 6.2–8.0, optimized at ±7.1 [29, 76]. In addition, other phenomena, such as the dissociation of important compounds (ammonia, sulfide, organic acids, etc.), are directly affected by pH [32]. Methanogenic bacteria are generally susceptible to pH and do not thrive at pH < 6.0 [77].

Homogeneous mixing within the digester improves the contact between the microorganisms substrate, improving the bacterial ability to obtain required nutrients. Also, by homogenization, scum formation and temperature increase within the digester are minimized. However, excessive mixing can disrupt the microorganisms; therefore, slow mixing is preferred [78]. According to Kossman et al. [79], with other parameters fixed, a well-agitated substrate can increase biogas production by 50%.

4. Various biomasses for biogas production

The anaerobic fermentation of manure for biogas production does not compromise the quality of the fertilizer supplement because the nitrogen and other substances remain in the treated sludge [80]. In the absence of appropriate disposal methods, animal dungs can cause various environmental and health problems, such as pathogenic contamination, odor pollution, and greenhouse gas emission [81]. Rain may flush these wastes into neighboring water bodies or percolate underground, springs,

S/N	Starting material	CH_4 yield	Substrate mixing ratio/ concentration	Biogas yield	Observation	Experimental condition	Ref.
WH only							
1	WH only	72.53%	5–30 g/L		Shoots only generated 6.86% extra methane	39°C substrate; concentration 25 g/L	[37]
2	WH only		C/N 35	202 L/ kg TS	TS not a significant factor	TS 1.59% ; 60 days	[38]
3	WH only	50–65%	C/N 16		No link between C/N ratio and production	51 days	[39]
4	WH only		1:4 (WH: Wate) C/N: 15	245 L/ kg VS		60 days; 30–37°C	[40]
5	WH only	65%		380 L/ kg VS	Pretreatment with NaOH yielded most methane composition (71%)	35°C; size 2 cm; 60 days	[41]
6	WH only		C/N: 25.9	75 L/kg TS		pH: 6.65; TS: 8%; size = 2 cm	[42]
7	WH only			360.1 L/ kg TS		40 days; Size. 1 cm; 45°C	[43]
8	WH only			221 L/kg TS	Increase of 75.61%	Size = 1 cm; pretreated at 60°C for 24 h	[44]
Co digested WH							
9	WH + cow dung		C/N : 32.0	108 L/ kg TS	Microbial consortium 7.26%	60 days; size = 15 cm; TS 5–10%	[45]
10	WH + MW[1]			16 L/kg		36-37°C ; pH 6.0–7.4	[46]
11	WH + MW	60.5%	4:1 (WH: waste)	230 L/ kg VS		TS = 4%; 15 days	[47]
12	WH + other biomass	68.67%		237.4 L; CH_4/kg VS		F:M = 1:1; 60 days	[48]
13	WH + other biomass		Duckweed: WH = 7:3; C/N = 16.4	20.55 L/ kg VS		8% TS size < 6 mm	[49]
14	WH + other biomass		Salivinia: WH = 0.5:1	406 L/ kgVS	VFA affected WH in 3:1		[50]
15	WH + other animal wastes	52.8%	1:2 (WH: Buffalo dung)	2.86 L/ day	Pretreatment increased biogas production by 102%; methane by 51%	Size: 6 mm	[51]

S/N	Starting material	CH_4 yield	Substrate mixing ratio/ concentration	Biogas yield	Observation	Experimental condition	Ref.
16	WH + poultry droppings		2:8 9 (WH: poultry manure)	34.65 L/kg		40 days	[52]
17	WH + cow dung	49–53% upgraded to 73%	3:1 (WH: cow dung)	3.2 L/kg		22.8–36.6°C	[53]
18	WH + cow dung	63.7%			Optimal OLR un-pretreated		[54]
19	WH + MW		F/M = 10.01 : 0.03	152 L/kg TS (daily)	TS = 6.76%		[55]
20	WH + cow dung	65%		270 L/ m^3	22%	Size = 5 cm; 10 days	[56]
21	WH + cow dung	56.4%	2:1 (WH: cow dung) C/N 10:1	3050 L/ day		40 days; size = 2–5 cm; 28–36.7°C; pH = 6.5–7.8	[57]
22	WH + pig waste	88.3%	1:3 (WH: Pig waste)			27–34°C	[58]
23	WH + pig waste	1.4 kg : 1L (piggery waste:WH) C/N = 30:1				TS = 14.02%; pH = 6.0–7.2; 12 days	[59]
24	WH + MAW[2]	64.9%	3:8:9 (pig dung: WH: Poultry droppings)	307 L/ kg		pH = 6.5; TS = 9.09%; 52 days	[60]
25	WH + MAW	62.14%	2:2:1 (WH: cow dung: poultry dropping)			0.02 m	[61]
26	WH + MAW			0.255 kg/m^3	OLR of 1.5 kg/m^3 yielded most biogas	36–37°C; TS = 9.98%; 60 days pH = 5.0–7.4	[62]
27	WH + animal waste	68%	3:7 (WH: animal waste)	14.09 L/ kg	Increasing temperature from 24 to 32 increased production by 186%	24°C	[63]
28	WH + animal waste + others		4:4:2 (waste WH: cow manure)	60 ppm CH_4 10,744 ppm		21 days; size = 2 cm	[64]
29	WH + animal waste		C/N = 20/1 5:3:2 (Prosopis juiflora pods: Duckweed: WH	96.6 L/ kg			[65]

S/N	Starting material	CH_4 yield	Substrate mixing ratio/ concentration	Biogas yield	Observation	Experimental condition	Ref.
30	WH + animal wastes + others			1 L/day		35°C; size = 3–5 cm; TS = 5.6% OLR[3] = 50 g/L	[66]
31	WH + animal wastes + others			273.3 L/kg		40 days	[67]
32	Phytoremediation		C/N 26.9	5195 m^3			[68]
33	Phytoremediation			23,650 cc/kg dry weight	Growing in 20% effluent increased production	35°C 21 days	[69]

[1]*MW = municipal waste.*
[2]*MAW = multiple animal wastes.*
[3]*OLR = organic load ratio.*

Table 1.
Recent studies on generating biogas (CH_4) from water hyacinth (WH) or water hyacinth-based (co-) digestion.

and wells used for sanitation and domestic purposes [82]. Poultry and livestock wastes often contain high concentrations of human pathogens, spilled feed, bedding materials, fur, wastewater, feed residues, feces, and urine. Therefore, the waste should be effectively managed to minimize environmental and public health risks. Such practices might result in acute gastrointestinal upset (e.g., nausea, diarrhea, and vomiting). Also, contact with affected surface waters during recreational activities can cause skin, ear, or eye infections.

4.1 Recent advances with WH only

Some researchers have investigated biogas generation from WH, either solely or co-digested with other waster materials (**Table 1**). Being tagged as a waterway menace, WH has been identified as a substrate for economically feasible biogas production [83, 84].

The effect of substrate concentration, particle size, and incubation period of dry WH shoots (WHS) and whole WH (WWH) plants on biogas production has been reported [37]. The CH_4 yield increased with substrate concentration till the sixth day (25 g/L) before declining. WHS consistently had a higher CH_4 yield than WWH, especially for every particle size. Recently, Syafrudin et al. [38] evaluated the optimization of biogas production using liquid anaerobic digestion. They used central composite and complete factorial design to determine the optimal values of enzyme concentration, C/N ratio, and total solid that generates the highest biogas volume. The optimum conditions for the C/N ratio were within 30–40 and 6% of the enzyme, with no significant effect of total solids. In another research, Patil et al. [40] chopped and ground WH to a fine paste and mixed with water in five ratios to evaluate the optimal level of dilution to produce the highest volume of methane. They reported that the slurry with WH 1:4 water had the highest volume of gas.

The effect of microwave pretreatment of fresh and dried WH on biogas production was also studied [42]. It was observed that the optimum condition is 560 W and 9 min contact time on fresh WH. However, in these conditions, CH_4 production is inhibited. In addition, Rozy et al. [43] studied how various parameters affected biogas production rate and volume. Their maximum biogas yield was achieved at 45°C and a pH of 7, with 1 cm particle size and 40% inoculum concentration and 0.2 mM of $MnCl_2$. Later, a study on the effect of organic (citric) acid pretreatment of WH on biogas production was carried out [44]. Later, response surface methodology was used to optimize the pretreatment parameters. About 76% increase in biogas yield was achieved. For optimum utilization of WH biomass, Hudakorn et al. [48] evaluated the production of biogas and biomass pellets from WH. They stated that, although biogas production commenced from the first day, flammable biogas didn't start to yield till the 10th day.

4.2 Recent advances with co-digested WH

4.2.1 WH + municipal wastes

Nugraha et al. [45] studied the effect of food to microbial (F/M) ratio on biogas yield from WH. They stated that the optimum F/M ratio and TS level were 10.0 and 6.76%, respectively. They concluded that biogas production reduces inversely with the F/M ratio. The same year, Ukwuaba [46] evaluated the performance of biogas yield from co-digesting kitchen wastes and WH. Temperature was identified as the optimal parameter. The highest and lowest gas pressure was observed at the 25th and 37th days, respectively. Previously, Hernandez-shez et al. [47] had investigated the potential of generating biogas from co-digesting WH with fruit and vegetable waste. They optimized the biogas production in terms of total solids concentration. Co-digestion increased the biogas produced. A total solid concentration of 80:20 (WH/food waste), corresponding to a C/N ratio of 20, was the optimal condition to avoid pH correction. However, for a continuous co-digestion, they recommended an organic loading rate of 2 kg VS m^{-3} d^{-1} and 15 days retention time.

4.2.2 WH + poultry droppings

Some co-digestion of WH with poultry droppings has been carried out. Ojo et al. [52] studied the best mix of WH with poultry manure (PM) that produces maximum biogas. The authors calculated for the optimum biogas production rate, a factor of the data collected using the following equation:

$$G^{1}{}_{max} = -abc[1 - c]^{c-1} \tag{7}$$

where $G^{1}{}_{max}$ = biogas production rate, a = ultimate biogas production, b = pseuso-biogas production velocity (rate constant), c = shape factor.

They observed that mixing WH and PM at 2:8 produced the highest volume of biogas. Furthermore, the volume of biogas produced increases slightly with temperature. Also, the highest biogas yield was observed on the 18th day. It was concluded that 2 WH: 8 PM is the best-aided WH digestion mix in daily biogas production, a cumulative volume of biogas produced, and a maximum biogas production rate. Elsewhere, Patil et al. [41] studied the effect of different pretreatment on biogas yield from WH. Notably, alkali treatment had no significant effect on the biogas produced from WH blended with poultry waste.

4.2.3 WH + cow dungs

Cow dung is a popular co-stock material for WH biodigestion is a popular one. Nugraha et al. [55] used response surface methodology (RSM) to study the optimization of biogas production by solid-state anaerobic digestion to discover the optimum total solids (TS), C/N, and microbial consortium (MC) for biogas production from a mixture of WH and cow dung. They then discovered that TS and MC had the most and least effect on biogas yield, respectively. The maximum biogas yield was obtained at TS concentration range 5–10%, C/N of 32.09, and MC of 6%. Somewhere else, Adegunloye et al. [58] evaluated the optimal ratio of variation of WH to pig dung to generate the maximum methane amount.

The ambient temperature affected the temperature in the digester as the temperature in the digester was higher than the ambient temperature by 1–3°C. The authors observed that 1:3 of WH to pig dung produced the highest amount of CH_4. In another report, Jayaweera et al. [39] evaluated the biogas production from WH grown under different nitrogen concentrations. The author carried out this study for four months at mesophilic temperatures using batch-fed anaerobic reactors. WH was grown in various folds of total nitrogen then co-digested with CD. They mentioned that WH roots contain high fiber and lignin content, thus making them unsuitable as a substrate. They recommended a retention time of 27–30 days for optimum results.

A process by which volatile fatty acids (VFAs) were extracted from WH and the VFAs laden slurry was developed [56]. The extracts were then used as a feed supplement to the conventional biogas digesters. The authors discussed that WH contains 60 g/kg of TS, requiring a large digester for significant biogas production. The VFAs were extracted by charging acid-phase reactors with a mixture of WH and cow dung slurry. The reactors were aerated and the pH kept within 5.5–7.0 to enhance acidogenic bacteria growth but that of methanogenic bacteria. **Figure 2** is a pictorial explanation of the experiment. For the same TS input, the VFA supplemented the feed, yielding 22% higher biogas amount. However, no significant changes happened to methane yield.

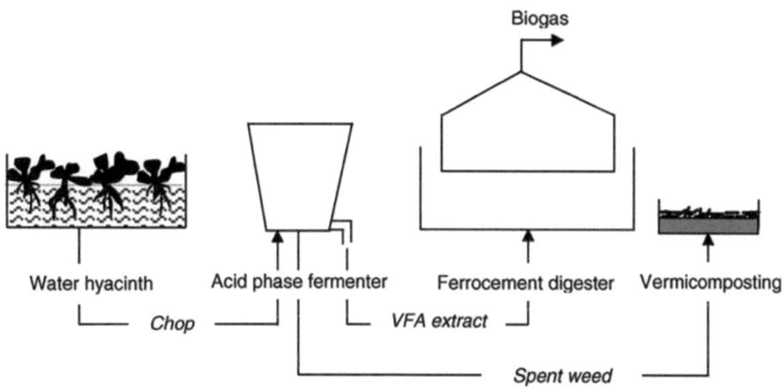

Figure 2.
Schematic representation of the process developed by the authors [56].

Eltawil et al. [85] studied the effect of stirring, dry oxidation, and water scrubbing processes on biogas quality from different substrates. Using five digesters equipped with handle stirrers (**Figure 3**), the gas produced from the digester was flushed through scrubbers to reduce H_2S and CO_2 concentrations of the biogas. They observed that stirring increased the biogas production rate by 45% for WH and cow dung mixtures but did not significantly impact the CH_4 volume of the biogas. We gathered that water scrubbing and dry oxidation removed 95% CO_2 and 97% H_2S. Therefore, this technology is recommended for developing countries where low-cost technology is needed.

Similarly, Akinnuli et al. [59] studied the performance of pig dung and WH for biogas production. The output gas was passed through KOH and anhydrous $CaCl_2$ to remove CO_2 and moisture, respectively. The authors stated that mixing pig dung and water hyacinth in the ratio 1.4:1 was optimum for biogas production. They recommended the digestion be carried out during the summer because low temperatures lead to low biogas generation. Elsewhere, a fixed dome digester was designed for biogas production using cow dung and WH (**Figure 4**) [57]. The digester is a semi-batch reactor composed of a fermentation chamber, feed and digestate pipes and, a

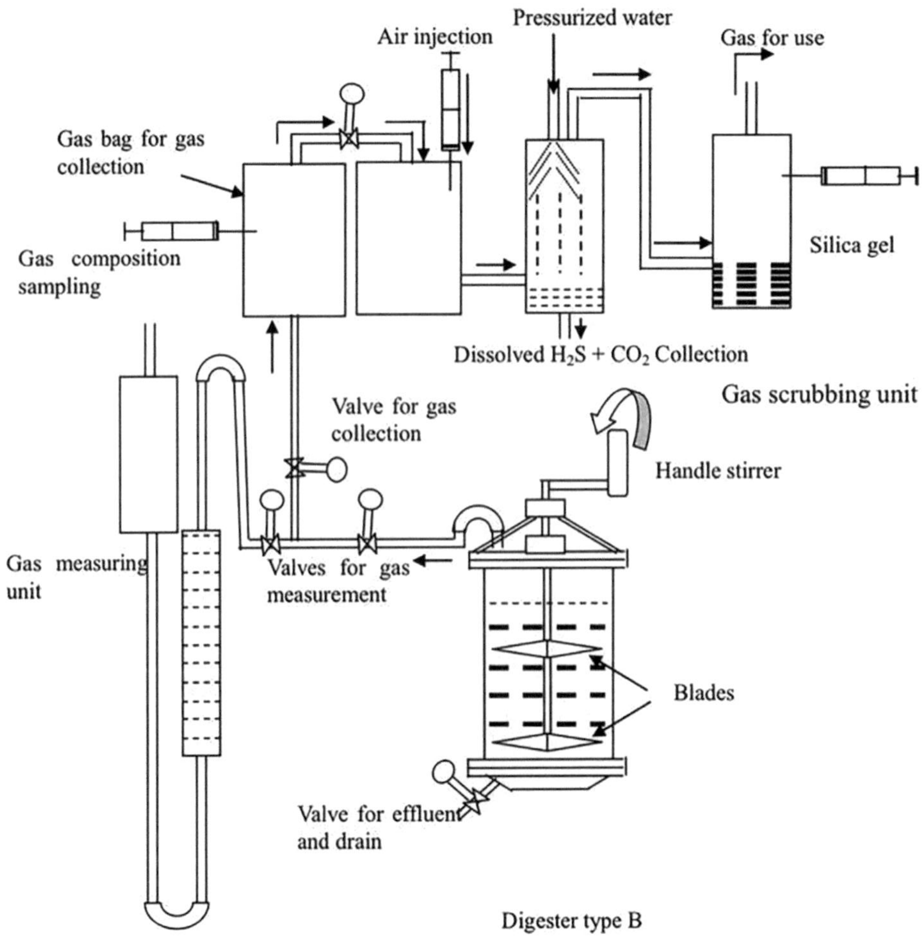

Figure 3.
Schematic diagram of the digester with stirring blade [85].

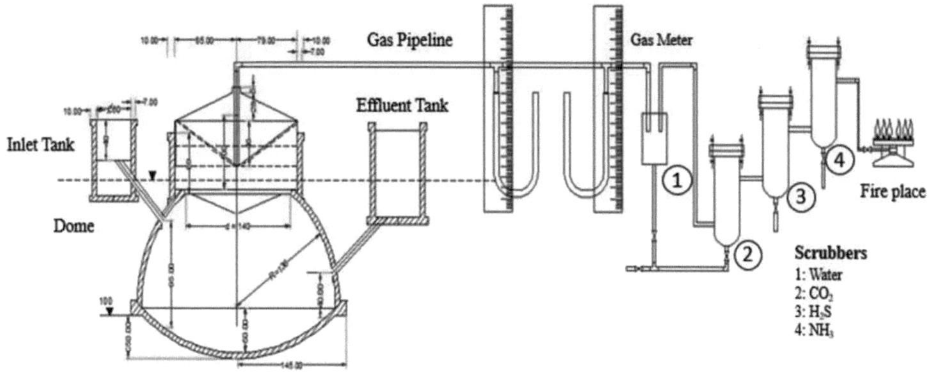

Figure 4.
Schematic diagram of fixed dome digester [57].

fixed dome on top for biogas storage. This configuration was recommended as a cheaper alternative for natural gas production.

4.2.4 WH + multiple animal wastes

Akindele et al. [60] reportedly co-digested pig dung, poultry droppings, and WH anaerobically. Lignocellulosic materials and animal manure co-digestion enhanced digestibility, biogas production, and equipment utilization. The respective mixing ratio of 3:9:8 was optimum for methane yield. Also, no gas production in the first four days as the enzymes adapt to a new environment. CH_4 production lapsed from 8th to 16th day. The biogas production increased with fermentation until the 40th day, with the highest biogas production observed on the 52th day.

Earlier, Fadairo et al. [61] co-digested WH with cow dung and poultry droppings. Their optimum mixing ratios were 2:2:1 and 1:1:0. The lower the WH dosage, the lower the biogas generated. However, the substrate containing WH alone produced the least biogas. Cow dung influenced biogas production than poultry droppings, attributed to the ammonium ions in the latter.

The effect of Organic Loading Rate (OLR) on biogas production systems has also been researched [62]. The rate of adding feedstock required alteration for optimal growth of methanogens, which directly influence biogas produced. The authors noted that direct charging above 1.5 kg/m^3 inhibits the growth of the methanogens. Recently, the co-digestion of (WH) biomass with ruminal slaughterhouse waste (RSW) was evaluated [63]. The highest and lowest biogas yields were with the substrate of solely slaughterhouse waste and WH, respectively. Also, the co-digestion of the waste with WH (5–50%) significantly reduced the retention time by 26 days, whereas if the proportion is >50%, no further impact on retention time will occur. The study recommended co-digestion of 30% waste and 70% WH at 32°C.

In some cases, WH and animal wastes are dosed with other waste materials. Sa'adiah et al. [64] evaluated biogas production from co-digesting Tofu waste, WH, and cow manure. They observed that adding more WH inhibited the production of biogas. They then recommended that mixing WH, tofu, and CD at 2:4:2 for optimal biogas production. Also, Prabhu et al. [65] investigated the anaerobic co-digestion of *Prosopis juliflora* pods with WH, dry leaves, and cow manure, modeled the biogas production kinetics using a modified Gompertz equation to examine the cumulative methane production (Eq. (8)).

$$Y = M * exp\left\{-exp\left[\frac{Rm * e}{M}(\lambda - 1) + 1\right]\right\} \quad (8)$$

where Y = cumulative methane production (L at time t), M = maximum methane production potential (L-CH_4), Rm = maximum methane production rate (L-CH_4/d), λ = Lag phase time (day), E = constant (2.71).

The authors noticed that methane composition was higher in biogas yielded by WH-rich mixtures than other mixtures. E.g, WH + dry (2:3) achieved the maximum methane yield of 80%. The coefficient of determination (R^2) between the experimental data and the model ranged as 0.991–0.999.

Moreover, Shah et al. [66] explored the potential of three plants (WH, giant reed, and maize) and poultry waste for biogas generation, using WH with 13% hemicellulose and poultry waste as inoculum. WH had the highest volatile solids, soluble solids and, a better C/N ratio. Thus, it was a relatively superior biogas substrate. The highest biogas yield occurred on the 11th day. From the four substrates, WH contributed the highest to biogas production. Likewise, Otaraku et al. [86] modeled the cumulative biogas produced from sawdust, cow dung, and WH. They concluded that the polynomial model best fitted the cumulative biogas production at any given day, with $R^2 > 0.9$. Similarly, the potential of biogas production from mixtures of WH, cassava peels, and cow dung using standard microbial techniques has been reported [67]. The highest total biogas yield from co-digesting the three substrates was noted. It was concluded that the prescribed treatment combinations could be facilitated with or without starter culture.

4.2.5 WH + other biomasses

To co-digest WH with other biomasses, Ogunwande et al. [49] constructed biodegradation and maximum biogas yield models based on first-order kinetics to describe and predict maximum biogas yields from the co-digestion of duckweed (DW) and WH. They made three assumptions: there was a correlation between the volatile solid and degradation of biogas yield at any time; a certain quantity of volatile solids in the substrates was recalcitrant to degradation within the retention time allowed; there was no lag time before the beginning of volatile solids degradation. They noted that biogas production started within the first day of digestion. The following biodegradation model was provided:

$$C_t = (C_0 - C_e)\, e^{-kt} + C_e \text{ at } 0 \leq t \quad (9)$$

where C_0 is VS concentration in the substrates at the beginning of the experiment (%, db), C_t is the VS concentration in the substrates at any moment (%, db), t is the time, k is the VS degradation rate constant based on the quantity of VS in the substrate (D^{-1}), C_e is the remnant VS concentration after retention time (%, db). Also, the researchers provided a biogas yield model equation as follows:

$$Y_t = Y_m\left(1 - e^{-kt}\right) \quad (10)$$

where Y_t is the biogas yield at time t and Y_m is the maximum biogas yield.

Elsewhere, Bhui et al. [50] explored the role of volatile fatty acids (VFAs) in WH and Salvinia plant digestion. The biogas production from both plants was highest at an inoculum to substrate ratio of 3:1. It was concluded that acetonic, propionic, and butyric acid were the common VFAs found in the plants that played a major role in biogas production. In the same year, the effects of hydrothermal pretreatment on biogas yield were investigated [51]. A dramatic surge in biochemical oxygen demand (BOD) occurred after the first 30 min of the pretreatment. The increasing BOD revealed that the microbes have larger access to cellulose, a substrate for biogas production. The biogas yield rate started to increase at 30 min of pretreatment, peaking 60 min. Longer hydrothermal pretreatment could reduce the methane yield. Also, the WH:buffalo dung had no significant effects on biogas yield without hydrothermal pretreatment. They recommended a 1:2 WH and cow dung mixing ratio for optimum biogas yield.

4.3 Phytoremediation with WH

Phytoremediation of polluted waster using WH has been researched widely [87]. However, some researchers have delved into adopting post-phytoremediation WH biomass for biogas generation. Singhal et al. [69] co-digested WH and channel grass used to phytoremediate paper mill and distillery factory effluents for biogas production. The plants grown in the effluents were chopped, sun-dried, and oven-dried at 60°C, before pulverizing to fine particles. It was then mixed with cow dung slurry and digested. The digester feed used for phytoremediation produced more biogas than pristine ones. Likewise, the effect of temperature and feedstock size on biogas production of WH used for phytoremediation was reported [88]. It was observed that improved biodegradation of organic matter occurred at high temperatures. Therefore, the digestion of WH should be done at thermophilic conditions with smaller particle size. Similarly, Kumar et al. [68] assessed the biogas production potential of WH that was initially used for phytoremediation of sugar mill effluent. They concluded that the biomass had high potential for biogas production than virgin counterparts.

5. Conclusions and recommendations

The invasive presence of water hyacinth (WH) on our waterways often hinders numerous socioeconomic, agricultural, and ecological processes, tagging the macrophyte an environmental nuisance. However, we have identified that it has some inherent benefits when exploited appropriately. WH is potential biomass for biogas production, and this fact has been adequately studied.

Biogas generation from WH is temperature-dependent, taking place between 25 and 50°C, provided the temperature is kept steady at ±2°C because methanogens are sensitive to abrupt temperature change. Also, pH can be used as a performance indicator due to acetogenesis bacteria's pH requirement of >5.0 pH, while methanogens require pH in the range of 6.2–8.0 to form CH_4. Overall, WH aids some other biomass biogas generation potential while it retards some others. By co-digesting Salivinia grass and WH at 1:0.5, the highest volume of biogas (406 L/kg VS) generated was reported, whereas co-digestion with pig dung generated the highest CH_4 content (88.3%).

Moreover, co-digesting WH with cow dung (most popularly researched) produced more biogas than with poultry droppings and buffalo dung. Although homogenization aids biogas yield by 50%, it showed no significant effect on CH_4 content. Also, in most cases, increasing C/N and F/M ratios inhibits CH_4 formation. Optimally performing wet scrubbers and dry oxidation for cleanup could remove up to 95% CO_2 and 97% H_2S from the raw biogas generated. It was also identified that digesting WH shoots without the roots could up biogas yield significantly.

Finally, because hydrothermal pretreatment before digestion has recently been identified to enhance biogas generation, we recommend that the effects of various pretreatments for methane generation from WH be further researched.

Author details

Adedeji A. Adelodun[1*], Temitope M. Olajire[1] and Ochuko Mary Ojo[2]

1 Department of Marine Science and Technology, The Federal University of Technology, Akure, Nigeria

2 Department of Civil and Environmental Engineering, The Federal University of Technology, Akure, Nigeria

*Address all correspondence to: aaadelodun@futa.edu.ng

References

[1] Demirbas A, Ozturk T. Anaerobic digestion of agricultural solid residues. International Journal of Green Energy. 2005;**1**:483-494

[2] Deublein D, Steinhauser A. Biogas from Waste and Renewable Resources: An Introduction. Weinheim, Germany: Wiley-VCH; 2008

[3] Maile OA, Tesfagiorgis HB, Muzenda E. Factors influencing chemical absorption of CO_2 and H_2S in biogas purification: A review. In: Proceedings of the World Congress on Engineering and Computer Science 2015 Vol II WCECS 2015. San Francisco, USA: Newswood Limited; October 21–23, 2015

[4] Lawal AK, Ajuebor FN, Ojosu JO. Characteristic of piggery wastes feeds stock for determination of design parameters to biogas digester plant. Nigerian Journal of Research and Review in Science. 2001;**2**:193-198

[5] Ortega M. Installation of a Low-Cost Polyethylene Biodigester. Inter-American Institute for Cooperation on Agriculture (IICA); 2009. Available from: http://www.iica.int

[6] Ofoefule AU. Investigation of the biogas production potentials of Bambara nut chaff (Vigna Subterranea). Advances in Applied Science Research. 2011;**2**(2):55-61

[7] Ilaboya IR, Asekhame FF, Ezugwu MO, Erameh AA, Omofuma FE. Studies on biogas generation from agricultural waste: Analysis of the effects of alkaline on gas generation. World Applied Sciences Journal. 2010;**9**(5): 537-545

[8] Rabah AB, Baki AS, Hassan IG, Musa M, Ibrahim AD. Production of biogas using abattoir waste at different retention time. Science World Journal. 2010;**5**(4):23-26

[9] Itodo IN, Agyo GE, Yusuf P. Performance evaluation of a biogas stove for cooking in Nigeria. Journal of Energy in Southern Africa. 2007;**18**(3):14-18

[10] Ezekoye VA, Okeke CE. Design, construction and performance evaluation of plastic bio-digester and the storage of biogas. The Pacific Journal Science and Technology. 2006;**7**:176-184

[11] Ilori MO, Adebusoye A, Lawal AK, Awotiwon OA. Production of biogas from banana and plantain peels. Advances in Environmental Biology. 2007;**1**:33-38

[12] Adeyanju AA. Effect of seeding of wood-ash on biogas production using pig waste and cassava peels. Journal of Engineering and Applied Science. 2008;**3**:242-245

[13] Babel S, Sae-Tang J, Pecharaply A. Anaerobic co-digestion of sewage and brewery sludge for biogas production and land application. International Journal of Environmental Science and. Technology. 2009;**6**:131-140

[14] Abbey AT. Biogas in Uganda: A new experience. Leisa. 2005;**21**(1):13-17

[15] AGAMA Energy. Employment Potential of Renewable Energy in South Africa. Study Commissioned by Sustainable Energy and Climate Change Partnership and Earthlife Africa. Zambia: AGAMA; 2003

[16] Omer AM, Fadalla Y. Biogas energy technology in Sudan: Technical note. Renewable Energy. 2003;**28**:499-507

[17] Ersahin MV, Gomec CV, Dereli RK, Arikan O, Ozturk I. Biomethane production as an alternative: Bioenergy source from codigesters treating municipal sludge and organic fraction of municipal solid wastes. Journal of Biomedicine and Biotechnology. 2011;**8**: 1-8

[18] De Vrije T, De Haas GG, Tan GB, Keijsers ERP, Claassen PAM. Pretreatment of Miscanthus for hydrogen production by *Thermotoga elfii*. International Journal of Hydrogen Energy. 2002;**27**(11-12):1381-1390

[19] Panagiotopoulos IA, Bakker RR, Budde MAW, De Vrije T, Claassen PAM, Koukios EG. Fermentative hydrogen production from pretreated biomass: A comparative study. Bioresource Technology. 2009;**100**(24):6331-6338

[20] Bekkering J, Broekhuis A, Van Gemert W. Optimisation of a green gas supply chain: A review. Bioresource Technology. 2010;**101**:450-456

[21] Holm-Nielsen J, Al Seadi T, Oleskowics-Popiel P. The future of anaerobic digestion and biogas utilization. Bioresource Technology. 2009;**100**:5478-5484

[22] Jantsch TG, Matttiason B. An automated spectrophotometric system for monitoring buffer capacity in anaerobic digestion processes. Water Research. 2004;**38**:3645-3650

[23] Nan LB, Trably E, Santa-Catalina G, Bernet N, Delgenes J-P, Escudie R. Biomethanation processes: New insignts on the effect of a high H_2 partial pressure on microbial communities. Biotechnology for Biofuels. 2020;**13**:141

[24] Li Y, Yan XL, Fan JP, Zhu JH, ZhouW B. Feasibility of biogas production from anaerobic co-digestion of herbal-extraction residues with swine manure. Bioresource Technology. 2011; **102**:6458-6463

[25] Sambusitu C. Physical, chemical and biological treatments to enhance biogas production from lignocellulosic substrates [PhD thesis]. Politecnico Di Milano: Department of Civil and Environmental Engineering; 2013

[26] Lozano CJ, Mendoza MV, de Arango MC, Monroy EF. Microbiological characterization and specific methanogenic activity of anaerobe sludges used in urban solid waste treatment. Waste Management. 2009; **29**(2):704-711

[27] Franke-Whittle IH, Goberna M, Pfister V, Insam H. Design and development of the ANAEROCHIP microarray for investigation of methanogenic communities. Journal of Microbiological Methods. 2009;**79**(3): 279-288

[28] Odeyemi O. Biogas production. In: Proceeding of the fifth Annual Conference of the Nigerian Society of Microbiology; 2nd–6th Dec, 2001; Ado Ekiti. pp. 17-24

[29] Poliafico M. Anaerobic digestion: Decision support software [masters thesis]. Cork Institute of Technology: Department of Civil, Structural and Environmental Engineering

[30] Nettmann E, Bergmann I, Klocke M. Methanogene Archaea in landwirtschaftlichen Biogasanlagen. In Bayerische Landesanstalt für Landtechnik editor. Internationale Wissenschaftstagung Biogas Science. Freising: ES-Druck; 2009. pp. 303-318

[31] Jenner M. The Biotown, USA Sourcebook of Biomass Energy (Internet). Indiana: Indiana State

Department of Agriculture and Reynolds; 2006. Available from: http://www.in.gov/biotownusa/pdf/Biotown_Sourcebook_040306.pdf [Accessed: March 31, 2015]

[32] Tingi K, Lee K, Worley J, Risse M. Anaerobic digestion of poultry litter: A review. Applied Engineering in Agriculture. 2010;**26**(4):677-688

[33] Erickson LE, Fayet E, Kakumanu BK, Davis LC. Anaerobic Digestion. National Agricultural Biosecurity Center, Kansas State University, Kansas; 2005

[34] Yen HW, Brune D. Anaerobic co-digestion of algal sludge and waste paper to produce methane. Bioresource Technology. 2004;**98**:130-134

[35] Liu X, Chen Y, Du G, Chen J. Effects of organic matter and initial carbon-nitrogen ratio on the bioconversion of volatile fatty acids from sewage sludge. Journal of Chemical Technology and Biotechnology. 2008;**83**:1049-1055

[36] Oyeleke SB, Onigbajo HO, Ibrahim K. Degradation of animal wastes (cattle dung) to produce methane (cooking gas). In: Proceedings of the eighth annual Conference of Animal Science Association of Nigeria (ASAN). Lagos: Journalijar; 2003. pp. 168-169

[37] Katima JHY. Production of biogas from water hyacinth: Effect of substrate concentration, particle size and incubation period. Tanzania Journal of Science. 2001;**27**:109-119

[38] Syafrudin, Nugraha WD, Matin HHA, Margaretha F, Budiyono. Optimization of biogas production from water hyacinth by liquid anaerobic digestion (L-AD) using response surface methodology. IOP Conference Series: Materials Science and Engineering. 2020;**845**:012046

[39] Mahesh W, Jayaweera JAT, Dilhani RKA, Kularatne SLJ, Wijeyekoo. Biogas production from water hyacinth (Eichhornia crassipes (Mart.) Solms) grown under different nitrogen concentrations. Journal of Environmental Science and Health, Part A: Toxic/Hazardous Substances and Environmental Engineering. 2007;**42**(7): 925-932

[40] Patil JH, Lourdu MA, AntonyRaj, Gavimath CC. Impact of dilution on biomethanation of fresh water hyacinth. International Journal of Chemical Sciences and Applications. 2011;**2**(1): 86-90

[41] Patil JH, Lourdu MA, AntonyRaj, Gavimath CC. Study on effect of pretreatment methods on biomethanation of water hyacinth. International Journal of Advanced Biotechnology and Research. 2011;**2**(1): 143-147

[42] Siswo S, Budiyono, Marian DT. The effect of microwave power and heating time pretreatment on biogas production from fresh and dried water hyacinth (Eichhornia Crassipes). American Institute of Physics. 2015;**1699**:050018

[43] Rozy R, Dar A, Urmila GR. Optimization of biogas production from water hyacinth (*Eichhornia crassipes*). Journal of Applied and Natural Science. 2017;**9**(4):2062-2067

[44] Tantayotai P, Mutrakulchareon P, Tawai A, Roddecha S, Sriariyanun M. Effect of organic acid pretreatment of water hyacinth on enzymatic hydrolysis and biogas and bioethanol production. IOP Conference Series: Earth and Environmental Science. 2019;**346**: 012004

[45] Nugraha WD, Syafrudin S, Pradita LL, Matin HHA, Budiyono.

Biogas production from water hyacinth (Eichhornia Crassipes): The effect of F/M ratio. IOP Conference Series: Earth and Environmental Science. 2018;**150**: 36-39

[46] Ukwuaba SI. Performance evaluation of biogas yields potential from co-digestion of water hyacinth and kitchen waste. European Journal of Engineering Research and Science. 2018;**3**:4

[47] Hernández-Shek MA, Cadavid-Rodríguez LS, Bolaños IV, Agudelo-Henao AC. Recovering biomethane and nutrients from anaerobic digestion of water hyacinth (Eichhornia crassipes) and its co-digestion with fruit and vegetable waste. Water Science & Technology. 2018;**73**:2

[48] Teerasak H, Noppong S. Biogas and biomass pellet production from water hyacinth. Energy Reports. 2020;**6**: 532-538

[49] Ogunwande GA, Adanikin BA, Adesanwo OO. Comparative evaluation and kinetics of biogas yield from duckweed (lemna minor) co-digested with water hyacinth (*Eichhornia crassipes*). Ife Journal of Science. 2018;**20**:3

[50] Mathew AK, Bhui I, Banerjee SN, Goswani R, Chakranorty AK, Shome A, et al. Biogas production from locally available aquatic weeds of Santiniketan through anaerobic digestion. Clean Technologies and Environmental Policy. 2015;**17**:1681-1688

[51] Yuhelsa P, Dewi M, Teguh K. Study of biogas production rate from water hyacinth by hydrothermal pretreatment with buffalo dung as a starter. Waste Technology. 2014;**2**(2):26-30

[52] Ochuko MO, Josiah OB, Adebisi OA, Olurinde L, Adelodun AA. Co-digestion of water Hyacinth and poultry manure for improved biogas yield. ABUAD Journal of Engineering Research and Development (AJERD). 2020;**2**(1):42-48

[53] Njogu P, Kinyua R, Muthoni P, Nemoto Y. Biogas production using water hyacinth (*Eicchornia crassipes*) for electricity generation in Kenya. Energy and Power Engineering. 2020;**7**:209-216

[54] Barua VB, Kalamdhad AS. Biogas production from water hyacinth in a novel anaerobic digester: A continuous study. Process Safety and Environmental Protection. 2019;**127**:82-89

[55] Nugraha WD, Syafrudin SAT, Abdul Matin HH, Budiyono. Optimization of biogas production by solid state anaerobic digestion (SS-AD) method from water hyacinth with response surface methodology (RSM). E3S Web of Conferences. 2018;**73**:01016

[56] Ganesh PS, Ramasamy EV, Gajalakshmi S, Abbasi SA. Extraction of volatile fatty acids (VFAs) from water hyacinth using inexpensive contraptions, and the use of the VFAs as feed supplement in conventional biogas digesters with concomitant final disposal of water hyacinth as vermicompost. Biochemical Engineering Journal. 2005; **27**:17-23

[57] Ajieh MU, Ogbomida TE, Onochie UP, Akingba O, Kubeyinje BF, Orerome OR, et al. Design and construction of fixed dome digester for biogas production using cow dung and water hyacinth. African Journal of Environmental Science and Technology. 2019;**14**(1):15-25

[58] Adegunloye DV, Olosunde SY, Omokanju AB. Evaluation of ratio variation of water hyacinth (*Eichhornia Crassipes*) on the production of pig dung biogas. International Research Journal of Biological Sciences. 2013;**2**(3):44-48

[59] Akinnuli BO, Olugbade TO. Development and performance evaluation of piggery and water hyacinth waste digester for biogas production. International Journal of Engineering and Innovative Technology (IJEIT). 2014;**3**(10):271-276

[60] Akindele OO, Olusola AA. Evaluation of biogas production from co-digestion of pig dung, water hyacinth and poultry droppings. Waste Disposal & Sustainable Energy. 2019;**1**(4):271-277. DOI: 10/1007/s42768-019-00018-8

[61] Fadairo AA, Fagbenle RO. Biogas production from water hyacinth blends. In: Proceedings from the 10th International Conference on Heat Transfer, Fluid Mechanics and Thermodynamics. Florida; 2014

[62] Orhorhoro EK, Ebunilo PO, Sadjere GE. Effect of organic loading rate (OLR) on biogas yield using a single and three-stages continuous anaerobic digestion reactors. International Journal of Engineering Research in Africa. 2018; **39**:147-155

[63] Omondi EA, Gikuma-Njuru P, Ndiba PK. Anaerobic co-digestion of water hyacinth (*E. crassipes*) with ruminal slaughterhouse waste for biogas production. International Journal of Renewable Energy Development. 2019; **8**(3):253-259

[64] Sa'diah S, Putra MD. Biogas production from wastes of Tofu Industry with effects of water hyacinth and cow manure additions. IOP Conference Series Materials Science and Engineering. 2019; **543**:012097

[65] Prabhu AV, Manimaran R, Raja SA, Jeba P. Biogas production from anaerobic co-digestion of Prosopis juliflora pods with water hyacinth, dry leaves, and cow manure. Energy Sources Part A: Recovery, Utilization, and Environmental Effects. 2019;**42**(3): 375-386

[66] Ali-Shah F, Mahmood Q, Rashid N, Pervez A, Iqbal A, Shah MM. Anaerobic digestion of water hyacinth, giant reed, maize and poultry waste for biogas generation. EC Agriculture. 2015;**2**(2): 277-284

[67] Asikong BE, Epoke JA, Bassey E, Antai EE, Eja ME. Potentials of biogas generation from mixture of three substrates, water hyacinth, cassava peels and cow dung- Wh+Cp+Cd. Chemical and Process Engineering Research. 2013;**17**:1-11

[68] Kumar V, Singh J, Kumar P. Adding benefits to phytoremediation of sugar mill effluent by growing water hyacinth (*Eichhornia crassipes*): Evaluation of biomass for biogas production. Archives of Agriculture and Environmental Science. 2018;**3**(3):275-288

[69] Singhal V, Rai JPN. Biogas production from water hyacinth and channel grass used for phytoremediation of industrial effluents. Bioresource Technology. 2003;**86**:221-225

[70] Parawira W, Mshandete AM. Biogas technology research in selected sub-Saharan African countries: A review. African Journal of Biotechnology. 2009; **8**:116-125

[71] Lethomäki A, Viinikainen TA, Rintala JA. Screening boeral energy crops and crop residues for methane biofuel production. Biomass and Bioenergy. 2008;**32**:541-550

[72] Carlsson M, Uldal M. Substrate handbook for biogas production. Report SGC. 2009;**200**:21

[73] Yadvika S, Sreekrishnan TR, Kohli S, Rana V. Enhancement of biogas

production from solid substrates using different techniques—A review. Bioresource Technology. 2004;**95**:1-10

[74] Thy S, Preston TR, Ly J. Effect of retention time on gas production and fertilizer value of biodigesters effluent. Livestock Research for Rural Development. 2003;**15**(7):1-24

[75] Arogo JO, Wen Z, Igno J, Bendfeldt E, Collins ER. Biomethane Technology. Virginia Polytechnic Institute and State University, Virginia: College of Agriculture and Life Sciences; 2009. pp. 442-881

[76] Gerardi MH. The Microbiology of Anaerobic Digesters. New Jersey: John Wiley & Sons, Inc; 2003

[77] Karki AB, Shrestha NJ, Bajgain S. Biogas as Renewable Energy. Malmoe: ETDEWEB; 2005

[78] Monnet F. An Introduction to Anaerobic Digestion of Organic Waste (Internet). Remade, Scotland 2003. Available from: http://www.remade.org.uk/media/9102/ [Accessed: August 30, 2013]

[79] Kossmann W, Pönitz U, Habermehl S. Biogas Biogas Digest. Boston: Paques; 2000

[80] Alvarez R, Lidén G. The effect of temperature variation on biomethanation at high altitude. Bioresource Technology. 2008;**99**: 7278-7284

[81] Harikishan SS. Cattle waste treatment and class a biosolid production using temperature phased anaerobic digester. Advances in Environmental Research. 2003;7:701-706

[82] Nwanta JA, Onunkwo J, Ezenduka E. Analysis of Nsukka metropolitan abattoir solid waste and its bacterial contents in south eastern Nigeria: Public health implication. Archives of Environmental & Occupational Health. 2010;**65**:21-26

[83] Akinbami JFK, Ilori MOM, Oyebisi TO, Akinwumi IO, Adeoti O. Biogas energy use in Nigeria: Current status, future prospects and policy implications. Renewable and Sustainable Energy Reviews. 2001;5:97-112

[84] Ubalua AO. Cassava wastes: Treatment options and value addition alternatives. African Journal of Biotechnology. 2008;**6**:2065-2073

[85] Eltawil MA, Belal EBA. Eval'uation and scrubbing of biogas generation from agricultural wastes and water hyacinth. Misr Journal of Agricultural Engineering. 2009;**26**(1):534-560

[86] Otaraku IJ, Anaele JV. Modelling the cumulative biogas produced from sawdust, cow dung and water hyacinth. International Journal of Advanced Engineering Research and Science. 2020; **7**(3):504-511

[87] Rezania S, Mohanadoss P, Talaiekhozani A, Mohamad S, Md Din MF, Mat Taib S, et al. Perspectives of phytoremediation using water hyacinth for removal of heavy metals, organic and inorganic pollutants in wastewater. Journal of Environmental Management. 2015;**163**:125-133

[88] Xu D. Effect of temperature and feedstock size on biogas production of water hyacinth used for phytoremediation of rural domestic wastewater in Shanghai. Key Project of Science and Technology Commission of Shanghai Municipality. 2010; No. 06DZ12311

Chapter 11

Role of Microcredit in Sustainable Rural Development

Muhammad Imran, Shamsheer Ul Haq and Orhan Ozcatalbas

Abstract

Around 1.7 billion adults have no access to transaction accounts in the world. The majority of those are poor and women in rural areas of two developing regions of the world (South Asia and Sub-Saharan Africa). Rural areas of these regions are home to the poor and poverty, hunger, unemployment/underemployment is widespread phenomenon. Access to financial services is crucial for economic development. However, poor and smallholder have been neglected by traditional banks for a long time. Microcredit a development model to provide loans to the poor who have no, or little collateral emerged in Bangladesh and has been adopted in many countries of the world. In this chapter, microcredit as a solution to much of the problems of the rural areas has been discussed. Over time there has been a shift in objectives of rural development. Rural development nowadays is about an overall improvement of the human quality of life in terms of economic, social, political, and environmental, issues. Access to microcredit has a positive impact on three dimensions of sustainable rural development; social, economic, and environmental. Microcredit helps in the alleviation of poverty, employment, entrepreneurship, higher productivity from agriculture, women empowerment, gender equality, reduced rural outmigration, better health and education, green entrepreneurship, and adoption of modern technology/inputs in agriculture.

Keywords: microcredit, poverty, rural areas, rural development, sustainability

1. Introduction

Widespread poverty, low standard of livings, and resultant depopulation of rural areas is the biggest development challenge for the world. Rural population accounts for 43.80% of the total world population and the ratio varies from region to region and country to country. South Asia has the highest percentage of its population (65.12%) living in rural areas, followed by Sub Sharan Africa (58.80%), East Asia and Pacific (39.30%), Central Europe and Baltics (37.50%), Middle East and North Africa (34%), Europe and Central Asia (27.39%), European Union (25%), Latin America and Caribbean (18.88%), and North America (17.45%). Similarly, countries like Papua New Guinea and Burundi have the largest percentage of the population (almost 87%) in rural areas, while countries like Qatar, Uruguay, Iceland have the lowest proportion of their population living in rural areas (1–4%) (World bank).

The urban-rural gap exists in the whole world and in some regions and countries it is consistently widening. Globally, rural areas have been facing endemic poverty. Estimates show that even in 2025 60% of the world's poor will be in rural areas [1]. Traditionally, the rural population has been engaged in agriculture for their livelihood. However, over the years agriculture has become less profitable and declining productivity is one of the main issues in regions like South Asia and Sub-Saharan Africa. Moreover, lack of access to essential services such as basic infrastructure, education, knowledge, markets (product and financial) is perpetuating rural poverty [2]. International Fund for Agriculture Development estimated that nearly 1 billion people in the world live on less than US$ 1/day and 75% of them live in rural areas. Almost 90% of the rural poor do not have access to financial services. The crucial role of financial services in reducing poverty is very well established. Access to financial services can enable poor people to move forward from hand-to-mouth survival to planning for the future, acquiring physical and financial assets, investing in better nutrition, health and education. Microfinance/microcredit has attracted the attention of the world as a tool to fight poverty since Muhammad Yunus and his Grameen Bank were awarded the Nobel peace prize in 2006. As microfinance services were developed around the world it also has attracted the attention of the researchers to examine the impacts of microcredit on the recipients, community, and society. The main objective of this chapter is to discuss the role and functions of microcredit in sustainable rural development.

2. Defining sustainable rural development

Rural and urban areas are different from each other in terms of socio-economic characteristics and other aspects. Therefore, development strategies aimed at rural areas should be different from those for urban areas. However, to achieve equality on welfare efforts should be made to minimize the differences in living standards of both areas. There is no universally accepted definition of a rural area or rural development. Rural areas can be described as "residential areas where the economy is mainly based on a natural resource such as agriculture & forestry, where face-to-face interaction is more common, rules of daily life are shaped by local customs, and social, economic, and cultural developments are relatively slow and delayed [3].

Similarly, rural development is a challenging notion, in theory, practice, and policy. World Bank [4] defined rural development as raising the level of rural income through agricultural modernization. In the '70s rural development was seen as a purely economic issue and rural development was seen synonymous with agricultural development. Later in 1980, World Bank redefined rural development as a strategy to improve the economic and social life of the rural people. Since then, rural development has been defined in various ways.

According to Kata [5] "rural development indicates the overall development of rural areas to improve the quality of life of the rural people".

Shepherd [6] defined rural development as "the set of activities and actions of diverse actors that taken together leads to progress in rural areas."

All these above definitions have many missing links and the authors failed to consider some important dimensions and aspects of rural development. From early 1900 to the 1990s the concept of rural development has changed significantly. In 1900s it was guided by profitability, however, in late 20th century it was directed to the sustainability of agriculture and other activities and based on a holistic, inclusive, and participatory local and territorial model [7, 8].

Madhu [9] defined rural development "as an activity concerned with the improvement of spatial and socioeconomic environments of rural areas to enhance the ability of the individuals to cater to and sustain their well-being".

"Rural development is people who live in rural areas and earn their living mainly from agriculture. with an inclusive approach to increase the welfare of people and families who provide such as nature, environment interaction, use of natural resources and technology, and agriculture-based industry. economic activities are carried out in harmony and aim to create an ecosystem" [3].

Rural development nowadays is about an overall improvement of the human quality of life in terms of economic social, political, environmental, and administrative issues [10]. Rural development has emerged from agricultural development to human, social, environmental development along with agriculture. Therefore, a shift from rural development to sustainable rural development has occurred in the recent past. Sustainable rural development is based on four basic pillars which are social, economic, environmental, and political.

According to a report on the 17th session of the Commission on Sustainable Development of the United Nations "sustainable rural development is vital to the economic, social, and environmental viability of nations. It is essential for poverty eradication since global poverty is overwhelmingly rural. The manifestation of poverty goes beyond the urban-rural divide; it has a sub-regional and regional context. It is therefore critical, and there is great value to be gained, by coordinating rural development initiatives that contribute to sustainable livelihoods through efforts at global, regional, national, and local levels, as appropriate. Strategies to deal with rural development should take into consideration the remoteness and potentials in rural areas and provide targeted differentiated approaches."

2.1 Importance of sustainable rural development

Rural development is concerned with improving the quality of life and social, economic wellbeing of the people living in the rural areas. Sustainable rural development is little different from rural development. It also focuses on the technical, socioeconomic, and environmental conditions of rural areas. Therefore, the daily basic needs of the rural population are covered by realistic public utilities combined with technical, socioeconomic, and environmental conditions to support regional economies and urban-rural linkages called sustainable rural development [11].

Rural development is concerned with the improvement in the different indicators like increasing productivity, raising employment opportunities, high adoption of modern technologies, low poverty, high income, and good infrastructure, etc. just focusing on these things without concerning environmental conditions is not sustainable development. Sustainable development links with social economic, and environmental sustainability of the rural areas which do not compromise the ability of the future generation to meet their own needs. For example, cultivation of crops by using high quantity of chemical fertilizer, chemical may contribute little to crop production, but it is not sustainable because the application of high chemical fertilizers degrades the agricultural land. Similarly, the zero application of fertilizer or pesticides is also not a concern of sustainable development which adversely affects the crop yield leading to low income of farmers. In this way, low income generates many socioeconomic problems in rural areas which adversely affect the current rural population. The main focus in sustainable rural development is meeting your daily needs without compromising the social, economic wellbeing of the current generation of rural population coupled with focusing on the wellbeing of future generations without compromising the environment quality.

3. Rural areas and their challenges

The rural-urban gap exists all over the world and eliminating inequality in access to services and opportunities is the main challenge for policymakers. The rural population is facing many challenges some of which are discussed below.

3.1 Home to poor

The rural areas are considered as a home for the poor. Generally, the rural population is poor as compared to the urban areas. The unequal distribution of assets, working facilities, and uneven land ownership cause high poverty in rural areas. Therefore, the poverty rate in the rural population is 17.2% three times more than in urban areas [12]. Endemic poverty has caused many negative consequences for the rural population. Especially rural population in countries of South Asia, Sub-Saharan Africa, and Latin America is more vulnerable to extreme poverty, climate change events (floods, heatwaves, disease, etc.), and social exclusion.

3.2 Gender inequality

Gender inequality can be observed more in the rural area as compared to urban areas. The women have limited access to services, capital, productive infrastructure, and technologies. Their limited freedom inhibits the women to perform their role actively in the development of the rural areas. According to International Fund for Agriculture Development [13], women spend 12–13 h in different daily activities, but they remain unpaid.

3.3 Limited employment opportunities

Rural areas are normally characterized by seasonal employment opportunities. Unemployment normally prevails more in rural areas. Moreover, underemployment is also a common problem in rural areas. The low productive jobs and low-waged activities discourage the rural youngsters which ultimately tend to migrate to the urban areas.

3.4 Subsistence economy

The majority of the rural population in many developing and even developed countries rely on agriculture as a source of livelihood. In developing countries, the majority of the farmers are small farmers, and the traditional economy is based on subsistence agriculture. Many researchers have estimated that there are around 500 million small farms (landholding up to 2 ha) in the world [14–16] which makes almost 70–80% of the total farms in the world [17]. Subsistence agriculture is volatile to extreme weather shocks, floods, droughts, degradation of resources.

3.4.1 Dependence on agriculture and related issues

The rural people are majorly depending on agriculture for their earnings. As a major source of earning, agriculture still faces many challenges and problems. The small landholding is one of the main problems impeding the development of rural areas. The 12% of the agricultural area is operated by small farmers having less than 2 acres and 75% of family farms cultivate 75% worlds' agricultural land [18]. Small farming leads to low adoption of modern technologies in agriculture. Low adoption of modern technologies is a constraint to improve productivity mainly caused by the low level of income of small farmers, low awareness, and low assets [19]. The small and family farms are mostly low adopters of modern technologies due to their limited knowledge, the limited desire for adoption, and low level of assets. Agricultural is commonly a profession of illiterate people. They are following the traditional farming practices and have no attention toward the adoption of different modern technologies. Deteriorating quality of water and land is also a growing problem of agriculture. The monoculture cropping system is commonly practiced in rural areas due to the limited availability of land. To attain a high level of yield, the small farmers try to apply more fertilizer, pesticides, and chemical inputs which degrades the health of the soil and also negatively affects the environment. It lowers agricultural productivity and affects the quality of life of the farmers [20].

Challenges	Description	Local Impacts	National Impacts
Quality and quantity of resource	Declining land holding land fragmentation, declining productivity of land, degradation of soil and water quality	• Low productivity • Making agriculture low-profit business • Loss of natural habitat • Rural outmigration	• Rising food insecurity • Urban sprawls • Urban poverty • Urban slums • Street crimes
Social challenges	Poverty, lack of access to educational, health, and energy services, poor infrastructure, absence of communication networks	• Social conflicts • Theft and crimes • Illiteracy • Poor health • Deforestation • Isolation • Low prices of output • Obliviousness	• Social unrest • Inequality • Illiteracy
Environmental/ climatic	Floods, heat waves, crop and animal diseases, rainfalls, droughts	• Low productivity • Diseases • Mortality • Crop and income loss • Famine	• Food insecurity • Emergence of new diseases • Increased health expenditures
Financial challenges	Lack of access to financial services	• Low productivity of crops and livestock • Illiteracy • Health problems	• Loss of national output • Food insecurity • Illiteracy • Low Human Development

Table 1.
Challenges of rural areas and their possible local and national impacts.

3.4.2 Lack of access to financial services

One important constraint discussed by many scholars is no or low access to financial services. Rural people with marginal demand for credit are often ignored by conventional financial institutions. Lack of access to financial services hampers the development of rural areas in many ways. Small (subsistence) farmers are resource scarce. Scarcity of financial resources causes problems in the adoption and use of modern technologies, seeds, and good quality inputs. Lack of financial capital also leads to the delayed application of inputs. This all results in low productivity>low income>low living standards. Therefore, subsistence farmers in many regions of the world are trapped in this cycle of low productivity and poverty (**Table 1**).

Though many of the problems in rural areas have an immediate impact on the local population. However, in long run, these negative impacts spread to the national level. Low productivity caused by natural hazards or financial constraints leads to national food insecurity. Lack of access to basic utilities in rural areas harms the human development of any country. Poor road, energy, and communication infrastructure cause low national output, obliviousness, environmental impacts, and inequality in the country.

Much of the challenges faced by the rural population stem from limited or no access to financial sources. Rural people are often neglected by the conventional lending institutions, especially, the small farmers. Small farmers have limited and consistent demand for credit, however, conventional banks often do not seem interested to fulfill their needs due to several reasons, such as low or no collateral, risk of recovery, high costs of operation, etc. In the background of all these financial challenges to the rural population, recently the world has embraced microcredit as an important strategic tool to combat challenges in rural areas. There has been increasing interest of policymakers, researchers, and government in the effectiveness of microcredit.

4. Microcredit and its history

Microcredit and microfinance are used interchangeably. Microcredit can be described as any credit or loan given by the bank or lending institutions for use by small enterprises such as smallholder farmers. Though the roots of modern microfinance are found in rural Bangladesh, however microcredit has existed in the world in a different form for centuries. The stems of informal lending and borrowing extend back for thousands of years in Asia. The term microcredit is new, and the term was invented by Muhammad Younas in the mid-1970s. The concept is to provide small loans to people with lower/weaker socioeconomic backgrounds. Though the concept of lending to people with lower socioeconomic background goes back to the period of the 1700s in Ireland. However, modern microfinance emerged in rural Bangladesh in the mid-1970s. Microcredit emerged with the establishment of Grameen Bank and BRAC (Bangladesh Rehabilitation Assistance Committee) in the 1970s with new models of lending. Muhammad Younas a Nobel Laureate played a key role in shaping the vision of the Grameen Banking model. He was looking for a practical solution to poverty in the rural areas of Bangladesh. The first-ever examples of microcredit originated from the group of 42 women who were making stools in Jorba village in Bangladesh. The women were earning very little profits of $0.02 on each bamboo stool because of the early repayment to suppliers. Muhammad Younas was shocked to find that the entire borrowing needs of 42 women which is equivalent to $ 27. He thought if women were provided with a loan, they could meet their business needs, sustain their business and get out of the poverty trap. The 42 women were lent $27 from his resources as an experiment and allowed them to sell their bamboo stools at reasonable prices and come out of this debt cycle. The experiment later led to the establishment of Grameen Bank. Later in 1983, Muhammad Younas decided to open a Grameen Bank in Bangladesh to realize his microcredit model.

The Grameen Bank Known as "Village Bank" came to existence and today works in more than eighty thousand villages across Bangladesh and serves more than six million active borrowers. Inspiring with the success of Grameen Bank, many new microfinance institutes came into existence around the world, many of them are started by several NGOs and funded by subsidies and grants from private and public sources. They signify/reveal that poor people could be relied on repaying their loans, even without collateral and microfinance is potentially a very feasible business. The Grameen Bank and its founder Muhammad Younas were awarded the Noble Prize in the Year 2006 for developing this development model and with time the model spread around the world with an estimated reach of 175 million people (Microcredit Summit Campaign). Since 2006, microcredit become a popular tool of economic development throughout the third world. Now the microcredit is widely propagated in many countries of the world.

The model of microcredit was adopted by many countries in the world for example in Pakistan there are specialized microfinance institutions, such as microfinance banks and microfinance institutions which provide loans to poor people. Several NGOs such as the Agha Khan Foundation, Akhuwat, etc. have also been involved in providing microloans to meet the credit needs of poor people. In the Philippines, to increase the income level of the poor government developed Microfinance Development Program (MDP) which provides easier access to credit. Evaluation report of MDP by Asian Development Bank (ADB) mentioned some important achievements, e.g., outreach increased from 1.3 million active borrowers in 2004 to 2.1 million buys 2008, increased loans to microenterprises created new jobs, and income of the households was increased. In India, a microcredit system called NABARD (National Bank for Agriculture and Rural Development) was developed by getting inspired by the Grameen bank model of Bangladesh. On the same lines, the self-employed women Association (SEWA) was developed in 1974 for microfinancing to rural and women. In Europe Microfinance has been growing steadily, a survey of 2015 revealed that there were 747,265 active borrowers with a gross microloan portfolio outstqand8ng of 2.5 billion euros. Similarly, microfinance has been widely developed and adopted in the African region where poverty is dominant. Many Institutions at different levels are providing microloans. Many microfinance programs have been initiated with the cooperation of the International Finance Corporation (IFC) across the African region. An extensive network of microfinance institutions is present in many developing countries. **Table 2** presents the microfinance network in three different countries of the world.

With the growing interest in microcredit as a major tool for poverty alleviation, the focus has been moved away from the NGOs models toward the sustainable microfinance industry by providing the microloans at the lowest prices and also making a reasonable return to the commercial investors. In such a way, many microcredit investment firms exist today. Many big banks also entered into the microfinancing industry such as Credit Suisse, Citigroup, and Deutsche. By the end of 2008, the US$15 billion of foreign investment had been channeled into microcredit institutions from private and commercial sources. Nowadays, microcredit has been the subject of

Bangladesh	
A. **Member-owned specialized institution**	Grameen Bank
B. **Non-Governmental Organizations**	BRAC, Proshika, ASA, BURO-Tangail, BEES, CODEC, SUS, TMSS, Action- Aid, etc.
C. **Commercial and Specialized Banks**	Krishi Bank (BKB), Rajshahi Krishi Unnayan Bank (RAKUB)
D. **Government-sponsored microfinance projects**	BRDB, Swanirvar Bangladesh, RD-12, etc.
Pakistan	
A. **Commercial and Specialized Banks**	First MicroFinance Bank, Advance Microfinance bank, Khushali Microfinance bank, Pak Oman Microfinance bank, NRSP Microfinance bank, Sindh Microfinance Bank, Finca Microfinance bank, Mobilink Microfinance bank, etc.
B. **Microfinance Institutions**	Akhuwat Islamic Microfinance, Dameen Support Program, Islamic Relief Pakistan, FFO support Program, CSC empowerment and inclusion Program, etc.
C. **Rural Support Programs**	National Rural Support Program, Punjab Rural Support Program, Sarhad Rural Support Program, etc.
Tanzania	
A. **Microfinance Institutions**	Adroit Financial Services Ltd., AJBE Microfinance Limited, Amani Microfinance Ltd., ASA Microfinance (Tanzania) Limited, BRAC Tanzania Finance Ltd., etc.
B. **Government Supported**	Zanzibar Economic Empowerment Fund (ZEEF)
C. **NGO's**	Tanzania Network of Religious Loaders Living with or Personally Affected by HIV and AIDS (TANERELA)
D. **Banks**	Akiba Commercial Bank (ACB), DCB Commercial Bank, Equity Bank Tanzania Limited, FINCA Microfinance Bank Tanzania (FINCA), MUCOBA BANK PLC, etc.

Table 2.
Microfinance networks in Bangladesh, Pakistan, and Tanzania.

various experiments and innovations. The huge mobile industry has also entered in sending and receiving microcredit loans.

5. Functions of microcredit

Microfinance/microcredit refers to financial services to the poor people and it's not limited to only credit but also provides services like money transfers and insurance. According to Murray and Boros [21] "Microfinance as a discipline has created financial products and services that are packaged in a way that enables low-income people to become clients of a banking intermediary. They enlisted the characteristics of microfinance products as follows:

- "Small amounts of loans and savings.
- Short loan terms (usually up to one year).
- Payment schedules featuring frequent installments (or frequent deposits).
- Installments made up from both interest and principal.
- High-interest rates on credit (higher than commercial bank rates but lower than loan-shark rates) reflect the labor-intensive work associated with making small loans and allowing the microfinance intermediary to become sustainable over time.
- Easy access to the microfinance intermediary, saving the client time and money, while also permitting the intermediary to better know the client in their home/business context Simple application forms which are easy to complete.
- Short processing periods (between the completion of the application and the disbursement of the loan).
- The availability of repeat loans in higher amounts for clients who pay on time.
- The use of tapered interest rates (decreasing interest rates over several loan cycles) as an incentive to repay on time. As larger-size loans are less costly to the MFI, some lenders charge lower interest rates: higher rates on small credit amounts and lower ones on larger credits.
- No collateral is required contrary to formal banking practices. To replace collateral (which low-income people generally do not have), microfinance intermediaries use alternative methods, such as the assessments of clients' repayment potential by running cash flow analyses based on the stream of cash flows generated by the activities for which loans are taken; all enterprise/household income and expenditure items; individual or group guarantees (solidarity groups); and compulsory savings schemes."

Microfinance has been growing rapidly over the last two decades into an important subfield of development studies and provides a place for researchers to research the causes and consequences of poverty.

The practice of microcredit includes lending small loans to poor people who do not have sufficient funds to meet their basic needs, to create employment opportunities for them. Microcredit refers to shorter loans offered to small and medium enterprises as well as to smallholder farmers for agricultural production. Such microloans are offered by thousands of financial institutions worldwide, ranging from government to non-government financial institutions like formal banks and public institutes. The beneficiaries of these financial programs are individuals and groups. In developing countries, the basic goal of microcredit is to alleviate poverty by providing job creation opportunities to needy people. Microcredit refers to the process of taking control over the acquisition and use of credit in the present time for a certain project, business activity, in exchange for a promise to repay in the future.

Commercial and Specialized Banks	
First MicroFinance Bank	
	a. Agriculture & Livestock (crop farming, purchase of small/large animals, construction of shed, etc.)
	b. Enterprise (purchase of inventories, small assets, common facility centers, etc.)
	c. General Purpose (Education, housing improvement, income-generating activities)
Microfinance Institutions	
Akhuwat Islamic Microfinance	
	a. Family enterprises loans
	b. Agriculture loans
	c. Liberation loans
	d. Housing, education, health, marriage loans
	e. Emergency loans
Rural Support Programs	
National Rural Support Program	
	a. Individual Loans (livestock, enterprise, house finance, etc).
	b. Groups Loans (Agriculture loans, livestock loans, women empowerment group loans)
	c. MSME (Micro Small Medium Enterprise Loans)

Table 3.
Microcredit institutions and their functions in Pakistan.

Recent trends have shown microcredit in a broader spectrum embracing housing, consumption, education loans, and loans to meet the basic needs of poor people such as electricity, water, and sanitation that are not provided by public financial institutes. For example in Pakistan, a variety of microcredit needs are met by loans from different microfinance providers. There are commercial and Specialized Banks such as First Microfinance Bank providing loans for crop farming, purchase of livestock, construction sheds to farmers, to the enterprise for purchase of inventories and assets, and also for education, health, housing improvements. Similarly, Akhuwat Islamic microfinance which provides interest-free loans covers different credit needs of households and small enterprises. Government-supported programs such as national and provincial rural support programs provide loans to individuals and groups for agriculture, MSMEs, housing, and women empowerment (**Table 3**).

Though eligibility criteria, tenure, and purpose of loans vary from country to country, however, the target group is the same (poor people).

5.1 Microcredit and family farming

Microcredit serves as a support system for family farms. Agriculture in developing and underdeveloped regions of the world is dominated by family farms. Family farms provide above 70% Of the global food supply [22]. Food and Agriculture organization-defined family farms as "An agricultural holding which is operated and managed by a household and where farm labor is mainly supplied by that household" [23]. According to Hazell et al. [24] a farm where the main purpose of growing staple food for the consumption of households and where the majority of the labor is supplied by household members. Family farms are different from other farms in many perspectives; freedom provided by self-employment, intergenerational continuity, residence on farms 9owner lives on the farm or in a nearby village, etc. In short family, farming is a lifestyle and tradition [22]. In recognition of its importance for humankind and the global economy, United Nation celebrated 2014 as a year of family farming.

6. Role of microcredit in sustainable rural development

Microcredit has been widely used as a development tool, especially in rural areas. Rural development is concerned with improving the quality of life and social, economic wellbeing of the people living in the rural areas. Sustainable rural development is little different from rural development. It also focuses on the environmental conditions of rural areas and the ability of future generations to meet their own needs. Therefore, the daily basic needs of the rural population are covered by realistic public utilities combined with technical, socioeconomic, and environmental conditions to support regional economies and urban-rural linkages [11]. The rural people must also develop some non-farming activities paired with farming systems, to counter the economic shocks and environmental challenges in the context of climate change. Based on the sustainable rural development aspects, the microcredit role cannot be ignored. Where the rural areas face many challenges especially in agriculture (major source of income) they need capital/financial support to manage their daily needs, but the main source of finance can be only microcredit because it is the loan available for the small farmers without collateral. Where the concern is sustainable rural development, the rural people's needs (especially small farmers) financial capital to overcome and timely manage their problems faced in daily life. Microcredit has a multidimensional and interlinked impact on rural areas.

Family farming—one important role played by microcredit is in the sustainability of family farms (discussed above). Farmers engaged in family farming solely rely on agriculture as a source of income, hence, their credit needs are higher compared to those with off-farm incomes. Large and commercial farming is posing a serious threat to the existence of family farms. Considering the socio-cultural, economic, and environmental contributions of family farming there is a greater need to protect and preserve it. In 2014, the importance of family farms for sustainable agriculture and rural development was greatly highlighted. Family farming is solving the problem of food security, nutrient supply, unemployment, and controlling rural outmigration. Hence, by providing timely and inexpensive credit to family farms, a tradition (family farming) can be preserved, and many issues faced by rural areas can be addressed.

Microcredit role in rural development is not limited to just family farming, it can play important role in capacity building of human resources, promoting cooperative culture, expanding production area, enhancing marketing capabilities of farmers, protecting and developing rural heritage and rural life, providing a safety net against climate shocks, controlling rural-outmigration, and reducing hidden or seasonal unemployment. Furthermore, the role of microcredit in the social, economic, and environmental development of rural areas is discussed below.

6.1 Economic development of rural areas

Among all other things, poverty and food security are the greatest threat to the existence of humans, and rural areas are home to a majority of poor people. Poverty is the situation in which the people lack the usual and socially acceptable amount of money or material possessions [25]. Lack of income and resources limited the capabilities of rural populations which leads to severe poverty. The provision of credit to the poor has been considered an important strategy to reduce poverty and promote rural entrepreneurship. Meehan [26] found in their study that, offering microcredit to rural people plays an important role in alleviating poverty and ensuring food security. Increased access to financial services enables the poor to smooth consumption (in case of adverse shocks), start or expand a business, cope with risk and diversify income. Microcredit distribution to the poor people of rural areas enables them to overcome their financial problems and limited resources. A small amount of loan or capital facilitates the rural people to generate small business opportunities generating enough income to feed their families, send their kids to school, and build suitable housing [27]. Shirazi and Khan [28] explored the positive role of microcredit in poverty alleviation and stated that microcredit reduced the poverty by 3.05 percentage points in the period of study under consideration. Moreover, access to credit empowers the rural poor by improving their access to production facilities. Microcredit enhances self-reliance, assists in the creation of employment opportunities, and engages women in economically productive activities. The provision of timely credit at an affordable rate increases the capacity of investment in a productive manner and may help to generate high income and improve the social and economic standards [29]. Kasali et al. [30] analyzed the role of microcredit in poverty alleviation in Nigeria. They stated the significant impact of microcredit on poverty reduction in the countries.

The role of SMEs in the development of any economy cannot be ignorable. Small and medium-sized business needs sudden financial capital to overcome the uneven changes in financial and market conditions. SMEs plays important role in generating a high income, and employment in areas that contribute to the economic growth of the country [31]. Rural areas SMEs are also very important in the development of the rural areas. SMEs plays role in the alleviation of poverty (social and economic sustainability), and improvement in the income of poor peoples (economic sustainability). Waliaula [32] described the strong and significant relationship between microcredit and SMEs in Kenya. Similarly, the SMEs owned by the women also need microcredit occasionally. The timely provision of microloans helps the women entrepreneurs to empower themselves and support their families which contributes to the sustainable development of the rural areas. Alhassan et al. [33] described the significant impact of microcredit on the average monthly profit earned by the women borrowers running SMEs themselves in Northern Ghana. Ruslan et al. [34] also stated the sale growth of SMEs with microcredit borrowers was having high sale growth and employment than non-microcredit borrowers.

Access to credit helps the rural economy in many ways; access to credit improves the ability of the households to meet their financial needs, enables farmers to purchase improved agricultural inputs and adopt modern technology, which in return increases the income of smallholders and break the perpetuity of poverty cycle they are entangled with. The positive impact of microcredit on uptake of improved inputs and increased agricultural productivity are extensively discussed in the literature. Zuberi [35] found that more than 70% of the microloans availed by smallholders are used in purchasing quality inputs. Rahman and Khandker [36], Khandker and Faruqee [37], Ahmad et al. [38] and Chandio et al. [39] found positive impacts of microcredit on the productivity of recipient farmers. Agriculture is an important source of income to rural households in most developing countries [40]. Agricultural productivity largely depends on the traditional farm technologies and land management practices which are labor and capital intensive. The poor rural people normally lack financial capital which limits their capacity to manage their farm at an efficient level. Access to microcredit has a positive impact on agricultural productivity. It provides the poor farmers with liquid capital to purchase timely farm inputs [41]. The timely purchase and application of critical farm inputs increase agricultural productivity. Ashaolu et al. [42] found that the user of microcredit was attaining higher profit than the non-user. Microcredit also increases the technical and economic efficiency of farmers by overcoming their financial constraints which affect their purchase of farm inputs on time. Moreover, it also allows the farmers to shift the most remunerative crops (**Figure 1**) [43].

6.2 Social development of rural areas

Microcredit stands to benefit the poor individual who lacks collateral, steady employment, verifiable credit history, or other requirements necessary to gain access to formal credit [44]. By putting money into the hands of poor families, and particularly poor women microcredit has the potential to increase households' health and education, empower women. Microcredit helps the rural poor to increase their productive capacity by bringing improvement in their human resources and financing [45]. The role of women in the rural economy area is significant and is not ignorable. They work as wage earners, farmers, and entrepreneurs. Empowering rural women has a significant impact on productivity and agricultural-led growth [46]. They are key agents for rural development by their agitator role toward gaining the transformational social, economic, and environmental changes necessary for sustainable development. Their limited access to credit, education, and health care facilities limit the capabilities of women in performing their active roles [47].

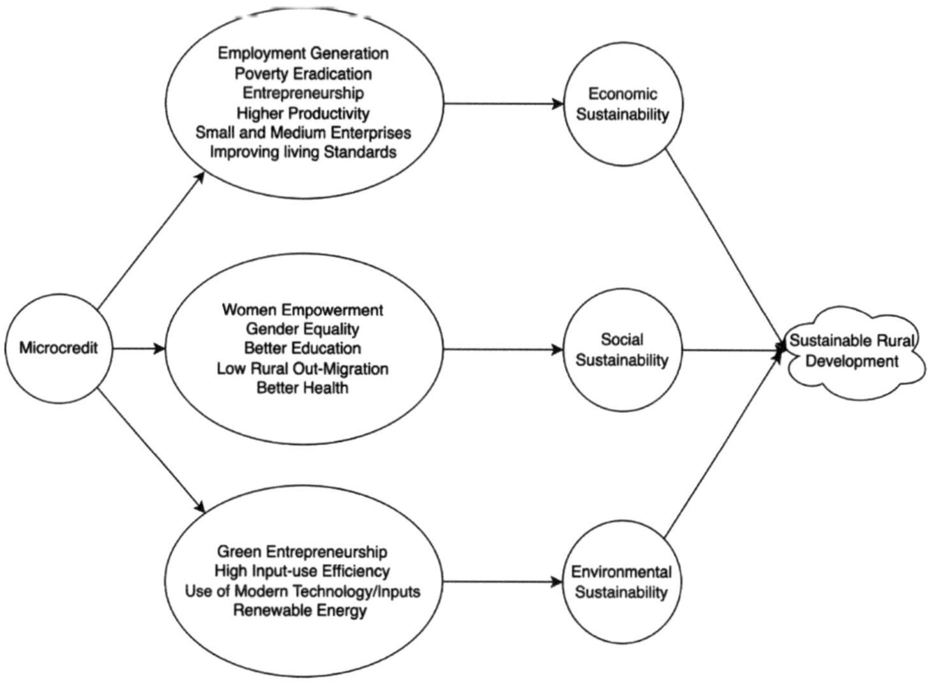

Figure 1.
Relationship between microcredit and sustainable rural development.

Microcredit can play a significant role in empowering poor women in rural areas. Ahmed et al. [48] described the role of microcredit in reducing the vulnerability of women living in rural areas. They assessed that the women with credit were generating high profits than those without credit access. They also stated that the borrower women obtained significantly high income which reduced their vulnerability effectively. Similarly, Shah and Butt [49] described the positive impact of microcredit on the socio-economic empowerment of female borrowers in Pakistan. Consequently, women's empowerment in rural areas increases the social and economic sustainability contributing to sustainable rural development. It is also evident that the women are more concerned about the environment and their role to manage the natural recourses of their families [50]. Microcredit also improves the well-being of women and their families. Nader [51] confirmed the higher children's education and availability of assets and the high-income level in the families of the women who borrowed the microcredit.

6.3 Environmental development of rural areas

The role of microcredit in the development of rural areas and economies around the world is not mysterious. Where microcredit became popular in uplifting productivity, poverty alleviation, generating a high income, and employment opportunities, its impact on the environment is also very important to assess. It is expected that microcredit plays important role in the adaptation of climate change. The timely availability of finance to the poor asset them to adopt the climate adaptation strategies such as irrigation technologies, harsh weather tolerant varieties, etc. Jordan [52] described the positive relation of microcredit and the adaptation of climate change strategies. Moreover, the microcredit to the poor in rural areas assists them in generating their high income and lowering their poverty level which requires the quality of the environment with time. They also have the chance of diversified income levels which reduce the risk of loss and have beneficial impacts on the environment [53]. The multidimensional impacts of microcredit in development can be sustainable, for example, microcredit helps the poor to alleviate their poverty which leads to uplifting their socio-economic sustainability, and once they are sustainable in their socio-economic conditions they require a quality environment. For this, microcredit institutions should focus the green entrepreneurship. The timely application of inputs for getting high yield and high income and becoming diversified in earning sources coupled with the environment care leads toward sustainable rural development.

6.4 Role of microcredit in achieving SDGs

Microcredit plays a vital role in achieving many sustainable development goals (SDGs) directly or indirectly. **Figure 2** describes the possible role and functions of microcredit for achieving the SDGs. The possible direct impact of microcredit on achieving the 1st SDG called "No poverty" has been extensively addressed in the literature. Microcredit is provided to the poor people who live on $1.25 or $2.00 per day. Poor families through microcredit can reduce the poverty, and ensure smooth consumption [54]. Malnutrition is also a very serious problem worldwide. To overcome this global problem, many studies have described the link between microcredit participation and the nutrition status of participants. Hamad and Fernald [55] reported the positive effect of microcredit on nutritional status and the food security

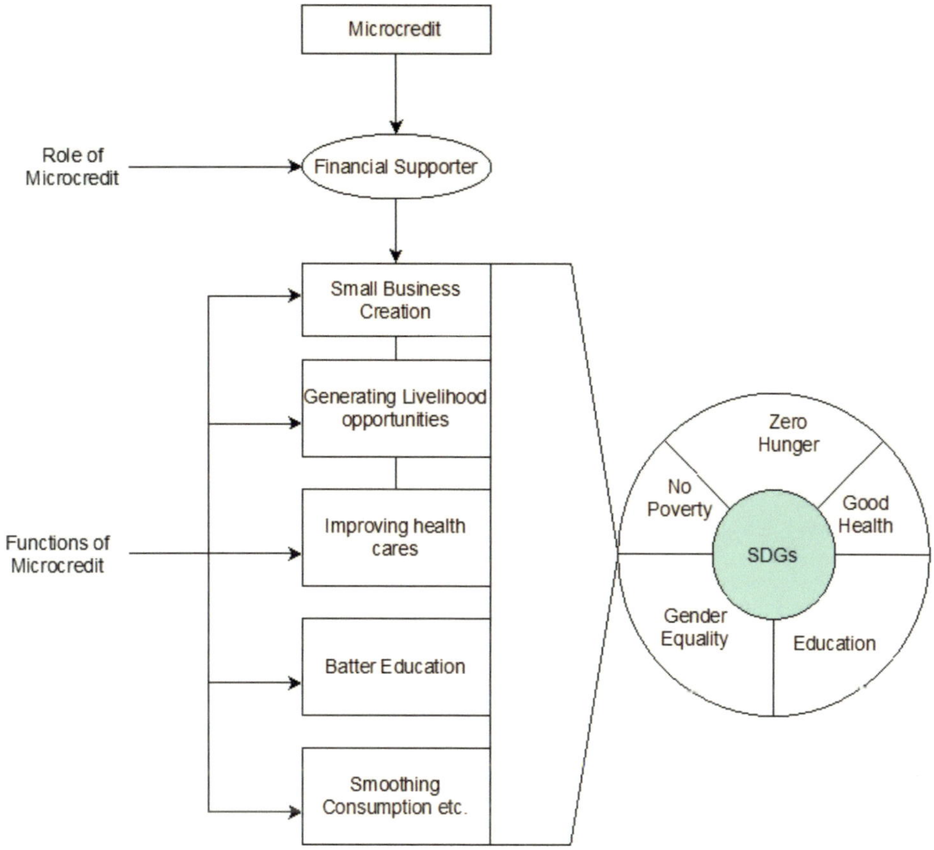

Figure 2.
Role and functions of microcredit to achieve SDGs.

of the participants. Similarly, the socioeconomic and health association has been widely unearthed. The poor people and those having low socio-economic status experience poor health. Microcredit provides the financial resources to the poor people to establish their small businesses, generate their income, and approach their self-sufficiency. In this way, the income and health link are widely acknowledged worldwide [56]. Poverty and health inequalities are indistinguishably linked, and microcredit is the option that can focus on multiple factors like no poverty, good health, low hunger, and better education, etc. Amin et al. [57] described that the poor families who joined the microcredit program tend to have better access to insurance. The 4th SDG "Education" is also approachable by the participation in microcredit programs. The poor families in the rural area mostly remain busy in farming and they keep their children working at the farm to stabilize their financial position. Consequently, their children remain absent from school. Microcredit helps them to stabilize their financial condition, and their children are more likely to attend school. Nader [51] find out a strong association between microcredit and children's education. Based on the discussion, microcredit has caused the change in income, enabled the participants to provide better health care, increase the provision of nutrition to the children, increase household consumption, enhance female empowerment, and enable the participants to save more [58].

6.5 Evidence from different countries

Microcredit has a positive impact on women's empowerment [59] which improves their family income and nutrition level. In Pakistan, it was s found that microcredit played important role in poverty alleviation and poverty reduced by 3.05% [28]. Similarly, in Bangladesh, every year, almost 1% of the total population is coming out from poverty in the country because microcredit played an important role in breaking out the vicious cycle [60]. In Zimbabwe, the microcredit borrowers experienced higher average growth in their business profit and family income [61]. In such a way, microcredit in terms of entrepreneurship is also very successful, especially in the case of women. In Malaysia, the microcredit entrepreneurs' profile was explored, and financing is one of the main variables that significantly affected the success of microcredit entrepreneurs [62]. Similarly, in Sri Lanka, the significant impact of microcredit on women entrepreneurship was found [63]. Moreover, the microcredit impact on nutrition can be explained by quoting the example from Malawi, which described that women's access to microcredit improved the young girls' long-term nutrition level [64].

7. Conclusion

Rural areas are facing problems of endemic poverty and hunger across the globe. Especially, in developing countries, rural economies are largely dependent on agriculture. A significant proportion of the population in these countries earn their livelihood from agriculture. Small landholdings, degrading soil and water quality, declining productivity, and low profits are making agriculture an unattractive business in these areas. Consequently, the world is facing challenges to achieve sustainable development in rural areas. Three pillars of sustainable rural development; economic, social, and environmental development depend to large extent on solving resource constraints of the rural population. Access to financial resources is one of the significant resource constraints for rural people. Microcredit which emerged from Bangladesh has the potential to ease the financial constraints of rural people. Microcredit offers small loans to poor people who have less or no collateral and are often ignored by the conventional banking system. Positive impacts of microcredit on poverty alleviation, employment generation, women empowerment, curbing rural outmigration, food security, better health and education, climate change adaptation, and green entrepreneurship are extensively documented in the literature. Hence, microcredit can be used as a tool to achieve sustainable rural development through the economic, social, and environmental development of these areas.

Author details

Muhammad Imran[1*], Shamsheer Ul Haq[2] and Orhan Ozcatalbas[3]

1 Department of Economics and Business Management, University of Veterinary and Animal Sciences, Lahore, Pakistan

2 Department of Economics and Business Administration, Division of Management and Administrative Science, University of Education, Lahore, Pakistan

3 Department of Agricultural Economics, Akdeniz University, Antalya, Turkey

*Address all correspondence to: maniuaf@yahoo.com

References

[1] IFAD. Rural Poverty Report 2001: The Challenge of Ending Rural Poverty. Oxford, UK: Oxford University Press for IFAD; 2001

[2] FAO. 2020. Available from: http://www.fao.org/policy-support/policy-themes/rural-poverty-reduction/en

[3] Özçatalbaş O. Rural development policies and strategies in Turkey (Türkiye'de Kırsal Kalkınma Politika ve Stratejiler). Journal of Yeni Türkiye. 113-114 Special Issue, Agricultural Policies Volume 2 Set, New Turkey. 2020

[4] World Bank. 1975. Available from: https://data.worldbank.org/indicator/SP.RUR.TOTL.ZS

[5] Kata S. Rural Development Principles and Policies and Management. New Delhi: Sage Publication; 1986

[6] Shepherd A. Sustainable Rural Development. New York, NY, USA: Macmillan International Higher Education; 1998

[7] Bryden J. Rural development indicators and diversity in the European Union. In: Proceedings of the Conference on Measuring Rural Diversity; 21-22 November 2002; Washington, DC, USA. Available from: http://citeseerx.ist.psu.edu/viewdoc/download?doi=10.1.1.199.4439&rep=rep1&type=pdf0

[8] Calatrava-Requena J. Origin and evolution of Rural Development concept and policies: From rural communities to territories. In: Serrão JV, Freire D, Fernández Prieto L, Santos R, editors. Proceedings of the Old and New Worlds: The Global Challenges of Rural History, International Conference; 27-30 January 2016; Lisbon, Portugal. Lisbon, Portugal: Lisbon, ISCTE-IUL; 2016. Available from: https://lisbon2016rh.files.wordpress.com/2015/12/onw-02461.pdf

[9] Madhu IA. Rural markets as a factor of rural development in N. Sukka region South-Eastern Nigeria (thesis). Nigeria Nsukka: Department of Geography University; 2000

[10] Wubayehu TZ. Conceptualizing rural development in the twenty-first century. International Journal of Agricultural Extension and Rural Development Studies. 2020;7(4):18-26

[11] Mihai FC, Iatu C. Sustainable Rural Development Under Agenda 2030. 2020

[12] UNSD. End Poverty in All Its Forms Everywhere. United Nations Statistics Divisions, Development Data and Outreach Branch; 2021. Available from: https://unstats.un.org/sdgs/report/2019/goal-01/ [Accessed: November 13, 2021]

[13] IFAD. Ten Things to Know about Gender Equality and Rural Poverty. 2021. Available from: https://www.ifad.org/en/web/latest/–/photo/ten-things-to-know-about-gender-equality-and-rural-poverty [Accessed: November 22, 2021]

[14] IFAD. Proceedings IFAD Conference on New Directions for Smallholder Agriculture; 24-25 January 2011. Rome: IFAD; 2011

[15] Hazell P, Poulton C, Wiggins S, Dorward A. The future of small farms: Trajectories and policy priorities. World Development. 2010;**38**(10):1349-1361

[16] Wiggins S, Kirsten J, Lambi L. The future of small farms. World Development. 2010;**38**(10):1341-1349

[17] HLPE. Investing in smallholder agriculture for food security. In: A Report by The High-Level Panel of Experts on Food Security and Nutrition. Vol. 6. Rome: FAO; 2013

[18] Lowder SK, Skoet J, Raney T. The number, size, and distribution of farms, smallholder farms, and family farms worldwide. World Development. 2016;**87**:16-29

[19] Muzari W, Gatsi W, Muvhunzi S. The impacts of technology adoption on smallholder agricultural productivity in sub-Saharan Africa: A review. Journal of Sustainable Development. 2012;**5**(8):69

[20] Eswaran H, Lal R, Reich PF. Land degradation: An overview. In: Response to Land Degradation. London, United Kingdom: CRC Press; 2019. pp. 20-35

[21] Murray U, Boros R. A Guide to Gender Sensitive Microfinance. The Socio-Economic and Gender Analysis (SEAGA) Programme. Rome, Italy: FAO; 2002

[22] Özçatalbaş O, Imran M. Current situation and importance of the family farming in agriculture of Turkey. Modern Agricultural Science and Technology. 2018;**3**(3-4):1-9

[23] FAO. Family Farming Year, World Food and Agriculture Organization. 2014. Available from: http://fao.org/familyfarming-2014/resources/audio-video/en/Rome

[24] Hazell P, Poulton C, Wiggins S, Dorward A. The future of small farms for poverty reduction and growth. In: International Food Policy Research Institute (IFPRI), Discussion Paper 42. Washington D.C.: IFPRI; 2007

[25] Michael Z. What's Class Got to Do with It?: American Society in the Twenty-First Century. Cornell University Press; 2004

[26] Mehan F. Usage and Impacts of Micro-credit Provision: A Case Study Based on the credit Operations of the Dedebit Credit and Saving Institutions (DECSI). Tigray: Mekelle University; 2001

[27] Karim MR, Tania S, Farazi MMR. Role of micro-credit in poverty alleviation of rural poor: Evidence from Laxmipur District of Bangladesh. Journal of Business and Technology (Dhaka). 2012;**7**(2):37-55

[28] Shirazi NS, Khan AU. Role of Pakistan poverty alleviation fund's micro credit in poverty alleviation: A case of Pakistan. Pakistan Economic and Social Review. 2009;**47**:215-228

[29] Muralimohan M, Devarajulu M. Role of micro credit in poverty eradication: A case study in Chittoor Rural Mandal of Andhra Pradesh. Asian Journal for Poverty Studies (AJPS). 2017;**3**(2):87-94

[30] Kasali TA, Ahmad SA, Lim HE. The role of microfinance in poverty alleviation: Empirical evidence from South-West Nigeria. Asian Social Science. 2015;**11**(21):183-192

[31] Christopher IF. Impact of microfinance on small and medium-sized enterprises in Nigeria. In: International Conference on Innovation and Management. Tokyo, Japan. 2010. Available from: https://www.pucsp.br/icim/ingles/downloads/papers_2010/part_9/7_Impact%20of%20Microfinance%20on%20Small%20and%20Medium-Sized%20Enterprises%20in.pdf [Accessed: December 05, 2021]

[32] Waliaula RN. Relationship between microcredit and the growth of small and

medium enterprises in Kenya (thesis). Nairobi, Kenya: University of Nairobi; 2013

[33] Alhassan EA, Hoedoafia MA, Braimah I. The effects of microcredit on profitability and the challenges of women owned SMES: Evidence from Northern Ghana. Journal of Entrepreneurship and Business Innovation. 2016;**3**(1):29-47

[34] Ruslan RAHM, Gan C, Hu B, Quang NTT. Impact of microcredit on SMEs performance in Malaysia. International Journal of Business and Economics. 2020;**19**(1):109-130

[35] Zuberi HA. Production function, institutional credit and agricultural development in Pakistan. The Pakistan Development Review. 1989;**28**(1):43-56

[36] Rahman RI, Khandker SR. Role of targeted credit programmes in promoting employment and the productivity of the poor in Bangladesh. Quarterly Journal of Bangladesh Institute of Development Studies. 1994;**22**(2/3):49-92

[37] Khandker SR, Faruqee RR. The impact of farm credit in Pakistan. Agricultural Economics. 2003;**28**:197-213

[38] Ahmad A, Jan I, Ullah S, Pervez S. Impact of agricultural credit on wheat productivity in district Jhang, Pakistan. Sarhad Journal of Agriculture. 2015; **31**(1):65-69

[39] Chandio A, Jiang Y, Wei F, Guangshun X. Effects of agricultural credit on wheat productivity of small farms in Sindh, Pakistan. Agricultural Finance Review. 2018;**78**(5):592-610

[40] Davis B, Winters P, Carletto G, Covarrubias K, Quinones E, Zezza A, et al. A cross-country comparison of rural income generating activities. World Development. 2010;**38**(1):48-63. DOI: 10.1016/j.worlddev.2009.01.003

[41] Guirkinger C, Boucher S. Credit constraints and productivity in Peruvian agriculture. Agricultural Economics. 2008;**39**:295-308. DOI: 10.1111/j. 1574-0862.2008.00334.x

[42] Ashaolu OF, Momoh S, Phillip BB, Tijani IA. Microcredit effect on agricultural productivity: A comparative analysis of rural farmers in Ogun State, Nigeria. International Journal of Applied Agriculture and Apiculture Research. 2011;7(1):23-35

[43] Morduch J, Haley B. Analysis of the Effects of Microfinance on Poverty Reduction (NYU Wagner Working Paper No. 1014). 2002. Available from: http://s3.amazonaws.com/academia.edu.documents/3470021/morduch_02_analysis_effects.pdf?AWSAccessKeyId=AKIAJ56TQJRTWSMTNPEA&Expires=1473291118&Signature=%2BLi5qJr7C lM%2FuN0zc3z5eUEFcPc%3D&respo nse-content-disposition=inline%3B%20-filename%3DAnalysis_of_the_effects_of_microfi nance.pdf [Accessed: November 29, 2021]

[44] Bauchet J, Marshall C, Starita L, Thomas J, Yalouris A. Latest Findings from Randomized Evaluations of Microfinance. Washington D.C.: World Bank; 2011

[45] Okurut FN. Access to Credit by the Poor in South Africa: Evidence from Household Survey Data 1995 and 2000. Stellenbosch: University of Stellenbosch; 2006

[46] ILO. Rural Women Need Equality Now. International Labor Organization.Available from: https://www.ilo.org/global/about-the-ilo/newsroom/statements-and-speeches/WCMS_621364/lang--en/index.htm; 2018 [Accessed: December 01, 2021]

[47] UNW. The role of women in rural development, food production and

poverty eradication. UN women. 2021. Available from: https://www.unwomen.org/en/news/in-focus/rural-women-day/2013 [Accessed: December 01, 2021]

[48] Ahmed F, Siwar C, Idris NAH, Mia MS. Role of microcredit in reducing vulnerability of the rural women: Evidence from Panchgarh district of Bangladesh. World Applied Sciences Journal. 2011;**15**(9):1325-1333

[49] Shah TH, Butt H. Income generating activities through microcredit and women's socio-economic empowerment: A study of district Kasur, Pakistan. Academic Research International. 2011;**1**(3):218

[50] BIIP. Women and the Environment, Global Women's Issues: Women in the World Today, Extended Version. 2021. Available from: https://opentextbc.ca/womenintheworld/chapter/chapter-11-women-and-the-environment/#:~:text=Women%20play%20a%20critical%20role,both%20forests%20and%20agricultural%20terrain [Accessed: December 03, 2021]

[51] Nader YF. Microcredit and the socio-economic wellbeing of women and their families in Cairo. The Journal of Socio-Economics. 2008;**37**(2):644-656

[52] Jordan JC. Climate shocks and adaptation strategies in coastal Bangladesh: Does microcredit have a part to play? Climate and Development. 2021;**13**(5):454-466

[53] Murali KS. Microfinance, social capital and natural resource management systems: Conceptual issues and empirical evidences. International Journal of Agricultural Resources, Governance and Ecology. 2006;**5**(4):327-337

[54] Pitt MM, Khandker SR. The impact of group-based credit programs on poor households in Bangladesh: Does the gender of participants matter? Journal of Political Economy. 1998;**106**(5):958-996

[55] Hamad R, Fernald LC. Microcredit participation and nutrition outcomes among women in Peru. Journal of Epidemiology and Community Health. 2012;**66**(6):e1-e1

[56] Salt R. Microcredit and the social determinants of health: A conceptual approach. Public Health Nursing. 2011;**28**(3):281-290

[57] Amin S, Rai A, Topa G. Does microcredit reach the poor and vulnerable? Evidence from Northern Bangladesh. Journal of Development Economics. 2003;**70**(1):59-82

[58] Okon DP, Etim NA, Offiong AA. Impact of micro credit scheme on rural farmers in Akwa Ibom State, Nigeria. Journal of Agriculture & Social Sciences. 2012;**8**(2):65-68

[59] Quibria MG. Microcredit and Poverty Alleviation: Can microcredit close the deal? (No. 2012/78). WIDER Working Paper. 2012

[60] Shukran K, Rahman F. A Grameen bank concept: Micro-credit and poverty alleviation program in Bangladesh. In: International Conference on Emerging Trends in Computer and Image Processing. 2011. pp. 47-51

[61] Copestake J, Bhalotra S, Johnson S. Assessing the impact of microcredit: A Zambian case study. Journal of Development Studies. 2001;**37**(4):81-100

[62] Fatimah-Salwa AH, Mohamad-Azahari A, Joni-Tamkin B. Success factors of successful microcredit entrepreneurs: Empirical evidence from Malaysia. International Journal of Business and Social Science. 2013;**4**(5):153-159

[63] Yogendrarajah R, Semasinghe D. Microcredit is a tool for women entrepreneurship development. Journal of Mathematics and System Science. 2015;5:385-390

[64] Hazarika G, Guha-Khasnobis B. Household Access to Microcredit and Children's Food Security in Rural Malawi: A Gender Perspective. Research Paper 2007/087. Helsinki: UNU-WIDER; 2007

Chapter 12

Sustainability of Soil Chemical Properties and Nutrient Relationships in Dairy and Beef Cattle in Antioquia, Colombia

Marisol Medina-Sierra, Mario Cerón-Muñoz and Luis Galeano-Vasco

Abstract

This chapter has been written with the purpose of increasing knowledge regarding the characteristics of soils dedicated to dairy and beef cattle farming in Antioquia, Colombia. Statistical analysis included several generalised additive models, with additive, smoothing, and tensor effects, such as geographic position and chemical parameters. Findings showed most farms belonged to small producers, 86.5% of cattle farms being family owned. Rotational grazing is the predominant system in 93% of farms; 58% of dairy farms and 94% of beef cattle farms do not fertilise their pastures. Results show high variability of soil chemical parameters. There are high levels of iron and low levels of sodium. Macronutrients, such as phosphorus and potassium show high levels in some dairy subregions and medium to low levels in others. Calcium (Ca) and magnesium levels are low for all subregions, excluding "Urabá" and "Occidente." Most subregions have organic matter (OM) levels below 13%. The distribution of some chemical parameters is related to geographical location, such as pH and Ca, which change according to latitude and longitude. Different correlations were found amongst OM, total nitrogen, Ca, and exchangeable aluminium. Due to the high variability of soil fertility parameters, management programmes should be implemented for each distinctive production system.

Keywords: available nutrients, farming systems, fertilisation programme, geographic information system, organic carbon, soil acidity, sustainable production

1. Introduction

Soils are the basis for sustainable production and the supply of nutrients to plants. The general aspects of soils and sustainability are as follows—soils are part of a fragile natural environment, so it is important to understand how they are formed and sustained, as well as how they relate to agriculture, forestry, ecology, conservation, and other areas of knowledge [1]. The variability of soil nutrients is affected to different degrees by soil formation factors. These factors, declared by Jenny are parent material, relief, climate, time, and potential biota [2]. Soil is considered an open and non-equilibrium system [2, 3]. The effect of humans on soil systems must be studied in the future to find new solutions to conserve the planet [2]. Furthermore, macronutrient stocks of nitrogen (N), phosphorus (P), and potassium (K) show significant spatial and temporal variability in soil [4].

Sustainability in pasture soils—environmental sustainability tends to reduce the inputs required for animal production by making more efficient use of internal resources. In this case, the correct management of pasture productivity may contribute [5] because the rational application of fertilisers and amendments helps in reducing nutrient leaching, and also, it could limit greenhouse gas emissions of CO_2, nitrogen, and others. In general terms, grassland soils can contribute to maintain the existence of carbon at least in the first cm of soil from the surface.

To contribute to sustainability, it is important to maintain grassland carrying capacities, use organic sources, and implement conservation practices—like to sow different kinds of trees or even to maintain the diversity of species, looking to contribute to conserve the soil is a key to any kind of system of production in agriculture. Sustainable soil management is key to achieving several SDGs (sustainable development goals) due to the dependence on plant production or different soil processes. Furthermore, agricultural research programmes should contribute to healthier soils [1].

Nutrient cycling in pastures: CH_4 and N_2O emissions need to be reduced from the livestock systems, and soil organic carbon (SOC) must be conserved [6]. Rational fertiliser programmes contribute to optimising grassland management systems to minimise environmental impacts and maximise pasture productivity.

Availability of nutrients in soils and fertilisation program—chemical elements for plant nutrition have high spatial variability, especially in agricultural soils [7]. It is essential to know the chemical properties of soils to be able to identify areas that require management practices.

Impact of cattle farming on soil fertility—international reports on the impact of cattle farming on soil nutrient conservation have shown that dairy farming extracts at least 2.5 times more nutrients, such as N, P, K, Ca, and S than beef farming [8]. Although cattle systems can provide some nutrients and contribute to maintaining soil fertility, which can reduce the use of amendments and fertilisers [9].

General description of soils in Antioquia—according to the general soil survey, soils in Antioquia are very variable in terms of their parent materials, relief and climate. There are soil orders with different degrees of paedogenetic evolution—soils with a deficiency in paedogenetic evolution such as Entisols, soils with a low degree of evolution specifically Inceptisols and several have mixed evolution including Andisols and Mollisols. The main factors limiting the use and management of soils in Antioquia are the slopes which range from slightly steep to very steep, erosion, extreme values of moisture content, acid reaction, high aluminium saturation, and low to very low fertility. Some alluvial valleys and piedmont landscape soils have moderate to high fertility [10].

2. Materials and methods

2.1 Location and sampling

The Department of Antioquia is located in the following coordinates—top right (8°52'25,748" N; 73°52'51,958" O), lower right (5°25'6,798" N; 73°52'51,958" O), top left (8°52'25,748" N; 77°7'40,239" O), lower left (5°25'6,798" N; 77°7'40,239" O).

Representative soil cartographic units were selected in areas dedicated to the cattle production system. The farms were sampled according to the access roads, and the consent of the producers to sample their farms, amongst other criteria. A total of 440 soil samples were collected from farms dedicated to milk and meat production in different subregions of the Department of Antioquia. Each sample was taken at a depth of 20 cm, making a zigzag path and avoiding taking subsamples at non-representative sites. A square-shaped hole was made and the subsample was taken from the central part of the walls where the soil on the edges was eliminated. About 10–15 subsamples were obtained and placed in a clean container, which was then mixed and homogenised and approximately a one-kilogram sample was taken. Subsequently, they were packed in boxes to prevent the entry of light and taken to the laboratory to perform chemical analysis. The analyses were carried out by wet chemistry using methods certified by the Colombian Corporation for Agricultural Research (AGROSAVIA) in Bogotá, Colombia.

2.2 Statistical model

Several generalised additive models were built. The models included various effects, such as longitude (from −76.82 to −74.62 W), latitude (from 5.71 to 8.76 N), altitude (from 0 to 2900 m above sea mean level), geoform (wavy, flat, and slope geoform) and also the combinations of the parameters soil chemicals. The analyses were performed with the MGCV library [11] of the statistical software R-project [12]. The Bayesian Information Criterion was used to select the best models.

3. Results and discussion

3.1 Overview of dairy and beef cattle farms in Antioquia

Cattle farms are located in mountain landscapes and are mostly small producers (**Figure 1**). Approximately 70% of farms visited have an area of less than 30 hectares, indicating that most belong to small producers. They are distributed throughout the department. Many of the producers use some amendments and fertilisers in the pasture fertilisation programmes (**Table 1**).

The size of the farms visited was between 2 and 20 ha for most of the dairy farms and between 2 and 100 ha for the beef cattle farms (**Figure 2**). In Antioquia, the predominant area of rural farms is less than 100 ha [13]. In Colombia, 83% of the Agricultural Production Units dedicated to beef cattle have an area of less than 20 ha, which indicates that the predominant area of beef cattle farms in Antioquia is larger than the average for the country [14].

According to the United Nations, the dairy sector can contribute to a selection of sustainable development goals, such as "No poverty in rural areas," as most dairy farms are small and located in developing countries [15]. Added to this, other authors [16] indicate that milk is mainly produced in mixed crop-livestock systems by small farmers in low-income countries. The authors also consider that dairy farms increase pressure on natural resources as most production systems tend to increase the size of their livestock to meet the growing demand for dairy food.

Most of the cattle farms in the study area were owned by farmers (**Figure 3**). This is encouraging as it contributes to the permanence of farming families in the Colombian countryside. Official data reported eight years ago indicated that 72% of dairy farms were owned by producers in the Department of Antioquia [14]. Studies in

(a) Typical dairy farms

(b) Typical beef farms

Figure 1.
Typical dairy and beef farms in Antioquia. (a) Author: Manuela Ortega Monsalve. (b) Author: María Mercedes Murgueitio.

other countries, such as Turkey, highlight the importance of the role of small families in dairy production; with farms averaging just 7.2 hectares and the contributions of all family members in activities related to animal production [17]. Women are involved in production, harvesting, processing, transport, and other important activities in agricultural production systems [18]. The participation of women in dairy farming systems assists towards their economic independence, which should be a blueprint for all women worldwide. It is worth noting that according to the SDGs 5 "Gender Equality," there have been achievements, but many challenges remain [19].

Rotational grazing is the predominant system with 93% of the cattle farms visited (**Figure 4**). In this grazing system, animals start from an initial paddock and then

Parameters	Dairy cattle	Beef cattle
Predominant area	2–20 ha	2–100 ha
Grasses	*Cenchrus clandestinus*, *Urochloa* spp., *Cynodon nlemfuensis*, *Axonopus* spp., *Paspalum* spp.	*Urochloa* spp., *Cynodon nlemfuensis*, *Hyparrhenia rufa*, *Ischaemum indicum*, *Axonopus* spp., *Paspalum* spp.
Predominant soil orders	Inceptisols, entisols, andisols	Inceptisols, entisols
Most common amendments	Calcium carbonate, composted and fresh organic fertilisers from different animal manures	Calcium carbonate, composted and fresh organic fertilisers from different animal manures
Most common fertilisers	Urea, compound fertilisers: 15–15–15 (15% N–15% P_2O_5–15% K_2O), 31–8–8 (31% N–8% P_2O_5–8% K_2O)	Urea, compound fertilisers: 15–15–15 (15% N–15% P_2O_5–15% K_2O), diammonium phosphate (18% N–46% P_2O_5–0% K_2O)

Table 1.
General information of the farms visited.

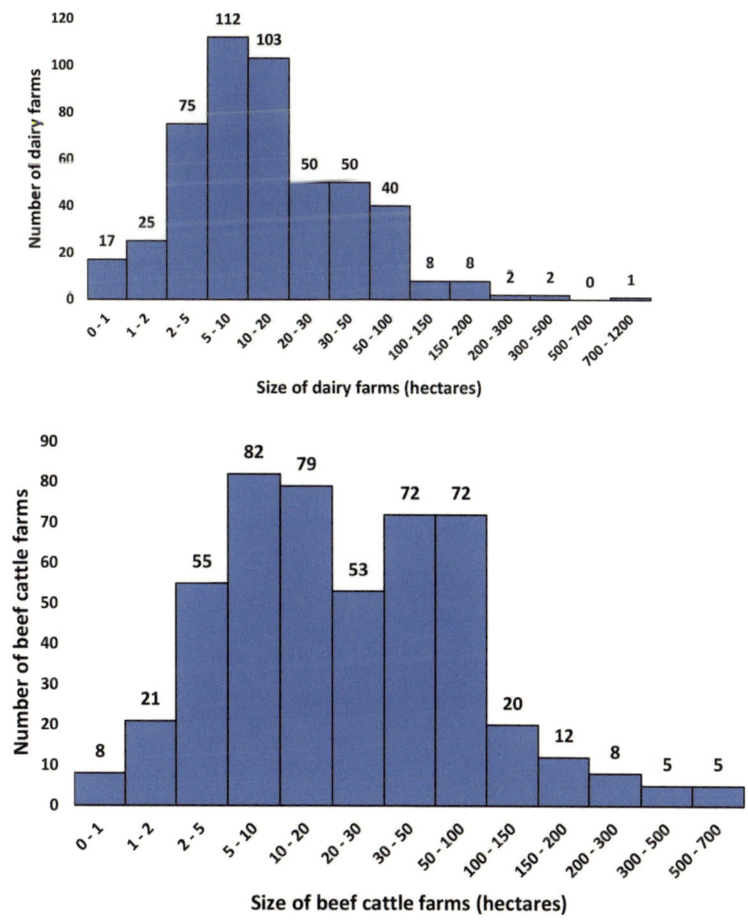

Figure 2.
Size of dairy and beef cattle farms in the subregions evaluated.

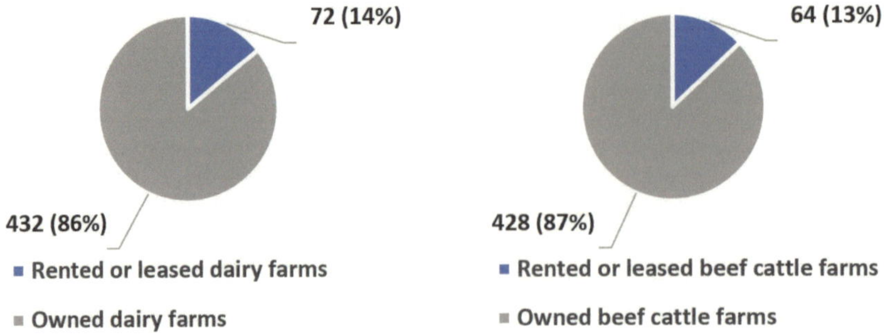

Figure 3.
Type of ownership of dairy and beef farms.

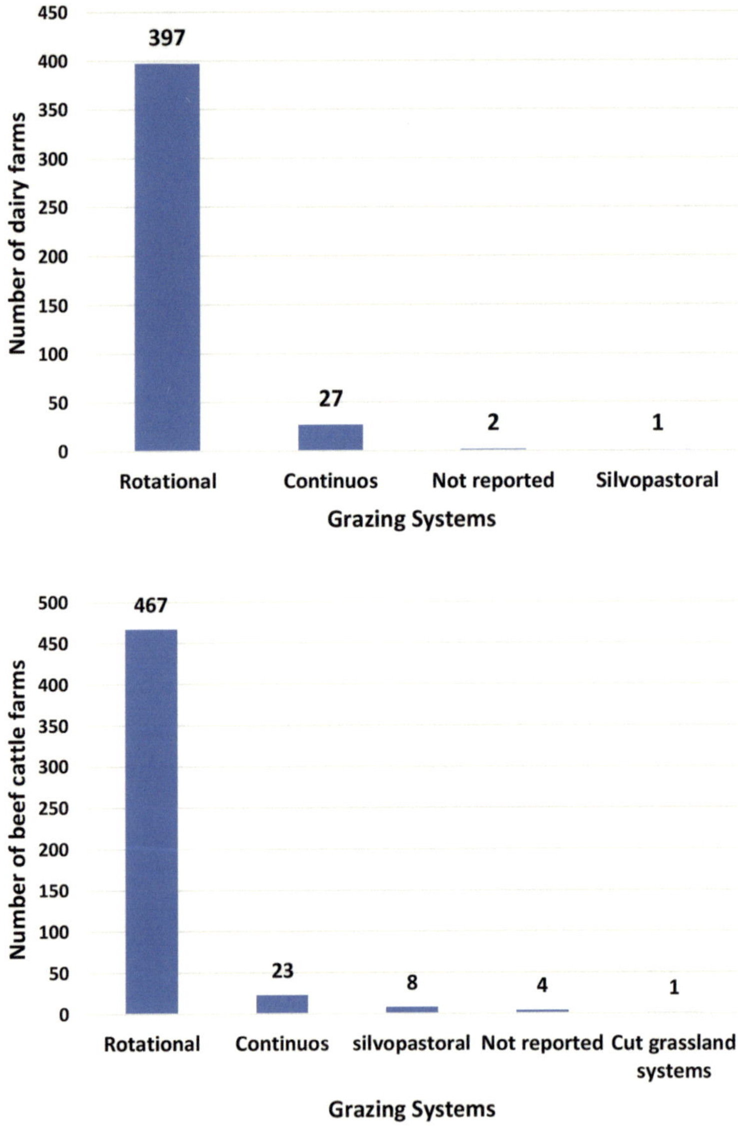

Figure 4.
Grazing systems used in the dairy and beef farms visited.

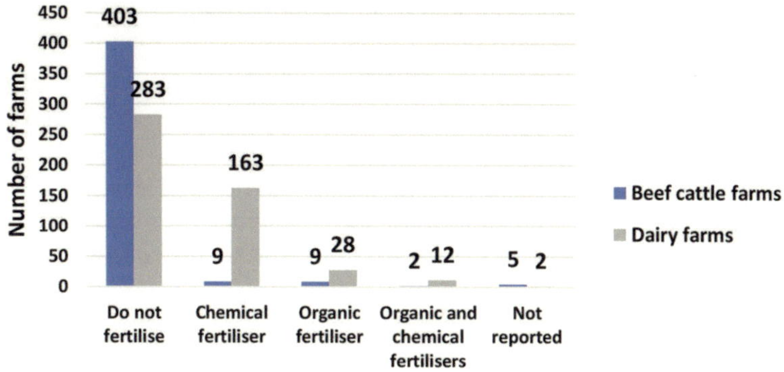

Figure 5.
Types of fertilisers used on dairy and beef cattle farms.

go through other paddocks until they return to the first when it is ready to be grazed again [20, 21]. In a study of beef cattle systems in Antioquia, using Criollo cattle breeds, it was found that 96% of producers use rotational grazing systems, and 4% use silvopastoral systems [22]. Silvopastoral systems include trees, shrubs, pastures, animals, and other crops, provide a good forage supply for animals, contribute to biodiversity conservation, contribute to nutrient recycling and soil fertility conservation, and provide shade, amongst other benefits [23, 24]. This indicates that amendments and fertilisers are required in most cattle production systems. However, there has been an increase in the establishment of silvopastoral systems in the country in recent years; these systems could positively contribute to the reduction of agrochemical application, and thus to environmental sustainability.

The type, amount, and form of application of fertilisers can affect the sustainability of pasture production. Of the 188 dairy farms in this study, 58% of farms do not have fertilised pastures, whilst the other 41.6% have fertilised pasture present. Approximately 33.4% of the farms use chemical fertilisers, 5.7% use organic fertilisers, 2.5% use a mixture of chemical and organic fertilisers and 0.4% of the dairy cattle farmers did not answer the survey question. In contrast, of the 428 beef cattle farms, it was found that 94.2% do not fertilise their pastures, while the remaining 4.7% do fertilise their pastures. 2.1% of the producers use organic fertilisers, 2.1% of the farms use chemical fertilisers, and 0.5% use a mixture of chemical and organic fertilisers. In addition, 1.2% of beef cattle producers did not answer the survey question (**Figure 5**). The above described demonstrates the low technological level of fertilisation management in beef cattle production systems compared to dairy cattle production systems.

According to the SDGs, implementing agricultural sustainability can help reduce poverty [25]. Although it was found that most farmers do not apply fertilisers, it is important to conserve the soil as the basis of agricultural production. For this reason, nutrient levels in soils must be determined in order to help implement rational management programmes according to the needs of the pastures. In this way, it contributes to the sustainability of livestock production in the conditions of the Colombian tropics.

3.2 Variability of the chemical parameters of the soils

There is a high variability of the chemical parameters of the soils in the Department of Antioquia. We show partial information on the descriptive statistics for some chemical parameters found for each subregion (**Table 2**). The maximum and minimum values found for most of the parameters presented very extreme values, which may indicate that in some farms the producers use high applications of fertilisers and amendments or it could also be that the taking of some soil samples for chemical analysis performed on recently fertilised paddocks. The general fertility parameters of the soils analysed are in accordance with that described by other authors [10]. Moreover, the levels found for soil nutrients are compared with the levels established for Colombia according to the ICA (Colombian Agricultural Institute) [26].

In general terms, the only nutrient with high levels in the soils studied is iron, the one with low levels in all soils is sodium. The electrical conductivity values (0.17–1.12 dS/m) and sodium levels (0.15–0.45 cmol(+)/kg) of the soils analysed indicate that they do not present salinity problems. In Colombia, soil sodium levels (Na < 1 cmol(+)/kg) are considered ideal values. Macronutrients, such as phosphorus, potassium, and sulphur, had high levels in some dairy areas (North and Valle de Aburrá subregions). These values were higher than 30 mg/kg, 0.4 cmol(+)/kg and 10 mg/kg for P, K, and S, respectively. In the other subregions, these nutrients had medium to low levels.

The Ca and Mg available bases are low for all subregions, except for the Urabá and "Occidente" subregions. However, the Ca:Mg ratio in these two subregions is low. This indicates that amendments containing calcium and magnesium need to be applied in all subregions, either to raise the levels of available bases or to improve the Ca:Mg ratio, which is recommended to be maintained at 3:1. The application of amendments would help to neutralise the exchangeable aluminium and as a consequence increase the pH, which tends to be acidic in most of the subregions of Antioquia.

Micronutrient levels are adequate for most of the soils analysed under the sampling conditions of this study. Cation exchange capacity (CEC < 10 cmol(+)/kg) is low in all subregions, except in the Urabá subregion (CEC = 15.83 cmol(+)/kg).

Although most areas have low to medium levels of organic matter (OM < 10%), the total soil nitrogen level is medium to high (TSN from 0.16 to 1%), which possibly indicates high mineralisation rates due to tropical conditions and in other cases could be due to excessive nitrogen application, mainly in dairy pastures. In the soils analysed, the loam textural class predominates, indicated by sand levels in the range of 36–52%, clay levels from 8 to 34%, and silt levels from 20 to 43%. The subregions of Urabá and Occidente present a clay loam textural class.

Previous reports on some dairy farms in the Oriente subregion were also characterised by acidic soils (low pH and exchangeable aluminium values of 1.2 cmol (+)/kg), similar Na values (0.02 cmol(+)/kg), and also low P levels (3 mg/kg). In contrast to this, there were low levels of available bases found for Ca, Mg, and K (0.09, 0.16 and 0.1 cmol(+)/kg respectively), while organic matter (OM > 22%) and boron (0.7 mg/kg) levels were high [27]. Another author [28] reported similar levels of P (9.4 mg/kg), K (0.29 cmol(+)/kg) and organic matter (OM = 5%) in pasture soils of the Occidente subregion. They also presented clay loam texture (34.7% sand, 34.5% clay, and 30.8% silt) similar to that found in this study. The other chemical parameters presented higher levels than those found on farms in the Occidente subregion, mainly pH 6.94 and available bases of 24.7 and 10.2 cmol(+)/kg for Ca and Mg, respectively [28]. In another study in the Norte subregion, similar levels of pH, Al, EC, and K and different values for organic matter and phosphorus were found [29].

Subregions of Antioquia								
Parameters*	Units	Nordeste	Norte	Occidente	Oriente	Suroeste	Urabá	Valle de Aburrá
Farms		72	118	31	57	28	131	3
MAMSL		906.94 ± 280.84 (833.5)**	2438.85 ± 240.52 (2500)	1586.58 ± 385.72 (1582)	1908.26 ± 511.45 (2136)	1932.04 ± 164.55 (1963)	80.27 ± 51.19 (70)	2550 ± 6.08 (2547)
pH	Units of pH	5.08 ± 0.36 (5.06)	5.03 ± 0.45 (4.99)	5.31 ± 0.47 (5.21)	5.1 ± 0.48 (5.04)	5.13 ± 0.4 (5.11)	6.16 ± 0.73 (6.22)	4.87 ± 0.43 (4.87)
Al	cmol(+)/kg	1.06 ± 0.87 (0.87)	1.12 ± 0.95 (0.95)	1.6 ± 1.59 (1.27)	1.06 ± 0.9 (0.89)	1.71 ± 1.58 (1.62)	0.38 ± 1.16 (0)	2.49 ± 2.07 (2.48)
EC	dS/m	0.17 ± 0.1 (0.14)	0.75 ± 0.55 (0.64)	0.2 ± 0.15 (0.16)	0.43 ± 0.35 (0.3)	0.32 ± 0.21 (0.26)	0.24 ± 0.24 (0.17)	1.12 ± 0.54 (1.2)
SOM	g/100 g	2.71 ± 1.36 (2.55)	8.81 ± 4.6 (8.57)	3.49 ± 1.84 (2.41)	9.12 ± 6.36 (6.76)	12.61 ± 5.68 (12.95)	1.66 ± 1.1 (1.28)	19.78 ± 3.28 (18.22)
P	mg/kg	3.17 ± 2.75 (2.07)	31.08 ± 39.81 (14.39)	8.66 ± 9.84 (5.96)	9.2 ± 10.79 (4.26)	6.56 ± 5.9 (4.66)	11.54 ± 13.84 (6.22)	38.85 ± 27.83 (24.46)
S	mg/kg	5.82 ± 3.54 (5.39)	24.21 ± 19.63 (22.42)	6.78 ± 7.15 (4.27)	8.58 ± 7.96 (6.68)	5.46 ± 4.48 (3.89)	7.98 ± 13.96 (3.28)	25.65 ± 8.27 (25.17)
Ca	cmol(+)/kg	2.13 ± 1 (1.83)	3.16 ± 2.12 (2.67)	5.85 ± 5.31 (3.93)	2.3 ± 2.23 (1.49)	2.06 ± 2.43 (1.42)	9.8 ± 5.56 (9.11)	3.26 ± 2.8 (1.7)
Mg	cmol(+)/kg	0.59 ± 0.41 (0.48)	0.89 ± 0.72 (0.68)	2.52 ± 3.19 (1.65)	0.85 ± 1.23 (0.4)	0.8 ± 0.99 (0.48)	5.8 ± 3.51 (5.15)	1.46 ± 1.36 (0.82)
K	cmol(+)/kg	0.23 ± 0.24 (0.13)	0.43 ± 0.42 (0.23)	0.35 ± 0.27 (0.25)	0.28 ± 0.24 (0.2)	0.3 ± 0.1 (0.28)	0.36 ± 0.24 (0.3)	0.41 ± 0.41 (0.17)
Na	cmol(+)/kg	0.15 ± 0.01 (0.15)	0.21 ± 0.09 (0.18)	NA***	0.3 ± 0.2 (0.23)	0.17 ± 0.03 (0.16)	0.45 ± 0.43 (0.26)	0.38 ± 0.21 (0.38)
CEC	cmol(+)/kg	3.3 ± 1.31 (3.04)	5.55 ± 2.81 (4.8)	8.18 ± 6.93 (5.7)	4.2 ± 2.62 (3.47)	5.28 ± 2.88 (4.56)	15.83 ± 8.31 (14.76)	8.5 ± 3.04 (8.24)

Subregions of Antioquia								
Parameters*	Units	Nordeste	Norte	Occidente	Oriente	Suroeste	Urabá	Valle de Aburrá
B	mg/kg	0.13 ± 0.05 (0.13)	0.22 ± 0.13 (0.22)	0.12 ± 0.07 (0.11)	0.19 ± 0.1 (0.2)	0.12 ± 0.07 (0.11)	0.23 ± 0.23 (0.19)	0.4 ± 0.07 (0.44)
Fe	mg/kg	186.75 ± 109.75 (159.24)	399.12 ± 267.64 (304.56)	263.98 ± 199.44 (205.13)	265.47 ± 212.74 (209.1)	298.05 ± 131.75 (318.5)	87.95 ± 89.13 (64.4)	541.22 ± 100.5 (489.2)
Cu	mg/kg	2.65 ± 2.03 (2.1)	3.22 ± 2.3 (2.25)	4.85 ± 3.29 (4.04)	2.97 ± 2.16 (2.02)	4.62 ± 3.1 (3.63)	3.51 ± 2 (2.84)	1.4 ± 0.19 (1.4)
Mn	mg/kg	5.56 ± 4.98 (3.66)	6.01 ± 6.14 (3.7)	10 ± 15.24 (4.38)	4.52 ± 4.73 (3)	5.34 ± 3.36 (4.43)	9.16 ± 13.08 (4.97)	6.12 ± 4.27 (4)
Zn	mg/kg	2.76 ± 3.12 (1.58)	9.32 ± 15.58 (4.14)	1.84 ± 0.66 (2.11)	4.67 ± 3.94 (3.72)	4.98 ± 6.51 (3.09)	2.62 ± 1.71 (2.13)	16.02 ± 4.09 (16.15)
OC	g/100 g	1.57 ± 0.79 (1.48)	5.11 ± 2.67 (4.97)	2.02 ± 1.07 (1.4)	5.29 ± 3.69 (3.92)	7.32 ± 3.29 (7.52)	0.97 ± 0.64 (0.74)	11.47 ± 1.9 (10.57)
TSN	g/100 g	0.2 ± 0.07 (0.19)	0.48 ± 0.25 (0.46)	0.27 ± 0.13 (0.21)	0.51 ± 0.31 (0.36)	0.83 ± 0.26 (0.88)	0.16 ± 0.09 (0.15)	1.08 ± 0.09 (1.04)
Sand	g/100 g	46.71 ± 10.86 (47.92)	52.21 ± 9.37 (52.38)	38.74 ± 14.6 (37.22)	51.15 ± 8.36 (51.58)	51.21 ± 9.64 (52.47)	36.21 ± 18.49 (33.44)	49.13 ± 2.39 (48.47)
Clay	g/100 g	33.79 ± 9.53 (33.26)	18.54 ± 7.46 (18.02)	32.32 ± 14.79 (29.98)	16.94 ± 10.72 (14.45)	14.19 ± 7.47 (12.04)	32.53 ± 13.71 (31.56)	8.18 ± 4.51 (7.39)
Silt	g/100 g	19.49 ± 6.08 (18.23)	29.25 ± 9.05 (28.74)	28.94 ± 5.78 (29.38)	31.91 ± 10 (30.43)	34.6 ± 6.67 (35.37)	31.27 ± 10.49 (29.03)	42.69 ± 5.36 (40.83)

Parameters: MAMSL = m above mean sea level, Al = exchangeable aluminium, EC = electrical conductivity, SOM = soil organic matter, P = available phosphorus, S = available sulphur, Ca = available calcium, Mg = available magnesium, K = available potassium, Na = available sodium, CEC = cation exchange capacity, B = available boron, Fe = available iron, Cu = available copper, Mn = available manganese, Zn = available zinc, OC = organic carbon, TSN = total soil nitrogen.
**Values in parentheses correspond to the median value.*
***NA = not available.*

Table 2.
Some physicochemical parameters of the soils in the subregions of Antioquia.

Some models were selected based on the verification of the significance of each effect and the lowest Bayesian Information Criterion (BIC). For the variables, soil organic matter and total soil nitrogen a value of $R^2 > 0.85$ was found, while the other variables showed a lower adjustment (**Table 3**).

The distribution of some chemical parameters in the department is related to the geographical location. It was found that parameters, such as pH and calcium level, showed homogeneous variations according to latitude and longitude, while some parameters, such as organic matter and phosphorus level, were highly variable and did not show any distribution according to geographical location (**Figure 6**).

3.3 Relationships amongst some of the chemical parameters of the soil

The soil parameters organic matter, total nitrogen, and calcium were positively correlated. Calcium levels showed a negative correlation with exchangeable aluminium, which is common for acidic soils in the tropics (**Figure 7**). Similarly, organic matter and calcium parameters also showed positive relationships for dairy cattle soils in the Norte subregion of Antioquia [29].

The altitude of the farms influences the pH values of the soil. Soils with acidic pH have low levels of calcium readily available (**Figure 8**). The low amount of available calcium is normal in tropical soils with naturally acidic conditions due to high rates of mineralisation and leaching of soil bases. In addition, it was found that medium to low levels of calcium and potassium are positively related to phosphorus levels (**Figure 8**). However, this relationship does not occur when phosphorus levels are high, due to the high applications of phosphorus fertilisers in some cattle farming areas, which also occurs in several crops in other areas of the country [30].

Independent variable	Smooth variables model	R square adjust	Deviance explained (%)	Number of observations
Soil organic matter	Latitude by longitude, total soil nitrogen by calcium	0.86	87.5	236
Total soil nitrogen	Latitude by longitude, soil organic matter	0.88	88.8	438
pH	Latitude by longitude, latitude, calcium	0.76	77.0	326
Calcium	Latitude by longitude, soil organic matter by aluminium, magnesium, potassium	0.70	73.6	285
Phosphorus	Latitude by longitude, calcium, potassium	0.54	58.4	219
Potassium	Latitude by longitude, magnesium, soil organic matter	0.31	33.8	346
Sodium	Latitude by longitude, soil organic matter by aluminium	0.27	30.5	126

Table 3.
Parameters of the selected models for the analysis of some variables.

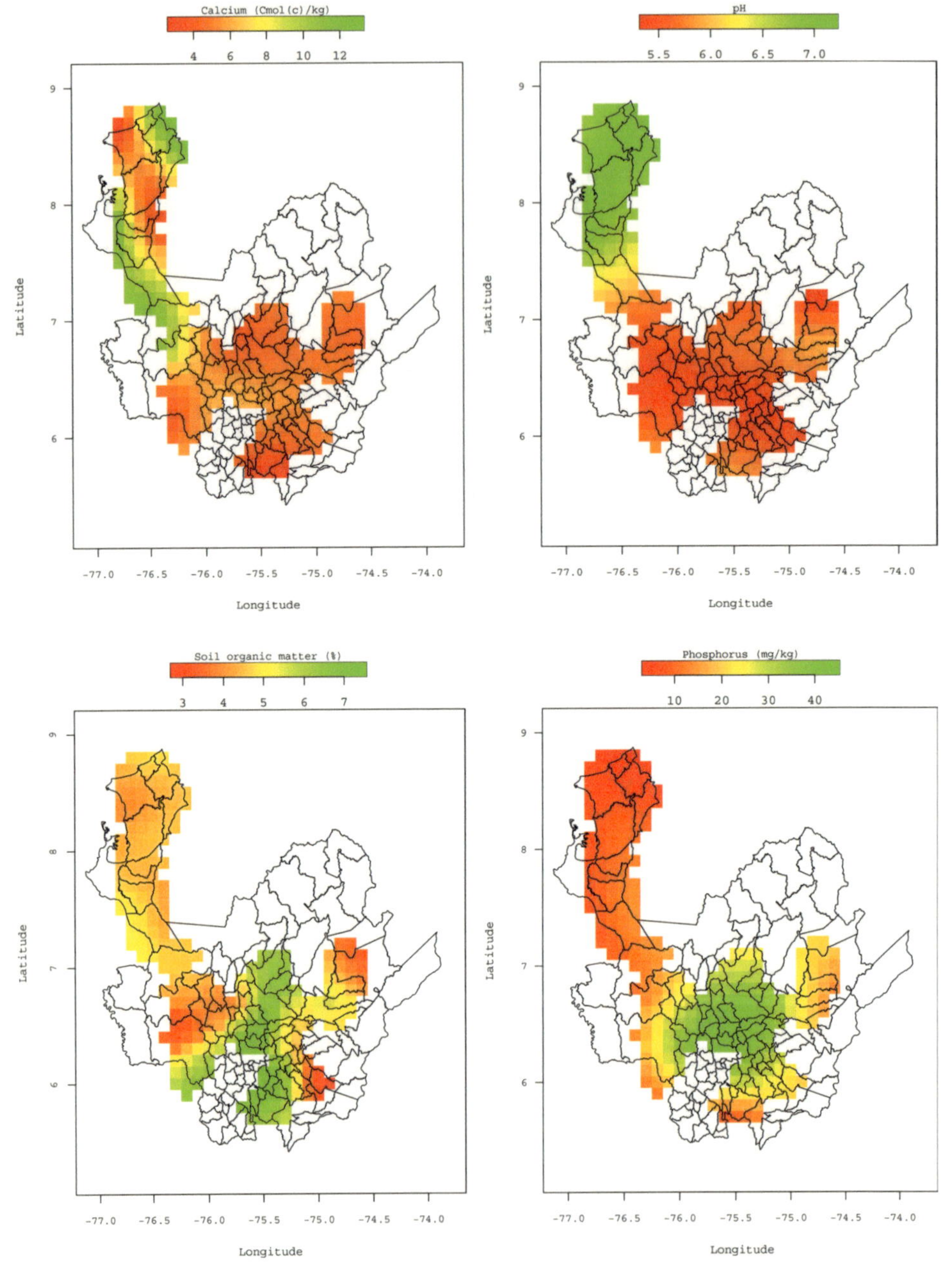

Figure 6.
The general behaviour of some chemical parameters of dairy and beef cattle soils in Antioquia, Colombia.

3.4 Integrated management of pasture fertilisation

Fertiliser applications to pastures without prior measurement of soil nutrient levels can affect soil health, the quality of food produced, and environmental sustainability. In the department of Antioquia, fertilisers are used in 50% of the Agricultural

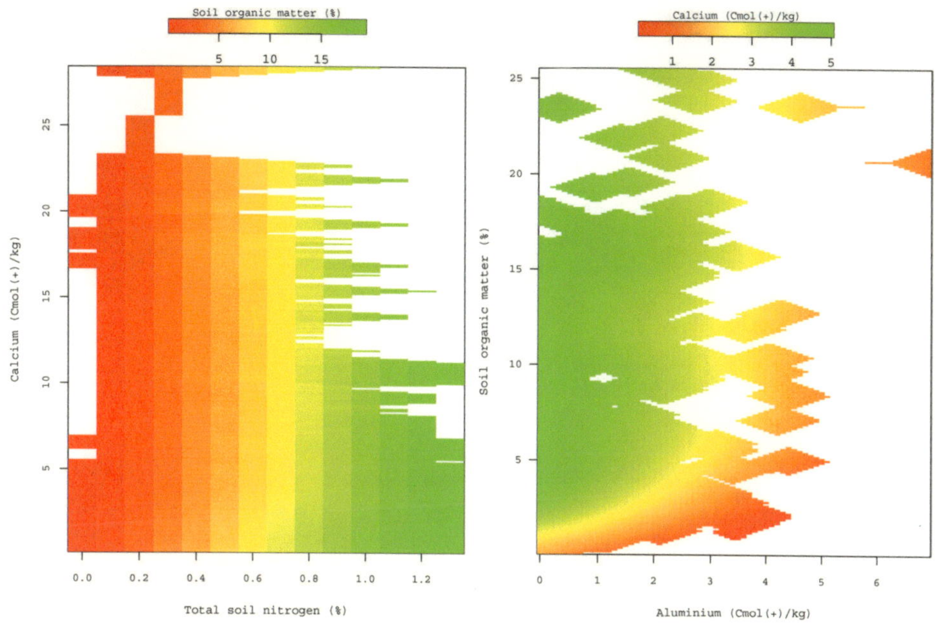

Figure 7.
Relationship amongst the levels of organic matter, total nitrogen, calcium, and aluminium.

Production Units [31]. Of the 185,000 tonnes of fertilisers consumed annually in Antioquia, 43% are compound fertilisers, 31% are simple fertilisers, and 25% are fertiliser mixtures [32]. Cattle systems use less fertiliser than reported for agricultural crops.

This study highlights the need to consider the fertility of soils and fertility programmes when we are looking for options to improve the efficient use of resources on smallholder farms. According to the results found, pasture fertilisation programmes in the analysed dairy and beef cattle subregions should be based on the application of calcium and magnesium amendments, the application of some macronutrients, and, in some areas, the application of some micronutrients, such as boron, is recommended. The low cation exchange capacity (CEC) in most subregions, except for the Urabá subregions, also highlights the importance of improving the levels of available bases, such as calcium, magnesium, and in some cases potassium, key elements for pasture nutrition.

We recommend the specific amendment and fertiliser programmes for each zone according to the soil analysis. It is important to make recommendations according to the botanical composition of the pastures, their biomass production, the type of grazing system, and the agronomic management implemented in each production system. To contribute to environmental sustainability, do not over-apply fertilisers, this will, therefore, conserve soil biota, contribute to the reduction of greenhouse gases, avoid contamination of water sources, and limit the loss of nutrients through leaching, contribute to animal health, amongst other beneficial effects. The measured application of nutrients and the implementation of appropriate agronomic practices contribute to sustainability in cattle production systems.

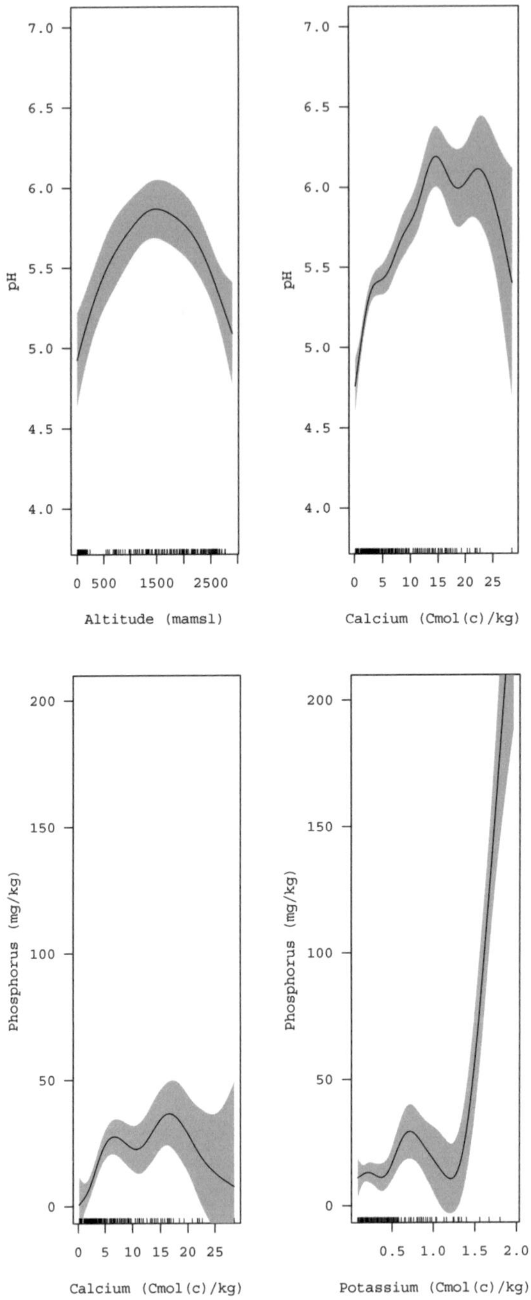

Figure 8.
Relationships of pH and phosphorus with other parameters.

4. Conclusions

Most of the farms visited belong to small producers, with 86% of the farms being owned by the producers and 70% of the farms being smaller than 30 ha. In 93% of the cattle farms, the continuous grazing system is most predominant, with fertilisation programmes implemented in 42% of the dairy farms and only in 5% of beef farms. The above indicates the low technological level of fertilisation management in beef cattle production systems compared to dairy cattle production systems in the subregions visited in the department of Antioquia.

The results of the soil analyses show high variability of chemical parameters in the studied subregions of the department. Micronutrient levels are adequate for most of the soils under the sampling conditions. Cation exchange capacity is low in all subregions except Urabá. Macronutrients, such as phosphorus and potassium, showed high levels in some dairy subregions and medium or low levels in the other areas. Other nutrients showed variable levels, possibly due to geographical location, soil and climatic conditions, grazing systems, botanical composition of pastures, and agronomic management; amongst other factors related to cattle production systems.

The distribution of some chemical parameters in the department is related to geographical location, such as pH and calcium, which had homogeneous variations according to latitude and longitude. Positive correlations were found amongst the parameters—organic matter, total nitrogen, and calcium; with negative correlations for calcium levels with exchangeable aluminium.

Pasture fertilisation programmes in the analysed dairy and beef subregions should be based on the application of calcium and magnesium amendments, some macronutrients, and some micronutrients, such as boron in some specific areas. Therefore, specific recommendations should be made for each farm based on the results of the soil analysis and the agronomic management in each distinctive production system. Thus, under the conditions evaluated, the impact of pasture management should be implemented to contribute to more sustainable dairy and beef farming.

Acknowledgements

The authors are grateful to the "Sistema General de Regalías" of the Colombian Government for the financial support for the project entitled "Desarrollo y Establecimiento del Centro de Desarrollo Agrobiotecnológico de Innovación e Integración Territorial, El Carmen de Viboral, Antioquia, Occidente (CEDAIT)," Expert System subcomponent, code BPIN 2016000100060.

Author details

Marisol Medina-Sierra*, Mario Cerón-Muñoz and Luis Galeano-Vasco
GAMMA Research Group, Faculty of Agricultural Sciences, Antioquia University, Medellin, Antioquia, Colombia

*Address all correspondence to: marisol.medina@udea.edu.co

References

[1] Lal R, Bouma J, Brevik E, Dawson L, Field DJ, Glaser B, et al. Soils and sustainable development goals of the United Nations: An International Union of Soil Sciences perspective. Geoderma Regional. 2021;**25**:e00398. DOI: 10.1016/j.geodrs.2021.e00398

[2] Amundson R. Factors of soil formation in the 21st century. Geoderma. 2021;**391**:114960. DOI: 10.1016/j.geoderma.2021.114960

[3] Rasmussen C, Troch PA, Chorover J, Brooks P, Pelletier J, Huxman TE. An open system framework for integrating critical zone structure and function. Biogeochemistry. 2011;**102**(1):15-29. DOI: 10.1007/s10533-010-9476-8

[4] He J, Dai Q, Xu F, Yan Y, Peng X. Variability in soil macronutrient Stocks across a Chronosequence of Masson Pine Plantations. Forests. 2022;**13**(1):17. DOI: 10.3390/f13010017

[5] Martin G, Barth K, Benoit M, Brock C, Destruel M, Dumont B, et al. Potential of multi-species livestock farming to improve the sustainability of livestock farms: A review. Agricultural Systems. 2020;**181**:102821. DOI: 10.1016/j.agsy.2020.102821

[6] Soussana JF, Lemaire G. Coupling carbon and nitrogen cycles for environmentally sustainable intensification of grasslands and crop-livestock systems. Agriculture, Ecosystems & Environment. 2014;**190**:9-17. DOI: 10.1016/j.agee.2013.10.012

[7] Bogunovic I, Pereira P, Brevik EC. Spatial distribution of soil chemical properties in an organic farm in Croatia. Science of the Total Environment. 2017;**584**:535-545. DOI: 10.1016/j.scitotenv.2017.01.062

[8] Abbona E, Presutti M, Vázquez M, Sarandón SJ. Los sistemas de producción de leche y carne bovina en la provincia de Buenos Aires ¿Conservan los nutrientes del suelo? Revista de la Facultad de Agronomía, La Plata. 2016;**115**(2):251-263. Available from: https://digital.cic.gba.gob.ar/bitstream/handle/11746/5254/744-2719-1-PB.pdf-PDFA.pdf?sequence=1

[9] Sadeghian S, Rivera JM, Gómez ME. Impacto de sistemas de ganadería sobre las características físicas, químicas y biológicas de suelos en los Andes de Colombia. In: Conferencia electrónica de la FAO sobre Agroforestería para la producción animal en Latinoamérica (FAO-CIPAV); 1 abril-1 septiembre 1998; Cali: CIPAV. 2000. pp. 77-95. Available from: https://www.fao.org/ag/aga/agap/FRG/AGROFOR1/Siavosh6.htm

[10] Instituto Geográfico Agustín Codazzi (IGAC). Estudio general de suelos y zonificación de tierras, Tomos 1, 2, 3. Bogotá: Departamento de Antioquia, Imprenta Nacional de Colombia; 2007 pp. 992

[11] Wood SN. Fast stable restricted maximum likelihood and marginal likelihood estimation of semiparametric generalized linear models. Journal of the Royal Statistical Society: Series B. 2011;**73**(1):3-36. DOI: 10.1111/j.1467-9868.2010.00749.x

[12] R Core Team. R: A Language and Environment for Statistical Computing. Vienna: R Foundation for Statistical Computing; 2022. Available from: https://www.R-project.org/

[13] Instituto Geográfico Agustín Codazzi (IGAC) [Internet]. 2011. Available from: https://sigot.igac.gov.co/sites/

sigot.igac.gov.co/files/sigot/Mapas%20 Tematicos/Departamentales/Antioquia/ Antioquia_Tam_Prom_Predios_Rurales_ V2_2012_01_18.pdf [Accessed: 30 January 2022]

[14] Departamento Administrativo Nacional de Estadística (DANE). Unidades productoras agropecuarias. [Internet]. 2014. Available from: https:// microdatos.dane.gov.co/index.php/ catalog/513/datafile/F13. [Accessed: 12 January 2022]

[15] Danish Dairy Board and Danish Agriculture & Food Council. Sustainable Development Goals. [Internet]. 2016. Available from: https://danishdairyboard. dk/products/sustainable-development-goals/ [Accessed: 05 March 2022]

[16] Tricarico JM, Kebreab E, Wattiaux MA. MILK Symposium review: Sustainability of dairy production and consumption in low-income countries with emphasis on productivity and environmental impact. Journal of Dairy Science. 2020;**103**(11):9791-9802. DOI: 10.3168/jds.2020-18269

[17] Özçatalbaş O, Akçaöz H, Firat M, Kutlar I. The analysis of socio-economic factors in the dairy farming of Antalya province of Turkey. Journal of Animal and Veterinary Advances. 2010;**9**(1):20-26. DOI: 10.3923/javaa.2010.20.26

[18] Özçatalbaş O, Sogué B. Improving women's role in agricultural production and food security promotion. In: Leal Filho W, Marisa Azul A, Brandli L, Lange Salvia A, Wall T, editors. Gender Equality. Encyclopedia of the UN Sustainable Development Goals. Cham: Springer; 2021. pp. 816-829. Available from: https://doi-org.udea.lookproxy. com/10.1007/978-3-319-95687-9_77

[19] United Nations Development Programme (UNDP). The SDGs in Action. [Internet]. 2015. Available from: https://www.undp.org/sustainable-development-goals [Accessed: 10 March 2022]

[20] Casaravilla NA. Sistemas de utilización de las pasturas. Sinopsis didáctica. Sitio Argentino de producción animal [Internet]. 2008. Available from: https://www.produccion-animal.com. ar/produccion_y_manejo_pasturas/ pastoreo%20sistemas/105-sistemas.pdf [Accessed: 20 January 2022]

[21] Senra AF. Principales sistemas de pastoreo para la producción de leche y su adecuación a las condiciones de Cuba. Revista Cubana de Ciencia Agrícola. 2005;**39**:415-426. Available from: https://www.redalyc.org/ pdf/1930/193017842004.pdf

[22] Parra-Cortés R, Magaña-Magaña M. Características técnico-económicas de sistemas de producción bovina de las razas criollas colombianas Romosinuano y Hartón del Valle. Revista MVZ Córdoba. 2021;**26**(2):e2079. DOI: 10.21897/rmvz.2079

[23] Zapata A, Silva BE. Sistemas silvopastoriles, aspectos teóricos y prácticos. 2nd ed. Cali: CIPAV; 2020 242 p. Available from: http://cipav.org.co/ wp-content/uploads/2020/08/sistemas-silvopastoriles-aspectos-teoricos-y-practicos.pdf

[24] Centro Internacional de Agricultura Tropical (CIAT). Establecimiento y manejo de sistemas silvopastoriles, Programa de gestión rural empresarial, sanidad y ambiente. [Internet]. 2015. Available from: http://ciat-library.ciat.cgiar.org/Articulos_ Ciat/biblioteca/Manual_Sistemas_ Silvopastporil_CRS_USDA_CIAT_2015.pdf [Accessed: 21 February 2022]

[25] Gil JDB, Reidsma P, Giller K, Todman L, Whitmore A, van Ittersum M. Sustainable development

goal 2: Improved targets and indicators for agriculture and food security. Ambio. 2019;**48**:685-698. DOI: 10.1007/s13280-018-1101-4

[26] Navarro ML, Urrego A, Morales A. Fertilización en diversos cultivos; Quinta aproximación. In: Manual de Asistencia Técnica No. 25. Tibaitatá: Instituto Colombiano Agropecuario (ICA), Subgerencia de Investigación, Sección Recursos Naturales; 1992 64 p. Available from: https://repository.agrosavia.co/bitstream/handle/20.500.12324/14124/27733_16902.pdf?sequence=1&isAllowed=y

[27] Echavarría NP, Jaramillo DF, Ruiz O, Parra L. Caracterización de un Andisol de la cuenca alta de la quebrada Santa Elena, Oriente Antioqueño, Colombia. Revista de la Facultad de Ciencias. 2017;**6**(1):24-38. DOI: 10.15446/rev.fac.cienc.v6n1.60628

[28] Jaramillo D. Variabilidad espacial de algunas propiedades de un Mollisol de clima cálido seco de Antioquia (Colombia). Revista Agronomía Universidad de Caldas. 2012;**20**(2):7-17. Available from: http://agronomia.ucaldas.edu.co/downloads/Agronomia20(2)_2.pdf

[29] Medina-Sierra M, Arismendi-Duque J, Cerón-Muñoz M. Descripción de las relaciones entre parámetros químicos en suelos de fincas lecheras del norte de Antioquia, Colombia. Livestock Research for Rural Development. 2019;**31** Article #102. Available from: http://www.lrrd.org/lrrd31/7/ceron31102.html

[30] Farfán MA, Forero SM, Avellaneda-Torres LM. Evaluation of impacts of potato crops and livestock farming in neotropical high Andean Páramo soils, Colombia. Acta Agronómica. 2020;**69**(2):106-116. DOI: 10.15446/acag.v69n2.82206

[31] Food and Land Use Coalition (FOLU). Diagnóstico Nueva Economía para la Alimentación y Uso del Suelo—FOLU Antioquia. 159 p. [Internet]. 2020. Available from: https://folucolombia.org/wp-content/uploads/2021/04/Diagonostico-FOLU-Antioquia.pdf [Accessed: 12 January 2022]

[32] Tovar JA. Estructura y poder del mercado en sector de agroquímicos en Colombia [Internet]. 2007. 105 p. Available from: minagricultura.gov: http://www.mamacoca.org/docs_de_base/Fumigas/Estudio_agroq_CEDE-UANDES_2007.pdf [Accessed: 23 January 2022]

Chapter 13

Community Collective Land Stewardship Contributions to Sustainable Rural Development: Lessons from Cubo, Mozambique

Simon M. Munthali, Jeremiah Machavi and Jonas Mongoè

Abstract

This chapter offers lessons on the ineffectiveness of community collective land stewardship as an enabling tool for local communities in semi-arid Africa to adopt biodiversity conservation to diversify their income and contribute to sustainable local-level rural development. While collective community stewardship of land could have transformed local land from an open accessed commodity into a collectively managed resource for community prosperity, and ensuring democratic decision-making, and permanent community benefits for generations, the Mozambique government's inability to effectively implement the statutes of its land law thwarted the Cubo community's dream to contribute to local sustainable development, due to competing land use. In the case of Mozambique, a number of factors contributed to the community's loss of its land to alternative use/agrofuel production, including the government's inadequate political will to enforce the land law's statutes; ineffective civil society to protect communities against the booming private interest in land for investment in agro-based businesses; blind loyalty of community members to their traditional leaders who are susceptible to corruption and manipulation by the private sector; illiteracy among community members, which renders them incapable of fully understanding their legal rights to land; and lack of financial capacity for the community to take legal recourse against the government's violation of its land law. We recommend that the new discourse on land tenure reform in Mozambique should: critically examine the effectiveness of how the government is enforcing its land tenure legislation; consolidate processes of accountable governance, transparency, and promotion of the rule of law. Additionally, Mozambique's civil society should: (a) proactively influence the government to prioritize implementation of existing laws and policies that promote devolved natural resources management to the local communities, and work on harmonizing cross-sectoral policies and legislation that improve management effectiveness of land and natural resources; (b) strongly advocate for implementation of Community-based Natural Resources Management models that strengthen locally accountable institutions for natural resource management and use—enabling local communities to protect their land and associated resources against foreign acquisitions; (c) improve transparency and effectiveness in enforcing the land law—to ensure that all its statutes are adequately implemented and enforced. The Cubo community's experience of losing its collectively secured land to alternative uses exemplifies one of the challenges faced in integrating local communities in biodiversity conservation and rural development programs in southern Africa, wherein some situations, power and money could easily trump laws and rules.

Keywords: community collective land stewardship, biodiversity conservation, sustainable rural development, ineffective land law, land alienation

1. Introduction

Rural development can be defined as a process that seeks social change and sustainable economic development for the rural community's ongoing progress. The goal is to improve the rural people's livelihoods and preserve the environment at the local level, where changes can be seen and felt in a more immediate manner, guarantee intergenerational equity, and ensure that the current generation must not compromise the ability of future generations to meet their material needs and enjoy a healthy environment [1]. In most African rural areas, agriculture (crop and livestock production) is often the dominant, and sometimes the exclusive economic sector that is considered as the driver of rural development. However, dependence on agriculture as the mainstay for livelihoods and rural development is severely constrained in the semi-arid areas, where soils are poor, and rainfall is insufficient. Consequently, rural people skewedly depend on natural resources (forests, wildlife, fish, grazing land, etc.) which are openly accessed by users, without individual or collective commitment to manage or regulate the use, which often leads to depletion, and environmental degradation.

To address this tragedy of the commons, the Cubo community in Massingir District, Mozambique, guided by the Land Law of 1997, chapter 3, Article 9 and Decree 66 of 1998 secured land rights to collectively manage and commit it to biodiversity conservation through the establishment of a Community Conservancy. The Conservancy was established as a mechanism through which the communities could partner with the private sector, and directly participate in the wildlife economy, through ecotourism marketing, selling of live wild animals, production of game meat to supply in the ever-growing venison market in Mozambique, and other associated benefits, such as employment in fencing and management of the Conservancy.

This collective community stewardship of land would have transformed local land from an open accessed commodity into a collectively managed resource for community prosperity. Under this arrangement, community members who have relationships with the land were expected to practice democratic decision making, and ensure permanent community benefits for generations. This was the basis through which communities themselves would have contributed to sustainable rural development in the Massingir District, Mozambique. This dream was however not realized due to competing land uses, and Mozambique's government's inability to effectively implement the statutes of the land law and this exemplifies one of the challenges of using collective land stewardship as a mechanism for promoting sustainable rural development in Africa.

This chapter highlights the fragility of community collective land stewardship as a tool for enabling sustainable rural development and addressing the problem of the tragedy of the commons. It elucidates the community's socioeconomic status, and its rationalization to integrate biodiversity management and sustainable use of wildlife into its land-use options; outlines the process the community had followed in acquiring land tenure security and its governance; highlights possible reasons for the Mozambique government's imposition of agrofuel production on a secured communal land, and provides lessons on Mozambique's non-committal to its own land law. Overall, this chapter exemplifies one of the challenges faced in integrating local communities in biodiversity conservation and sustainable rural development in Africa where power and money could easily trump laws and rules.

2. Cubo community's socioeconomic profile and rationalization to adopt collective land stewardship for biodiversity conservation as a strategy to improve their livelihoods, and contribute to rural development

The Cubo community comprises three adjacent villages (Cubo, Chivovo and Mbidzo), collectively referred to in this chapter as the Cubo community, inhabited by at least 2500 people of the Shangaan tribe. Based on a socioeconomic sample survey of 152 households, carried out by Munthali, et al. [2], the Cubo community occurs in a predominantly savannah woodland, typified by poor soils, and low and erratic rainfall (300-400 mm/annum). Consequently, conventional agriculture is generally unsustainable. Hence, people's livelihoods largely depend on livestock, forestry, and wildlife resources. At least 51 plant species are being utilized for food, medicinal purposes, timber, and firewood [2]. In addition, 21 wild animal species are being used for food leading to the extinction of almost all large mammals outside the protected areas [2].

As the population of this community is predominantly of middle age (49 ± 4.0 years), with a fairly large family size (5 ± 2 children per household), the pressure on natural resources can be expected to escalate with time, aggravating land degradation and poverty, which is already characterized by unacceptable indices, such as high illiteracy rate (56.2% among males and 83.8% among females); low employment (≈ 6% of the population); and high food insecurity (with 65% of male-headed and 50% of female-headed households) running out of food within 6 months of the year; and scanty household possessions [2]. Additionally, social amenities, such as schools and health facilities are scarce, and community members have poor access to clean drinking water. Household earnings averaged about US$0.56/day [2], which was below the US$2/day threshold recommended by the United Nations [3].

In recognition of the high levels of social dissatisfaction, the Cubo community opted to adopt biodiversity conservation, production of wildlife, and ecotourism development adjacent to the Great Limpopo Transfrontier Park (**Figure 1**). In southern Africa, local communities usually adopt biodiversity conservation through programs commonly known as Community-based Natural Resources Management (CBNRM). CBNRM has been variously defined as "a broad rubric encompassing a wide range of resource management programs that share a recognition of the participation of people who live near or interconnected with natural resources [4], or as a broad spectrum of new management arrangements and benefits sharing partnerships for the involvement by people who are not agents of the state, but who, by virtue of collective location and activities are well placed to enhance the present and future status of natural resources, and their own well-being [5]. The approach is a community based because the communities managing the resources have the legal rights, the local institutions, and the economic incentives to take substantial responsibility for sustained management and use of these resources. CBNRM has been firmly rooted in wildlife management with income earned from tourism and trophy hunting providing the main economic incentive for rural communities to invest in wildlife as a form

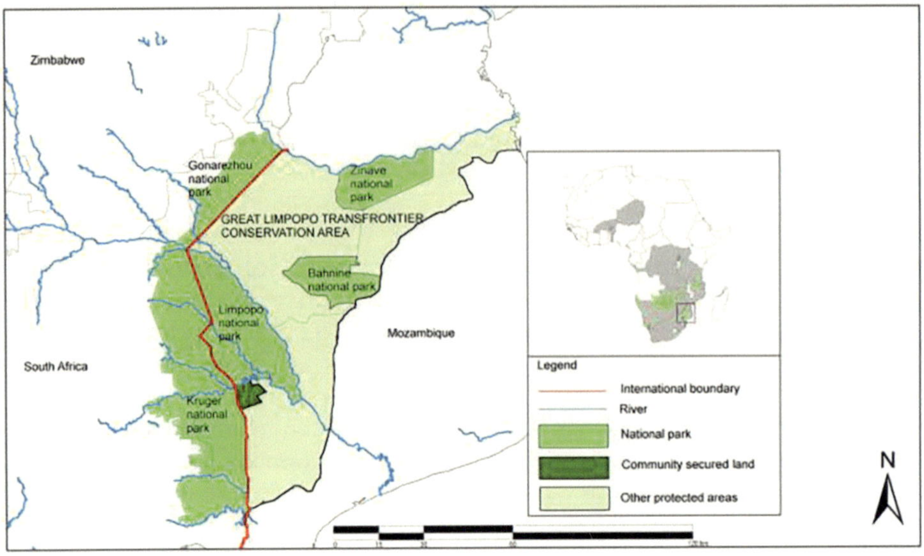

Figure 1.
Positioning of the secured community land (Cubo, Chivovo and Mbindzo) relative to the Kruger and Limpopo National Parks (map drawn by Gordon Ringani).

of land use, improving local economic options, and extending the amount of land used for wildlife across the region to communal areas.

At the core of CBNRM initiatives is widespread recognition among policymakers that for wildlife to persist outside state protected areas, and private and communal lands, it must be an economically competitive land-use option for landholders [6]. This perception has led to a proliferation of CBNRM initiatives, all with a common agenda—integration of biodiversity conservation and improvement of rural livelihoods, with the wildlife economy providing multiple private sectors and community partnerships opportunities in the live wildlife animal sales, ecotourism supply chain and game meat production for local consumption as well as commercial trade in supermarkets and urban restaurants, thus being the trigger for improved household incomes of the rural poor, who usually have very limited economic capital assets. Additionally, several compelling reasons have forced governments to adopt CBNRM as an operational tool for their national biodiversity conservation programs. Notable among these being (i) a realization that protected areas (a system widely adopted to safeguard representative examples of ecosystems and biodiversity worldwide) are expensive to maintain without the support of rural communities, and are rarely financially sustainable in the face of competing demands on dwindling government budgets [7, 8]; (ii) the growing realization both from the conservation movement, starting with the 1980 World Conservation Strategy [9] and within the rural development theory of the importance of understanding the needs and perspectives of local people; and (iii) the Convention on Biological Diversity, which emphasizes three equally important objectives: conservation, sustainable use and equitable sharing of benefits—has reinforced the role of local people in nature conservation and management.

For the Cubo community, their quest to venture into CBNRM, through the establishment of a Community Conservancy, under a community-private sector partnership was triggered by an opportunity arising from its location adjacent to the Great

Limpopo Transfrontier Conservation Area (GLTFCA). The Conservancy was strategically positioned where the Kruger and Limpopo National Parks meet (**Figure 1**); close to the existing tourism markets of the southern end of Kruger National Park (a world's renowned wildlife park, which prior to the Covid-19 pandemic attracted more than a million tourists per annum), and near an airstrip, and the Massingir Dam; thus, making the Conservancy quite attractive for investments in wildlife production, and tourism development and marketing—enabling community members to earn additional income from fencing and management of the Conservancy, and services, such as the supply of food to lodges, laundry, maintenance of the Conservancy fence, and waste disposal. Communities would have been shareholders through the allocation of their collectively owned and secured land to biodiversity conservation, wildlife production, and, tourism development and would have benefited from profit dividends.

For the Cubo community securing a collective tenure of its communal land, as described below, was essential because the provision of security of tenure is a prerequisite for better natural resources management and sustainable development [10]. Rural people generally need both secure individual rights to farm plots and secure collective rights to common-pool resources, such as flora and fauna upon which they depend. These are also preconditions for sustainable rural development, which aims to improve the rural people's livelihoods and preserve the environment at the local level, as well as guarantee intergenerational equity, and ensure that the current generation must not compromise the ability of future generations to meet their material needs and enjoy a healthy environment [1].

3. Securing land rights: process and governance

3.1 Process

Land issues and their relation to poverty have gained great importance throughout eastern and southern Africa. It has become a high-profile issue in virtually every country in these regions, more particularly in response to the scramble for land in the context of privatization and a search for foreign investment. Land held under various forms of communal tenure has particularly come under serious threats. Consequently, debates on land reform and provision of secure land tenure systems to the often-disenfranchised local communities have taken center stage by the governments, donors, civil society, and NGOs. This need was emphasized at a Conference on Land, Labour and Food Security in Southern Africa held in Johannesburg in 1997. During this conference, a Charter was drafted demanding governments, among other things to acknowledge that equal access to, and ownership of land is a basic human right, and that land reform policy should: (i) break the monopoly of landholding by landlords and commercial farmers, and give equal and secure ownership of land to those who live and work on it; (ii) be developed with full participation and input of the landless and rural poor, with emphasis given to the interest of rural women and the youth; (iii) should be driven by the principles of social justice and basic human needs as opposed to market forces; and (iv) that people who have been displaced from their land by conflict or unjust policies should have the right to claim their rights to land [11].

The argument is in favor of consolidating communal land tenure is that customary tenure which is most prevalent in rural Africa is insecure for the smallholder farmers and provides no incentive for land improvements, prevents land from being used as collateral for credit and that it prevents the transfer of land from inefficient uses to efficient ones [10]. Provision of communal land tenure security is, therefore, seen as a precondition for intensifying agricultural production and is increasingly stressed as a prerequisite for better natural resource management and sustainable development. In response to these needs, over the past three decades, many countries in eastern and southern Africa have been addressing the issues of inequitable access to land.

To address the problem of inequitable access to land, the Mozambique government enacted a land law in 1997, which is supposed to accommodate the new political, economic and social context and guarantee access and secure tenure to land, both for the Mozambican local communities and the national and foreign investors [12]. Under this law (Article 1), a 'local community' is defined as "a grouping of families and individuals, living in a territorial area that is at the level of a locality or smaller, for the purpose of safeguarding their communal interests through the protection of traditional areas, agricultural areas, whether cultivated or lying fallow, forests, places of cultural importance, pastures, water sources and areas for expansion". According to this Law (Lei de Terra) of 1997, tenure security for local communities is supposed to be protected in three ways:

i. By recognizing that the right to use land can be acquired by formal request (Article 12.c of the land law), by occupation by individuals and local communities according to customary norms and practices (Article 12.a), or by national individuals who have already utilized the land in good faith for at least ten years (Article 12.b), and the law explicitly states that the absence of title and/or registration does not hinder the right acquired by occupation (Articles 12; 13.2 and 14.2).

ii. Giving local communities an important place in the new legislation. First, they are explicitly recognized as subjects for the right to use land (Article 10.1). Second, they are assigned an active role to play in the management of natural resources, in the resolution of conflicts, and in the titling process to ascertain if the land is vacant (Articles 13 and 24).

iii. Allowing the proof of land use rights based on testimony by members of local communities, as well as through the presentation of the title document (Article 15).

Thus, in terms of land, the range of rights protection under this law is extremely broad, encompassing all the major categories of land use among rural communities, now and in the future. This is re-enforced by: (a) recognition of customary land rights; (b) granting greater leasehold security to smallholder and commercial interests—thus strengthening smallholders' chances to defend their rights in the face of growing competition for land from commercial interests; and (c) granting women land rights [13, 14]. Based on these statutes, Mozambique's land law theoretically provides an enabling legal framework upon which local communities can harness collective land stewardship and tangibly participate and benefit from biodiversity conservation through joint management partnerships with private investors, or with the state. It was on this basis that the Cubo community secured its rights to land on which they intended to establish a Conservancy.

Guided by the Land Law of 1997, Chapter 3, Article 9 and Decree 66 of 1998, the African Wildlife Foundation (AWF) used a step-wise approach in facilitating the titling and securing of the Cubo community land. This involved identification and delimitation of the Cubo communal land. The community was the major player in the identification of its rights and the delimitation of its territory. AWF simply facilitated the process of self-determination by providing the community with a participatory methodology that translated indigenous knowledge into a form that was registered in a modern registry system—the National Cadastre. The approach was participatory, allowing the community and its neighbors to delineate and reach a consensus on the boundary of what they perceived to be their land. The delineated boundaries coincided with the historical lineage territory over which the Cubo community has jurisdiction over land and other natural resources, such as wildlife, forests, water, and pasture. A series of validation mechanisms were integral to the land delimitation process, whereby the Cadastral Department consulted the neighboring communities to verify (i) if they were consulted in the process of land delimitation and (ii) if the delineated boundaries were correct. In addition, the administrator of the Massingir District ensured that the registration process was transparent and that it considered various interests of those desiring to secure their land, including the interests of neighboring communities and other stakeholders, such as the authorities of the adjacent Limpopo National Park, and the private concessionaires. The outcome of this process included:

- The consensus among different interest groups within the community and its neighbors, *i.e.*, state-protected area and private concessionaires on the delimitated land for Cubo;

- Official mapping of the Cubo communal land (101,000 ha) by the Provincial Cadastral Department;

- Issuing of *Certificates of Land Use and Benefit Rights* to Cubo community by the Provincial Governor of Gaza. Thus, offering them a co-title for their communal land. Co-titling refers to the registration of land that collectively belongs to the community, and not to individuals within that community. It confirms the existing rights of the community to their land with the same degree of security as a land title for a private concession.

- Participatory zoning of the secured land, allowing the Cubo community to accommodate within its secured land various uses, such as allocating land for (i) the establishment of a Conservancy for biodiversity conservation and tapping into the wildlife economic opportunities, (ii) settlement and agriculture, and (iii) livestock grazing, which included areas where community members could extract forestry resources (fruits, timber, medicinal plants, etc.) (**Figure 2**). Of the secured 101,000 ha, the community assigned 41,000 ha (about 41%) of nearly pristine land to the establishment of a Conservancy, with the potential for natural dispersal of wildlife from the Kruger and Limpopo National Parks to the Community Conservancy. The Conservancy's wildlife carrying capacity, in terms of biomass, was estimated at 49,200 kg ha^{-1} [15]. Infrastructure and management costs for the conservancy were to be initially paid by AWF and in-kind donations from donors for a period of at least 8 years after which period these costs would have been taken over by the community and paid for by dividends from wildlife-based enterprises. The assumption was that by the 8th year, the Conservancy through conservation enterprises, such as ecotourism, sales of live wild animals and game meat, etc., would have been financially viable and profitable to pay for the Conservancy's management costs.

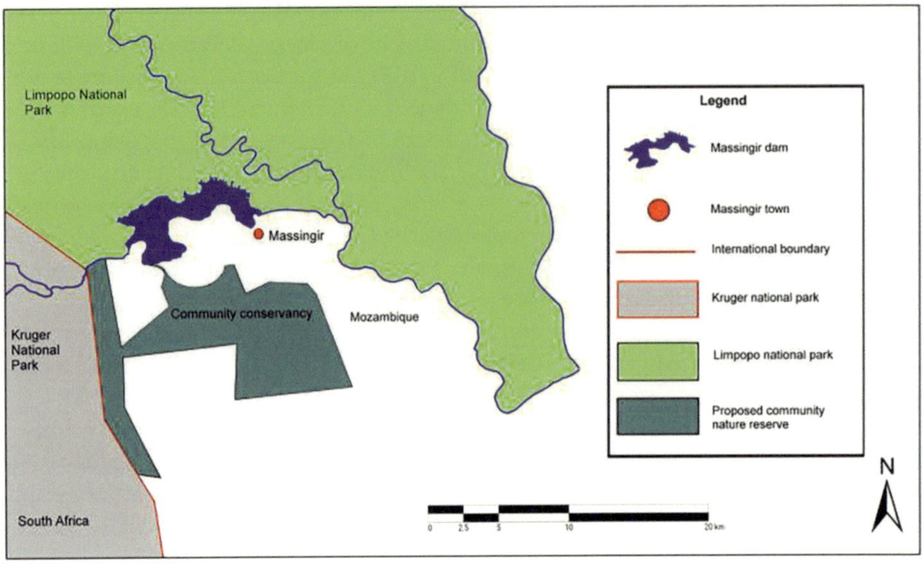

Figure 2.
Cubo secured communal land allocated to biodiversity conservation (map drawn by map drawn by Gordon Ringani).

Securing and leveraging communal land for biodiversity conservation and tapping into the wildlife economy for the Cubo community was expected to have several benefits, including:

i. Changing from an open-access resource use regime, where overexploitation is inevitable to a legally secured common property regime in which communal right holders would have managed their natural resources (forests and wildlife) exclusively to preserve and enhance their long-term productive capacity for the benefits of current and future members of the community. This was especially important for the Cubo community, where deforestation is rampant due to charcoal production.

ii. Shifting from subsistence use of forestry and wildlife resources to market-orientated production and marketing would have led to the diversification of household income for the community holding communal rights to the Conservancy.

iii. Using the secured land/the Conservancy as collateral for the community to attract extra support (*e.g.*, financial grants, or soft loans) from non-governmental organizations and donors for investment in wildlife production, conservation, and ecotourism development. Better access to land resources, with greater security, is a necessary condition for the community to negotiate joint partnerships with the private sector in the development and management of the conservancy.

iv. Providing the community with an opportunity to lease its natural capital assets (the Conservancy) to the private investor, and this would have been consistent with successful CBNRM programs in southern Africa, which follow an empowerment and enterprise approach, where community-based organizations (CBOs) are registered and contract directly with the private sector [8]; and

v. The community's quest to establish a Conservancy conformed to the resolutions of the 5th IUCN's World Parks Congress of 2003, and the Convention on Biological Diversity 7th Conference of 2004, which recognized and recommended the establishment of "Community Conserved Areas" as the fundamental mechanism in support to biodiversity conservation, and equitable sharing of benefits arising from biodiversity conservation [16]. It was also in agreement with Mozambique's own legislation on land (*Lei de Terra* of 1997), and forestry and wildlife legislation (Mozambique government 1997).

3.2 Governance of the secured land and conservancy

In southern Africa, the legal personality for governing CBNRM is either a trust, Associações (as is the case in Mozambique), communal property association, or community-based organization (CBO) which is empowered to govern and represent the communities' interests in the management of the natural resources. In addition, the empowerment of communities over their land and associated natural resources, such as wildlife and forests establishes them as prospective partners in any commercial joint venture deal with private investors, and/or state agencies.

For the Community Conservancy, an Associação was established comprising three females and seven men elected by the Cubo community. The inclusion of women was consistent with Mozambique's constitution, in which men and women are equal before the law in all aspects of political, economic, social, and cultural life. As equal members within the community, women representatives are supposed to fully represent the interest of women within their communities in land management decision-making processes and other aspects of socioeconomic development. The Associação was established to reinforce local accountability—a prerequisite for them to gain direct benefits from investments in conservation-based business and/or in-kind donations from effective local-level land and biodiversity management. The roles of the Associação were to:

- Consolidate collective rights to land and biodiversity assets;

- Garner collective coercion among the community members in setting rules for the governance/management and use of their secured land, forest, and wildlife;

- Represent their members in negotiating partnership arrangements with private investors in the management of the Conservancy, and conservation enterprise development;

- Promote collectiveness in harnessing equitable sharing of benefits from investments on the Cubo community land;

- Represent the community's interests in various developmental endeavors, such as social amenities (health, schools, roads, water supply), livelihood security (food production, livestock husbandry, etc.), land use conflict resolution, and other aspects depending on the community's needs; and

- Defend the community's rights to its secured land tenure.

To fulfill the Associação's mandate, its members were trained by AWF in various aspects (resource management, conflict resolution, financial management, etc.) to ensure that it appropriately served the community's interests and expectations.

4. Vulnerability of community collective stewardship of the land as a strategy for promoting conservation and sustainable rural development

4.1 Corporate's grabbing of the community secured land

Despite the Cubo community's success in legally securing its land rights and unconditional support from AWF and donors to develop a Conservancy, as a tool for diversifying its income opportunities and contribute to rural development, the Mozambique government through its local administration in Massingir District rescinded its decision to allow the community to develop 41,000 ha of its secured communal land for biodiversity conservation and wildlife-based enterprise development. The state instead signed a contract with a London-based Central African Mining Company (CAMEC), operating locally in partnership with ProCana for agrofuel production, with an estimated output of 120 million liters of ethanol [17]. This was intended to be produced from sugarcane that was to be planted over an area of 30,000 ha of the community land without any compensation for the community's loss of its land.

Sugarcane production was preferred despite speculation of serious water availability problems (http://www.osisa.org/node/10517) and the inability by ProCana to secure the requisite investment capital for ethanol production [17]. It is estimated that a liter of agro ethanol produced from sugarcane can use as much as 4000 l of water [18]. Therefore, for ProCana to produce its pledged volume of 120 million liters of ethanol, it would use at least 480 bn liters of water, which would deplete the available surface and underground water supply in Massingir District; hence from an environmental perspective, agrofuel production is not a sustainable venture in Massingir District. Additionally, the rationale that sugarcane growing, and ethanol production could create about 7000 local jobs was questionable, especially as similar ventures in Mozambique have failed to improve rural livelihoods. There is some evidence that levels of pay offered by agrofuel production ventures in Mozambique are so low that those employed are not any better off [19]. Furthermore, most agrofuel crops require little labor, mainly in the form of short-term work clearing the land to make way for the plantation and some work at harvest time, thus, there are few long-term jobs for local communities [20]. It is estimated that one permanent job is created for every 100 ha of agrofuel planted, and where mechanized farming methods are used, employment levels are even lower; for instance, a harvesting machine can replace 100 jobs [21]. Therefore, adopting agrofuel production based on pledges of improved employment opportunities needs scrutiny by governments in Africa.

For the community of Cubo, which lost its land and the opportunity to diversify its livelihood strategies through the development of a Conservancy and tapping into the wildlife economy, its dream was shuttered. The multiple benefits that could have accrued from non-governmental organizations and donor subsidized investments in biodiversity conservation, such as improved productivity of their savannah woodland through wildlife production, and marketing, would have increased opportunity for entrepreneurship/small and medium enterprise development by increased potential for meat production from wildlife sustainable harvesting scheme; community capacity building and skills development in wildlife management and its associated businesses (e.g., participation in the ecotourism supply chain, and marketing with many spin-off benefits) were completely ignored by the state in weighing options for local economic development in Massingir District. The community's long-term benefits were ignored in favor of the unproven pledge of 7000 jobs that the proposed sugarcane and its associated ethanol production may have created. Considering the high illiteracy levels in Massingir District (56.2% among males and 83.8% among females [2], the benefits from the ethanol industry were unlikely to be substantial for the community members because they could only benefit from low-paid manual work, on an average earning about $0.7/day, which is below the UN recommended threshold of US$2/day [3].

Besides poor wages, the use of large areas of land for agrofuel production may result in a range of detrimental environmental impacts, which can include deforestation and loss of habitat, soil degradation because of inappropriate farming methods, water pollution from pesticide and fertilizer use, and the depletion of water resources—threatening biodiversity, carbon stocks, and land and water resources. While converting forests and rangelands to mono-cropping—a common phenomenon in the production of agrofuels reduces diversity in flora, fauna, and agrobiodiversity, as well as aboveground and subsurface carbon stocks [18].

Irrespective of all these negative social and potential environmental impacts, the Mozambique government denied the Cubo community an opportunity to directly benefit from its secured land and wildlife resources in favor of agrofuel production, which never materialized. 15 years later (in 2021) the land which had been alienated from the community proposed conservancy continues to be degraded by deforestation due to charcoal production and overgrazing. In addition, the Massingir District harbors some of the most notorious poachers in southern Africa, who have been involved in decimating the population of elephants (*Loxodonta africana*) and rhinoceros, both black rhino (*Diceros bicornis*) and white rhino (*Ceratotherium simum*) in the Great Limpopo Transfrontier Conservation Area. These are the environmental ills that the communities themselves wanted to mitigate.

The indifference and lack of community support to biodiversity conservation in the transfrontier conservation area (though not confirmed by any research), can in part be linked to Mozambique's government's denial of the community's opportunity to establish, develop, and manage its own nature Conservancy for the collective socio-economic benefit of the community. The case of the Cubo community exemplifies the existence of high competition for land and other natural resources between local people and outsiders in Mozambique, and elsewhere in Africa, especially in areas perceived to be valuable for agricultural production or tourism development. The area where the Cubo community wanted to establish a Conservancy is one such area. Being part of the land, which is adjacent to the Great Limpopo Transfrontier Conservation Area, it has high value as a productive asset, attracting a wide array of investors, including unscrupulous ones, both from within and outside Mozambique, who bypass or cursorily pay attention to legally prescribed procedures in accessing land for investment in agriculture production. Most of these land seekers are speculators who use their connections with senior political and government officials to secure land rights over large areas but many lack the financial resources to develop their concessions. It is unfortunate that the government disregards its own people's interest in favor of the private sector, irrespective of the latter's credibility.

5. Lessons learned from Cubo

The Mozambique government's denial of the Cubo community's desire to develop a nature Conservancy in favor of agrofuel production contradicts the primary purpose of its land law (Lei de Terra of 1997) which is to secure customary land rights, thereby helping to manage and reduce conflict over land tenure and promote rural development through among others, conservation and communities' participation in the wildlife economy as an option to diversify livelihood strategies in rural areas, especially where conventional agriculture is not sustainable, such as in Massingir district. Similarly, there is a special recognition in the land law, of the rights and interests of local communities, including mandatory requirements for community consultations and hearings when land is transferred to new uses and/or users; hence it is quite puzzling that the implementation of these positive legal and institutional frameworks has been ignored by the government in addressing the Cubo community's lost opportunity to adopt collective land stewardship as a strategy to improve its rural livelihoods and contribute to local level sustainable development.

Although the Mozambique government's contempt for its own land law has not been fully assessed, it may be due to the following interrelated factors.

5.1 Ineffective enforcement of the land law statutes

There is a gap between the law-making rhetoric and on the ground enforcement of the land law due to weak institutional capacity, poor governance, lack of political will to enforce the law, and vested state's interests—contrary to communities' needs. These shortfalls are being amply taken advantage of by the district authorities, who under the pretext of promoting national economic priorities, promote the interests of investors over local communities' needs. The ease with which ProCana secured a leasehold concession on the community's legally secured land is subject to wild speculations, but it appears the process may have been aided by powerful Mozambicans who had a stake in the venture and bribing of traditional leaders by ProCana [17]. Traditional leaders have since colonial times been susceptible to corruption and in almost all southern African countries; tenure insecurity is sometimes caused by the exploitative behavior of traditional leaders and rent-seeking government officials [10]. Contrary to the statutes of the land law, the Cubo community members were cursorily consulted about the transfer of its land to the private sector. The district authorities focused their land acquisition consultations on community elders/village headmen, who granted permission unwittingly without the full involvement of the wider community [17]. Such agreements are made in exchange with promises from the investor to improve employment and facilities in the area, but such pledges are hardly fulfilled, leading to resentment from the community [19].

The Cubo case demonstrates that legal rights to land are not enough. They need to be set in the broader historical and political context of the country. Nelson and Agrawal [22] observed that institutional reforms that devolve rights to the local level have been relatively successful in countries where public institutions are efficient and the rule of law operates—for example, in Namibia, Botswana, and pre-crisis Zimbabwe. This, they argue, is not the norm across sub-Saharan

Africa—where in general state institutions are characterized by patrimonial relationships and weak rule of law. The partial devolution of valuable natural resources such as wildlife to the local level (as observed in Mozambique) appears to be fundamentally at odds with the interests and incentives that dominate governance processes [22].

5.2 Illiteracy and "blind" loyalty to traditional leaders and local government officials

The Cubo community's "blind" loyalty to traditional leadership and local government officials may have contributed to its lack of effort to challenge the alienation of its land by ProCana. The community, through its Associação could easily have used the extant land law to challenge the grabbing of its land, but it has not done so, and this was compounded by a lack of funds to hire a lawyer to challenge the state in the court of law. Illiteracy, which is prevalent among the community members, may have contributed to a lack of fully understanding of the statutes of the land law. This is being aggravated by the lack of effective civil society's support in Mozambique to assist communities like Cubo to take legal recourse against the government's violation of its land law.

5.3 Susceptible communal tenure security

The sources of land tenure insecurity in Mozambique are more complex than generally acknowledged. The statutory mechanisms for securing land tenure rights are insufficiently effective to protect the full range of land interests in modern and globalized economic circumstances. Those with the least status, knowledge, or means, such as local communities are least well served. The State itself is a source of insecurity due to the way it easily transfers communal land tenure rights to the private sector—a process that appears to have strong economic incentives for political elites and central bureaucracies to consolidate their control over natural resources at the expense of local communities who are custodial owners of these resources.

5.4 Lack of political will in support of CBNRM programs

CBNRM programs are greatly constrained by a lack of political will to enforce the extant enabling policies and legislation, regardless of international NGOs and donor support. Attempts to set up CBNRM projects against the backdrop of legislation, such as Mozambique's land law, that is not being effectively enforced, wastes donor funds, derails the morale of the supporting NGOs, and erodes the community's capital assets—notably the natural capital (e.g., land & its associated stocks of natural resources and environmental services); social capital (natural resources governance associations, norms, trust and disposition to work for a common good for biodiversity conservation); and loss of potential by the communities to diversify their income generation from conservation enterprises and the related spinoff businesses. International NGOs spearheading CBNRM efforts are poorly positioned, in a political sense, to address the problem of ineffective enforcement of the land law statutes.

5.5 Ineffective civil society

Mozambique's civil society has so far been very ineffective in protecting communities like Cubo from losing their land to private interests. Civil society faces the challenge of balancing the different expectations of the government and the local communities, especially as the functions of government agencies in implementing and enforcing the land law are constrained by weak processes of accountable governance, limited financial resources, and scarce capacity. The national civil society needs capacity building so that it should be able to encourage the government to prioritize implementation of existing laws and policies that already promote devolved natural resources management and work on harmonizing cross-sectoral policy and legislation that improve the management effectiveness of land and natural resources. This process should be reinforced by strong civil society's advocacy for CBNRM models that strengthen locally accountable institutions for natural resource management and use, enabling local communities to protect their land and resources against foreign acquisitions. Civil society should also focus more on improving transparency and effectiveness in enforcing the land law, to ensure that all its statutes are adequately implemented and enforced.

The civil society should also strongly advocate for Mozambique to implement various international conventions, such as Convention on Biodiversity (CBD) and the United Nations Convention on Combating Desertification (UNCCD), among many others. While national sovereignty is paramount for any country, these Conventions have been widely embraced by many countries in the world in promoting and implementing biodiversity conservation programs, and Mozambique is a party to these conventions. The CBD, for example, specifically recognizes the potential role of local communities in biodiversity conservation through Articles 8 (j), 10 (c), 10 (d), and 11 [23].

Additionally, UNCCD also recognizes the importance of secure land and resource tenure and forms of decentralization. The UNCCD places considerable emphasis on promoting the sustainable use of natural resources [Article 3 (b)], alternative livelihoods [Article 10. 4], and capacity building of local communities for sustainable land and resource management [Article 19]. The strong convergence between the key principles of the CBD, UNCCD, and the generic approach to CBNRM in southern Africa should provide sufficient impetus for the civil society in Mozambique to encourage the government to improve its effort in implementing policies and legislative mechanisms that safeguard community's land rights, prevents alienation of their land without their collective consent, and guarantees their participation in conservation-based enterprises as a means of diversifying community livelihood strategies, and contribution to sustainable rural development.

6. Conclusion

Despite its contemporariness, Mozambique's Land Law is failing to guarantee secure tenure to land for its local communities, especially in areas perceived to be valuable for agriculture or biofuel production. This is primarily because of the government's inadequate political will to enforce the land law's statutes; ineffective civil society to protect communities against the booming private interest in land for investment in agribased businesses; blind loyalty of community members to their traditional leaders who are susceptible to corruption and manipulation by the private sector; illiteracy among community members, which renders them incapable of fully understanding their legal rights to land; and lack of financial capacity for the communities to take legal recourse against the government's violation of its land law. The discourse on land tenure reform in the past two decades in Mozambique has dominantly focused on land-law formulation and institutional restructuring for implementation, but as witnessed by the Cubo community, there is a need to critically examine the effectiveness of how the government is enforcing its land tenure policy and legislation and find out why these enabling frameworks are being superficially implemented. Additionally, the new discourse on land tenure security should aim at consolidating processes of accountable governance, transparency, promoting the rule of law, and identifying sustainable mechanisms for mobilizing resources to enhance the government's capacity to effectively enforce its land law statutes.

Mozambique's civil society should: (a) proactively influence the government to prioritize implementation of existing laws and policies that already promote devolved natural resources management to the local communities, and work on harmonizing cross-sectoral policies and legislation that improve management effectiveness of land and natural resources; (b) strongly advocacy for CBNRM models that strengthen locally accountable institutions for natural resource management and use—enabling local communities to protect their land and associated resources against foreign acquisitions; (c) improve transparency and effectiveness in enforcing the land law—to ensure that all its statues are adequately implemented and enforced; and (d) ensure that adoption of monoculture ventures, such as biofuel production is guided by objective assessments of their social and environmental impacts on the rural communities, so that such undertakings do not erode communities' natural and social capital assets, and denial them the opportunity to adopt collective land stewardship to pursue their locally rural development agenda.

Author details

Simon M. Munthali[1*], Jeremiah Machavi[2] and Jonas Mongoè[3]

1 Development and Environmental Education, Lilongwe, Malawi

2 African Wildlife Foundation, White River, Mpumalanga, South Africa

3 Cubo Community Association, Massingir, Mozambique

*Address all correspondence to: muchina.munthali@gmail.com

References

[1] Batie SS. Sustainable development: Challenges to the profession of agricultural economics. American Journal of Agricultural Economics. 1989;**71**(5):1083-1101

[2] Munthali S, Machavi J, Mongoe J. The Socioeconomic Status of Cubo, Chivovo and Mbindzo Communities: Setting Baseline Indices for Monitoring Change Triggered by Adoption of Biodiversity Conservation and Ecotourism Development. White River, South Africa: AWF, Limpopo Heartland; 2006

[3] UNDP. Monitoring Human Development. Human Development Report. New York: United Nations Development Programme; 2004. Available from: http://www.hdr.undp.orgstatistics/data

[4] Lyons A. An Effective Monitoring Framework for Community Based Natural Resource Management: A Case Study of the ADMADE Programme in Zambia. Msc. Thesis. Florida, USA: University of Florida; 2000

[5] Metcalfe. In: Western D, Wright M, Strum S, editors. CAMPFIRE: Zimbabwe's Communal Areas Management Programme for Indigenous Resources in Natural Connections: Perspectives in Community-based Conservation. Washington D.C.: Island Press; 1994

[6] Child B. Parks in Transition: Biodiversity, Rural Development and the Bottom Line. Earthscan: London, UK; 2004

[7] Hanks J. Financing Africa's Protected Areas, Southern African Wildlife Association. Transfrontier Conservation Initiatives in Southern Africa. Cape Town, South Africa: Conservation International; 2001

[8] Josserand HP. Community-Based Natural Resource Management in Africa—A Review. USA: ARD-RAISE Consortium; 2001

[9] IUCN. The World Conservation Strategy: Living Resource Conservation for Sustainable Development Gland. IUCN/UNEP/WWF: Switzerland; 1980

[10] UNDP. Land tenure Systems and Sustainable Development in Southern Africa. ECA/SA/EGM.Land/2003/2. South Africa: United Nations Southern Africa Office; 2003

[11] Palmer R. Contested Lands in Southern and East Africa. A Literature Survey. UK: Oxfam; 1997

[12] Republic of Mozambique. Lei de Terras, No. 19/97 de Octubro. Boletin da Republica, 1 Serie—numero 40. Government of Mozambique; 1997

[13] De Wit P Land Reform in Mozambique: Acquired Values and Needs for Consolidation. 2001. Available from: http://www.fao.org [Accessed on October 5, 2003]

[14] Tanner C. The Reform and Implementation of Land Policy in Mozambique—A case Study of FAO support. 2001. Available from http://www.fao.org [Accessed on October 5, 2003]

[15] AWF. Management and Business Plan for the Cubo proposed Nature Reserve. White River, South Africa: AWF Limpopo Heartland; 2006

[16] CBD. The Vth IUCn World Parks Congress was held in Durban, South Africa in 2003. 2004. Available from: http://www.iucn.org/about/union/commissions/wcpa/wcpa_puball/

wcpa_parksmag/?2137/2003-Durban-World-Parks-Congress

[17] Welz A. Mozambique's Sugar Lush: A Land War Threatens our Neighbours' Hard-won Democracy. Claremont, California, USA: Noseweek; 2009

[18] Friends of the Earth. 2009. Africa: Up for Grabs: The Scale and Impact of Land Grabbing for Agrofuels. Available from: www.foeeurope.org

[19] Ambiental J, Mozambique U. Jatropha! A Socio-economic Pitfall for Mozambique. Brasil: National Union of Students (União Nacional dos Estudantes); 2009 Available from: www.swissaid.ch/global/PDF/entwicklungspolitik/agrotreibstoffe/Report_Jatropha_JA_and_UNAC.pdf

[20] von Braun J, Meinzen-Dick R. "Land Grabbing" by Foreign Investors in Developing Countries: Risks and Opportunities. Washington, D.C., USA: IFPRI; 2009

[21] Ustulin EJ, Severo JR. Cana-de-Açúcar: Proteger o ambiente e continuargerando empregos. 2001. Available from: NDP

[22] Nelson F, Agrawal A. Patronage or participation? Community-based natural resource management reform in sub-Saharan Africa. Development and Change. 2008;**39**:557-585

[23] Roe D, Nelson F, Sandbrook C. Community Management of Natural Resources in Africa: Impacts, Experiences and Future Directions, Natural Resource Issues No. 18. London, UK: International Institute for Environment and Development; 2009

Chapter 14

A View of Sub-Saharan Africa from the Perspective of Food Security and Gender

Eriola Marius Charlot Adenidji and Orhan Özçatalbaş

Abstract

Nowadays, in most households, food and nutrition affairs are delegated to women. However, in the sub-Saharan regions, like the most undeveloped countries, food security is still in danger. Such a situation is not observed in developed countries. Why so? Is women's positive contribution to food security only reserved for developed countries? How can women in less developed countries intervene in food security? This paper analyses food security from a gender perspective and focuses on women's participation and empowerment as means to eradicate food insecurity.

Keywords: sub-Saharan Africa, SSA, food security, food safety, equality, gender

1. Introduction

Although food security is seen as a primary problem in Africa and Asia, it is essentially a problem for the whole world. However, besides food security, the issue of food safety is also in the field of interest of every country, especially in industrialized countries. The issue of ending poverty (no poverty) and hunger (zero hunger) on a global scale is among the top priorities of the UN SDGs. Unfortunately, this global problem could not be ended in the last century and is among the priority issues of the twenty-first century. Therefore, ensuring food security is a very important goal for the UN. According to the UN FAO, 'Universal access to safe food is a key requirement for the 2030 Agenda for the Sustainable Development Goals'. Despite this, an estimated 600 million people get sick each year from eating unsafe food, and 420,000 of them die [1], Safe food is also critical for economic development and the international food trade. It should also be noted that, unfortunately, due to the negative effects of the COVID-19 pandemic, it will not be possible for the UN to achieve its goals in this area until 2030 [2].

In our patriarchal society, roles and perceptions are categorized at a very early age according to gender. Therefore, throughout life, responsibilities will be assigned, preferably according to sex [3]. But with economic progress, society is becoming more liberal. Additionally, more voices are challenging the pre-existing gender perception by calling for gender equality. This demand is reflected in the fifth goal of the seventeenth sustainable development goal (SDG5) adopted by the United Nations General Assembly: Achieve gender equality through the empowerment of women and girls [4].

Thus, it is encouraged to highlight the potential of women and girls and to give them greater exposure.

More visibility, indeed, in a society that has always tended to stifle and control the female gender in almost everything. In such a society, until now, it has often been observed that for the same qualifications as men, women are paid less. In Germany, for example, the first power in the European Union, 21.5% of women are paid less than men [5]. The primordial roles that our so-called modern civilizations offer and accept for women without amalgamation are those related to the housewife position; in other words, the role of the housewife. As it is stated by Eurostat, the percentage of women engaged in domestic chores is 97.3% in Finland; even in Belgium, the capital of the European Union, 95.6% of women handle the home's well-being [6]. Overall, the share of women who take care of children, housework and cooking is much higher than that of men. In 2016, the number of women aged 25–49 (with children under 18) who cared for their children every day stood at 93% in the EU, compared with 69% of men. In addition, 78% of women cooked or (and) did daily household chores, compared with 32% of men [7]. However, women's involvement in household welfare is unavoidable. Moreover, any situation that disrupts women's conditions may impact food security within the household.

Even though agriculture accounts for three-quarters of women's income in sub-Saharan Africa [8], and women make up a significant portion of the agricultural labour force [9], the region is experiencing an increase in the prevalence of food insecurity. In such a situation, would it not be helpful if chefs in the households had detailed nutritional knowledge? Should more attention not be paid to the training and education of the female gender to achieve sustainable development goals, especially in reducing hunger and promoting food security? This study aims to explain and find the means of combating the scourge of malnutrition and hunger in sub-Saharan Africa.

2. Definition of sustainable food security

To examine the role of women's actions in the battle against food insecurity, it is important to have an explanation of what this concept conveys. For this reason, considering the wording of [10], food security

- Corresponds to the ability of all people to have physical and economic access to basic food needs at all times [11]. Furthermore, a national food security strategy cannot be envisaged without ensuring food security at the household level [12].

- Is the capacity to ensure that the food system provides the entire population with a nutritionally adequate food supply over the long term [13].

- Is ensured when the viability of the household defined as a unit of production and reproduction is not threatened by a food deficit [14].

Furthermore, when considering the statement of the World Food Summit (1996) about the concept as it is given, 'when all people, at all times, have access to sufficient, safe, nutritious food to maintain a healthy and active life', the need for sufficient availability of nutrient-rich foods seems to define food security. Hence, in FAO's logic, clearly, 03 elaborated conditions must exist to define food security. These are

- *food availability*: sufficient quantities of food available consistently;
- *food access*: sufficient resources to obtain appropriate foods for a nutritious diet; and
- *food use*: appropriate use based on knowledge of basic nutrition and care, as well as adequate water and sanitation.

Thus, as seen above, food security is any system that provides the consuming entity with enough food to meet nutritional needs. When this system can be long-term [13], the concept of sustainability has its legitimacy. Thus, in sustainable food security, reference is made to any system that can provide the consumer necessary quantity and quality of goods over time. Quality refers to the nutritional value of the food. As for the notion of time, it refers to future generations. However, it is essential to point out that there is no exact definition of the 'food security' concept because quality requirements are progressive over time.

After examining what is hidden behind the 'food security' concept, the positioning of the present work is to clarify how women can provide support in the three pillars that define food security as defined by the FAO. The present work also investigates if women's support can contribute to the achievement of food security objectives. To do so, the light will first be shed on the skills of women in different sectors of activity.

3. Definition of sustainable food safety

Food and nutrition literacy is a newly emerged concept to connect food-related knowledge and skills to healthy diets [15]. Nutrition literacy is the combination of access to, analysis and evaluation of the information related to food and nutrition, making and implementing good decisions, maintaining healthy eating, choosing and consuming a proper amount of healthy food, and the motivation, knowledge, skill, attitude, behaviour and abilities required for the evaluation of the motivation needed to ensure food security and the working of the food system. The recognition and dissemination of food and nutrition literacy in society will aid in the improvement of well-being, healthy nutrition maintenance and healthy food selection [16]. It would be useful if all stakeholders were to cooperate in the planning and implementation of food literacy initiatives where effective, theory-based training methods focusing on knowledge, skills and behaviour are used.

Activities that involve conscious behavioural change such as education and extension, as well as practices that increase food education and training, can enhance diet quality by reducing food insecurity [17, 18]. Of course, here along with limited food literacy, access to food and the ability to prepare healthy foods are also important. According to this, all stakeholders should work in cooperation for food and nutrition literacy.

In this context, it is important to establish an effective food control system. Because an effective food control system provides assurances to governments and the public that the available food is safe for human consumption and can be sold or traded. It, therefore, aims to [19].

(1) protect public health by reducing FDB risks; (2) protect consumers from fraudulent practices including mislabelling and adulteration; and (3) support economic development by ensuring the quality and safety of products sold and or traded [20].

Components of a national food control system include an enabling legislative framework, a food control management system, food inspection, laboratories for monitoring hazards and surveillance, information, education, communication and training of value chain operators and consumers.

These data and determinations show that there is a need for a comprehensive food safety policy on a national and global scale in any case. In many African countries, the food safety mandate is spread over many agencies and authorities, with unclear responsibilities leading to inaction and duplication. In most cases, countries lack effective national coordination mechanisms [19].

4. Gender definition

Gender empirically encompasses both males and females. From a cultural and discursive perspective [21], gender ideology refers to the beliefs and values held concerning what is right for men and women and determines the patterns by which a particular society judges or evaluates the appropriate conduct of a man or woman. The following lines illustrate gender concepts for a better understanding [22, 23].

- *Gender analysis* is the systematic gathering of data and information about gender relations to identify, understand and correct inequalities.

- *Gender discrimination* is the unwelcome treatment of people due to their gender, which denies them the opportunities, rights and resources they deserve.

- *Gender division of labour,* how specific ideas based on socially determined roles define what is appropriate for women and men.

- *Gender equality and equity*, Gender equality involves a woman being treated the same as a man in terms of accessing public life. Gender equity denotes equalization of life outcomes between men and women, recognizing their diversity in interests and needs, and requires redistribution of power and resources.

- *Gender mainstreaming*, is an organizational strategy designed to bring gender equity to all aspects of an organization's work, through building capacity and accountability.

- *Gender needs*, are shared and prioritized needs identified by women that arise from their common experiences as a gender.

- *Gender planning,* the technical and political processes and procedures necessary to implement gender-sensitive policy.

- *Gender relations,* hierarchical relations of power between women and men that tend to disadvantage women.

- *Gender training, a facilitated* process of developing awareness and capacity on the issue of gender, to bring about personal or organizational change towards full equality.

- *Gender violence, is* any act or threat by male-dominated institutions that inflicts physical, sexual or psychological harm on women or girls.

The paper considers gender in an original binary logic (masculine/feminine) to finally focus exclusively on the female gender. It inspects the contribution of the female gender to different sectors to identify its potentialities that can be beneficial in the fight for food security.

5. Women's role in the households and society

Gender relation underlies the hierarchical aspect existing between both genders in the family. And without a doubt, these relations are widely considered to be dominated by men within households. Furthermore, few women have the responsibility of being heads of families. But when they do have such responsibility, it is usually the synonym of being single, divorced, separated mothers, widows or when the husband is completely disabled or has a serious illness issue that makes him infirm. Therefore, this is the position that, despite being assimilated into cultural posturing, the female gender consistently finds itself in, according to studies. However, others studies have demonstrated that despite their vulnerability, women are excellent household leaders. It has been shown that households headed by women manage better deprivation [24]. Namely, they make a better distribution or allocation of the scarce resources at their disposal. Besides, the distribution of time allocated to household chores is generally in conflict with the status of non-household workers. This means that often in the family, there is a dilemma for women to just do childcare, housework or go for a job outside the home. Thus, in the presence of the man, the woman is much more willing to engage in housekeeping. Then, about the hierarchical paradigm of societal assignment of tasks, the predominant cultural mentality forces them to stay at home.

Socially, roles generally assigned as females are almost less valued than those designated as males. Women are expected to fulfil the reproductive role of carrying and raising children, caring for other family members and household management tasks, as well as home production. Men, on the other hand, tend to be much more associated with productive roles, specifically paid work, and commodity production. In the labour market, although the overall engagement rate of women is growing, they tend to be restrained in a relatively narrow range of occupations or clustered at lower levels than men and generally earn less [25].

Despite some wives making a greater contribution to household income than men, society does not recognize women's equal participation in household management. Their wealth is not a sufficient justification to not take on domestic tasks. Even if these things happen, it is not well perceived. Further, as far as discrimination and domestic violence are concerned, the female gender is the most vulnerable. Likewise, poverty is common among rural housewives because they depend on their husbands to provide them with economic help [26–28].

In order to best describe the contemporary place of women in society, it has been important to understand where women are and how they're doing in that society. Therefore, to achieve a sustainable, update gender-based social solution for a problem, prior women's roles in the family circle and the community must be recognized.

6. Food security in sub-Saharan Africa

If agriculture crop production had met 80% of the continent's food requirements in 2013, whereas animal production was not growing quickly enough to meet the demand [29], the situation has recently deteriorated due to the pandemic in the years 2019–2020. In fact, by 2020, 281.6 million Africans had been malnourished, up 89.1 million from 2014. East Africa accounts for 44.4% of the continent's undernourished

Country	2020 Score	Rank / 113
South Africa	57.8	69
Botswana	55.5	74
Ghana	53.0	77
Mali	52.7	79
Côte d'Ivoire	51.0	82
Kenya	49.0	86
Niger	47.6	87
Burkina Faso	47.4	88
Tanzania	47.1	89
Senegal	46.4	=90
Benin	46.2	92
Togo	44.9	93
Cameroon	44.7	94
Uganda	42.9	95
Angola	42.1	97
Congo (Dem. Rep.)	40.7	98
Mozambique	40.6	99
Nigeria	40.1	100
Guinea	39.5	102
Chad	39.4	103
Rwanda	38.8	104
Madagascar	37.5	106
Burundi	37.1	107
Ethiopia	37.0	108
Sierra Leone	37.0	108
Malawi	36.7	110
Zambia	36.6	111
Sudan	36.0	112

Source: Computed by the author and retrieved from FAO, IFAD, UNICEF, WFP and WHO, 2020. To be representative of sub-Saharan African states in terms of food insecurity prevalence, this table includes a sample of countries from each of sub-Saharan Africa's four sub-regions: the western, eastern, central and southern regions.

Table 1.
Food security index of selected sub-Saharan African countries.

people, while West Africa accounts for 26.7%, Central Africa accounts for 20.3%, North Africa accounts for 6.2% and Southern Africa accounts for 2.4%. Along with the 346.4 million Africans who are food insecure severely, another 452 million are food insecure moderately [30]. Global standards for food security are far from being met. Food security scores in sub-Saharan Africa are designed to accurately reflect this situation (**Table 1**).

The tables below illustrate African food security, under-malnourished people (**Table 2**) and the food production stat (**Table 3**).

Food production in sub-Saharan in late 1980 was estimated to be $876 billion, but in 2010, the estimation was $196 billion (**Table 3**). Besides this growth, food security is still a challenge. According to [2], it is estimated that 75% of low-income children in Sub-Saharan Africa are malnourished; only 25% are not low-income food deficient. Furthermore, this condition is exacerbated by climate change, civil unrest, armed conflicts [31] and the current widespread pandemics, which make the poorest populations even more vulnerable. Consequently, malnutrition statistics in the continent are among the highest in the world. Additionally, the food insecurity situation in sub-Saharan Africa today could be illustrated by the state of malnutrition and newborns' weight and height. These indicators reflect whether mothers fed themselves

009–2011 2014–2016 2016–2018	2000–2002	2004–2006	2009–2011	2014–2016	2016–2018	2017–2019	2018–2020
WORLD	819.2	804.0	652.3	613.8	622.7	632.9	683.9
Africa	199.8	198.4	194.8	201.4	217.1	224.9	248.0
Central Africa	39.7	41.2	40.1	44.4	47.8	49.7	53.2
Eastern Africa	102.3	101.0	99.9	97.5	104.8	107.7	115.3
Northern Africa	15.0	15.5	15.0	13.6	14.8	15.2	16.0
Southern Africa	2.9	2.8	3.6	4.7	4.9	4.9	5.6
Western Africa	39.9	37.9	36.2	41.2	44.9	47.4	57.8

Source: FAO, retrieved from FAO, CEA, and CUA, 2021. https://doi.org/10.4060/cb7496fr. Undernourishment is defined as the condition of an individual whose usual food intake is insufficient to provide, on average, the amount of dietary energy required for a normal, active and healthy life. The relative indicator is called 'prevalence of undernourishment', which is an estimate of the percentage of individuals in the total population who are undernourished. Source: FAOSTAT.

Table 2.
Number of undernourished people (millions).

African sub-region	Food production (livestock and crop) (%)
Northern	23
West	33
East	23
Central	7
Southern	17
Africa (US billions $)	196

Source: Self-computation from NEPAD, African agriculture, 2013.

Table 3.
Food production in Africa.

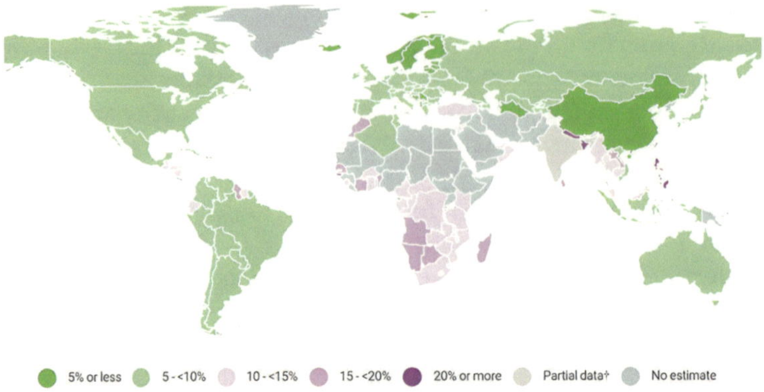

Figure 1.
Low birth weight prevalence, by country and region, 2015. Source: UNICEF-WHO low birth weight estimates, 2019.

well during pregnancy to ensure their good health and that of their infants. In other words, statistics on malnutrition reflect regardless of whether a country is safe from food insecurity, reiterating that the birth weight of a newborn is an important indicator of maternal and foetal health and nutrition (**Figure 1**).

7. Food accessibility in sub-Saharan Africa

In rural areas, people have less access to safe food than people in urban areas. It is the rural counties that supply the urban cities with food. Perhaps rural people do not have enough cash to afford sufficient nutritious meals. Or, due to lack of resources,

Country	Urban area (%)	Rural area (%)	Year
Benin	88	66	2003
Burkina Faso	93	51	2003
Cameroon	93	58	2007
Côte d'Ivoire	92	51	2002
Ghana	100	74	2006
Guinea	95	49	2007
Mali	97	56	2006
Mauritania	100	88	2004
Niger	99	50	2007
Nigeria	82	64	2003
Senegal	100	87	2005
Sierra Leone	99	58	2003
Chad	98	58	2003
Togo	100	72	2006

Source: CIRAD, Afrisat, AFD AGRA, 2020 https://agra.org/wp-content/uploads/2020/09/AASR-2020-Feeding-African-Cities.pdf.

Table 4.
Food accessibility in the selected African regions.

they are unable to buy the required food supplements. As noticed by [32], even in abundant regions, food shortages can happen, mostly due to poor conservation techniques or post-harvest losses. In fact, at the sub-regional level, sub-Saharan Africa (SSA) is the region with the highest prevalence of hunger with one person in four being undernourished [33]. Additionally, owing to SSA's population growth, the demand for food will still rise, and persist, thereby leading to the push for stronger interventions to boost food security and nutrition, as well as to end hunger. Agriculture should become a priority as it accounts for approximately 30% of African economies and comprises over 60% of the workforce in these nations [34, 35].

The above table illustrates food accessibility in some African countries (**Table 4**).

8. Female gender and food security in sub-Saharan

8.1 Women and crop production

Smallholder farmers produce 80% of agricultural production in sub-Saharan Africa [36, 37]. In subsistence farming, as practised in Africa, women are the primary sources of food production and little from men is necessary [37–41]. Sub-Saharan African women have the highest average agricultural labour force participation rate in the world, at 62.5% in 2012, compared with 36.4% globally. In Ghana, for instance, women produce 70% of the country's food crops, offer 52% of the agricultural labour force, provide 85% of the labour for agro-processing operations and 86% for food supply. Women made up 43.8% of the economically active population in sub-Saharan Africa as a whole in 2010, and 65% of those women worked in agriculture, with women delivering 70% of farm labour in Kenya [9, 42]. The importance of female agricultural workers in the world is unparalleled, even though life for rural women is not easy. Women do not have the same rights as men and must often balance domestic duties with agricultural work, such as sowing, weeding, harvesting and collecting firewood and water. Additionally, it can be more difficult for women to produce as much as men do in the farm setting, due to poor access to land, agricultural extension services and technologies.

8.2 Women and livestock farming in SSA

In less developing countries such as SSA, women generally can rear small animal husbandry (poultry, sheep, etc.), but men take care of the big head. As relayed by [43, 44], women are seen as key stakeholders in backyard poultry farming systems, and successful engagement in this sector requires gender-appropriate methods. Research has proven that women's poultry farming initiatives are more likely to lead to successful nutritional outcomes [33, 45–50]. In fact, the Food and Agriculture Organization enhances that increasing the productive resources available to women would reduce the number of hungry people in the world by about 12–17% [51]. Furthermore, it is undeniable that women play a fundamental role in food and nutrition protection at all food chain stages [52].

8.3 Women and food transformation in SSA

African women's involvement in every stage of the food system is crucial [53]. In SSA, most of the food processing infrastructures are traditional and women once again are the key point of such a system of food processing. They are present from production, to distribution and nutrition. In fact, based on traditional knowledge, rural women process food and pass this knowledge to the next generation through their daughters. It is rare to have a modern infrastructure for food processing in urban or rural areas. Most modern processed foods are imported. And for the rare locally modern processed food, the price is high for the local market; as result, a large number of citizens cannot afford it. Therefore, it opts for foreign markets. In such decor, the only possibility which remains for these regions is to empower local actors like women with knowledge to create healthy and wholesome foods. So, then they could supply the market with the appropriate dishes fitting their social reality [52, 53].

As noticed, previously and with the current food insecurity situation in the underdeveloped world and peculiarly in SSA, empowering women is still the most suitable way for a rapid effective sustainable food security promotion.

9. Comparing food security in two countries: Benin and Turkiye

Benin is a country in sub-Saharan Africa that has regional recognition for its culture, democracy and prowess in cotton production [54]. But a lot remains to be done when it comes to food security (accessibility, affordability, quality, and safety). Attempts have been made at the national level in recent years. In particular, the multiplication of school canteens in schools to enable all pupils to have a decent and full meal throughout the country, but there is still a great deal to be done. According to the latest 2020 report on the global food security situation, the Global Food Security Index Ranking, the country has recently been in a moderate position

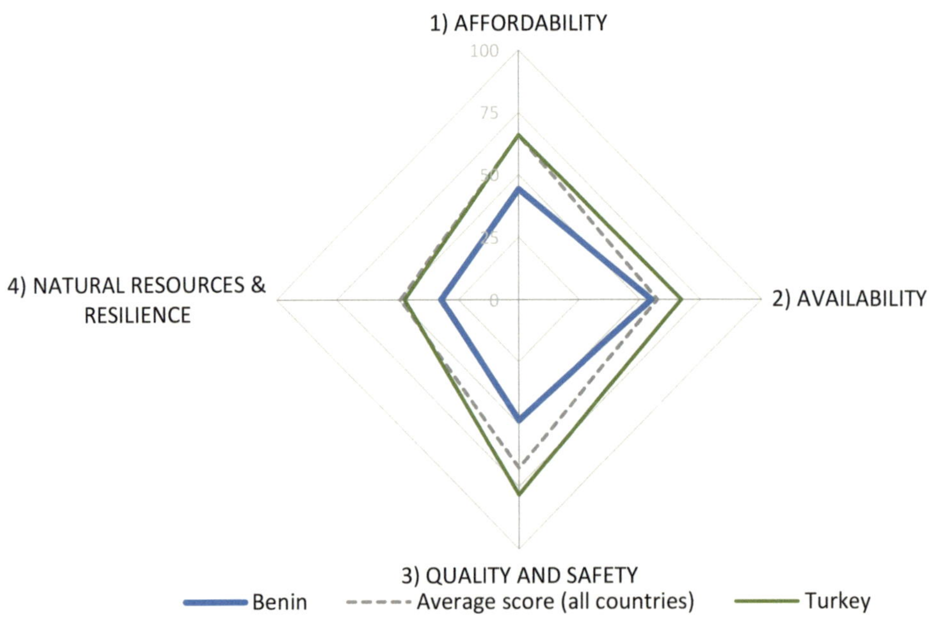

Figure 2.
Comparing food security in two countries: Benin and Turkiye. Source: GFSI (2020) computation from https://foodsecurityindex.eiu.com.

Series	Benin				Turkiye				Average Score(All Countries)
	Score	D	Rank	D	Score	D	Rank	D	
Overall food security environment	46.2	0.1	92	—	65.3	−0.7	47	—	60.4
Affordability	44.2	0	91	−1	66.0	−4.6	65	−4	65.9
Availability	54.8	0.2	68	+2	67.2	0.2	24	1	57.3
Quality and safety	48.3	−0.1	92	−1	78.3	3.6	43	5	67.6
Natural resources and resilience	32.2	0.4	113	—	47.4	0.1	-53	5	49.1

Source: GFSI (2020) computation from https://foodsecurityindex.eiu.com.

Table 5.
Comparing food security in two countries: Benin and Turkiye.

[55]. Since 2019, the country's global food security situation in 2020 had slightly increased by 0.1 points compared with 2019, but its place in the ranking remains unchanged (as the values in the table illustrate). On the other hand, relative to Benin, Turkiye, a country ranked 54th in the most recent HDI 2020 study (Benin 158), suffered a 0.7 decline in overall food security. Despite this downturn, the nation is well ahead of Benin and other underdeveloped countries. Moreover, at the global level, the country's food security skills are above average (the global average score is 60.4) (**Figure 2** and **Table 5**).

This fantastic above-average performance was achieved through several initiatives in the food sector, particularly targeted towards women. Additionally, the initial programme was focused on education. Similarly, data from the World Bank show that 93% of Turkish women are literate, as compared with the world average of 82.65%, while in Benin only 31% of women have this level of literacy (see **Table 6** below). In general, the growth of literacy for Turkish women has been 1.16% per year [56]. Similarly, the school enrolment for women is reportedly at 99.61% [57].

Moreover, as [58] points out on the issue of food security, he revealed that special attention has been paid to the matter. Thus, sectoral policy reforms in the Ministry of Agriculture and Rural Affairs have resulted. Therefore, these reforms have given a substantial boost to the public extension services in agriculture. Indeed, there have been many agricultural initiative programmes implemented in Turkiye, as well as programmes spearheaded by the Turkish Ministry of Agriculture and associated institutions that aim to allow (especially for) women entrepreneurs to enter into agricultural activities [59]. Most importantly, these initiatives involve:

a. Supporting Female Entrepreneurship in farming Programmes Project for 18–40-year-year-old women entrepreneurs in the Agriculture sector.

b. Young in Agriculture Entrepreneurial Women Empowerment Project area for young women aged 18–40 years).

Selected countries	Most recent year	Most recent value
World	2019	83
Turkiye	2017	93
Benin	2018	31
Burkina Faso	2018	33
Burundi	2017	61
Central African Rep	2018	26
Ghana	2018	74
Kenya	2018	78
Niger	2018	27
Somalia	1972	4
Côte D'Ivoire	2018	40
Sierra Leon	2018	35
Chad	2016	14
Togo	2015	51

Source: World Bank Group. Available from: https://data.worldbank.org/indicator/SE.ADT.LITR.FE.ZS?locations=TR-BJ-1W-KE-GH-BF-SO-DJ-BI-NE-CF.

Table 6.
Literacy rate, adult female (percent of females ages 15 and above).

c. Young Farmers Project, which is given to women and men on equal terms.

d. 'Agricultural Extension Project for Women Farmers', which was initiated by the Ministry of Agriculture in 2000 and has increased in intensity since 2015.

However, in less developed countries such as Benin, vocational training programmes can hardly help women who work in agriculture. The interests and welfare of the household take precedence over other responsibilities when a woman has a major decision-making role. This should indicate in underdeveloped countries that every well-educated woman has the power to prevent hunger. Therefore, in those countries, more women should be encouraged to start businesses (in agriculture-related fields too) in urban and rural areas where malnutrition (particularly) is prevalent.

10. Female gender empowerment in nutrition and food security

In most households, men handle finances while women take care of young children, food and nutrition. That's why without their effective implication, food security and the achievement of sustainable development goals will be jeopardized. Also, in undeveloped countries where food security is threatened, the participation of rural women in agriculture is considered necessary for its sustainability. For this reason, women must have maximum control over the production cycle of staple goods, food processing and importation and food storage. As the increase in the above-mentioned aspect, the availability of food will also increase. Furthermore, since food availability is one of the first pillars of food security [10], women's empowerment in food availability must be effective by:

10.1 Training them on agricultural production techniques

In rural areas, women's participation in production activities is not yet proven [60]. However, even if it does exist, it is relegated to the background because women generally do not have equal access to production resources as men do. This observed inequality makes the economic situation of women in non-urban areas very vulnerable. Speaking of the vulnerability of rural women, [61] points out that in the majority of least developed countries, rural women are not only poor, but this unfair treatment is very detrimental to women. Women's participation, even if it is not at the forefront, is either helping or growing staple foods on the land for the family's needs. Rarely do women produce cash crops.

To this end, to facilitate food availability for food self-sufficiency, the existing production system can be adjusted in ways that women might be more efficient in staple crops' production. In rural areas produced staple crops are almost the only ones feeding the household. In this regard, it would be through the awareness programmes and production policies of the states affected by the scourge of food insecurity. At present, many of the countries concerned are the least developed, i.e. about 12% of the countries in the world [62]. In this process, the empowerment of women in producer groups or cooperatives is more than vital. This empowerment itself will be achieved through:

**Easy access to production resources:* Production resources are capital, labour, land and managerial capacity. At first glance, women's conditions of access to land resources should be reviewed without discrimination. This means that they can freely exploit the land as they see fit. They would produce differently, increase agricultural productivity and respond more effectively to household needs. Needs sometimes in developing countries are different from the needs of the market. In these countries, the majority of cash crop (cotton, coffee, cocoa, etc.) production is for export [29]. If given the necessary tools, women who involve highly in the household may be able to strike a balance between cash crops and subsistence nutrient food production.

The second approach is to support them through targeted financing mechanisms to take up more profitable and productive production initiatives. This aspect should not be neglected. In the production process, inputs have a cost, as does production as a whole. In addition, access to the fund will enable women to empower themselves with skilled labour and enhance the value of women's labour in the different sectors in which they are involved. When women have a source of income, they have a positive impact on the household situation. They invest their funds in the household and the children. Finally, for women and groups of women producers, regular training to raise the level of women heads of agricultural production enterprises must be repeated. This will make it possible to maintain dynamic production. This may require the involvement of agricultural advisers and agricultural policies [63–65].

It is important to remember that increasing agricultural production alone is not enough to reduce food insecurity [51, 66, 67]. As proof of this, most of the countries where the disease is rife are those where nearly 70% of the active population is in agriculture. Therefore, initiatives must be taken, and habits must be introduced to ensure that production meets the need for food security because the situation persists.

Encouragement and support for women who are already heads of agricultural enterprises. Awareness raising of the need and necessity to produce according to nutritional needs. This necessarily involves education to this end. Again, academicians, non-governmental agencies, farmers' organizations, agricultural advisers and agriculture ministries, all have a part to play. These institutions need to support women by educating them in the use of technologies that can advance their businesses. Balakrishnan and Fairbairn-Dunlop [68] states that: 'Training should build the capacity of rural women according to their multi-segment production tasks, and new information and communication technologies should be harnessed to improve rural women's access to technical information and public sector support services. The existing models of Farmer-field-school and farmer-to-farmer learning approaches may have ignored gender biases that prevent women from taking advantage of such technology transfer approaches'.

Resources	The gender gap	Closing the gap
Land	Women make up 10 to 20% of landowners across the world, but this varies by country, and even in the same region. However, female minority landowners are primarily located in Africa.	To ensure equitable access to land and other agricultural assets, reform is required. A process must be put in place to hold officials and community leaders accountable for upholding the law. It aims to empower women in several ways including awareness and the ability to claim their rights.
Labour Markets	Farms run by female-headed households are likely to be smaller and have fewer members working on the farm. Women have heavy household tasks and are therefore more likely to take time away from productive activities.	Women's participation in and access to rural labour markets requires freeing women's time through labour saving technologies and the provision of public services. It also entails raising women's human capital through education, eliminating discriminatory employment practices, and capitalizing on public work programmes.
Financial Services	The ability to access credit and insurance is necessary to accumulate assets. Smallholders often have less access to financial services, but due to the types of assets they hold, women do not have the same access as men.	Closing the gender gap in financial services involves legal and institutional reforms that will meet the needs and constraints of women, as well as efforts to enhance their financial capacity. Innovative delivery channels and social media can reduce the cost of financial services that were previously unavailable to rural women.
Education	While there have been improvements in gender parity at the national level, women and girls continue to lag in most societies. The gender gap in education is greatest in rural areas, where female household heads sometimes do not have as much education as male household heads.	Organizations and other forms of collective action can be a useful measure of building relations and networks and addressing gender gaps in other areas of life. Groups of female leaders can positively impact social change by helping eliminate gender inequality.
Technology	Women utilize less purchased inputs, and materials and implement less technology than men. In many countries, women use fewer fertilizers than men. One of the main reasons for the status quo is a lack of available credit.	Improving women's access to agricultural technologies can be facilitated through participatory gender-inclusive research and technology development programmes, the provision of gender-sensitive extension services

Sources Training Guide – Gender and Climate Change Research in Agriculture and Food Security for Rural Development [51, 69].

Table 7.
Women empowerment in agriculture through access to resources.

In school curricula too, *gardening programmes,* learning, educating schoolchildren on good nutrient practices can be introduced. In this way, a more aware generation will be able to take up the challenge of food security because no change is possible if youth are not involved. The hope of tomorrow is the youth. Regardless of the sector, a youth who masters the stakes will know how to take up the challenge (**Table 7**).

10.2 Training and awareness raising on good nutrition practices

It is essential to be knowledgeable about the good and bad eating habits. The healthiest foods provide the best nutrients [70]. Since food is needed for the body to function properly, people must eat the right types of food. Therefore, public awareness [71] campaigns can successfully guide diet recommendations when they are supported by national policy and women's organizations in remote areas. If adequate resources are available and policies are implemented with the participation of NGOs and professional extension agents, rural areas would greatly benefit from the use of such an approach.

Once consumers understand basic food composition, they can acquire the specific food ingredients they want through a variety of channels.

10.3 Accessibility to food

Increasing the purchasing power of the population and of women, in particular, will not only lead to a better life but will also enable households to fill the basket qualitatively so that food security objectives can be achieved. When women's economic level is improved, they will be able to easily obtain the food and ingredients needed for consumption. Because of this, a policy of 'affirmative action' is very important to help women get jobs they could not get.

The other aspect that is important to remember is the multiplication of infrastructures. Infrastructures facilitate access to food wherever it is found. For example, rail, ships, road networks, etc. can be used to provide access to food. If it is in the South that there is easy access to this type of production, citizens in the interior should have no difficulty in obtaining it. Women traders will easily be able to act as intermediaries in the transport of the goods concerned. According to [72], infrastructure must be implemented immediately to make landlocked areas accessible.

10.4 Food utilization

Women must be able to process and offer new foods that meet quality and hygiene standards in such a way as to meet the nutritional expectations of consumers. These expectations are sometimes influenced by social and religious norms in each geographical area. The use of food in the context of food safety must not only be integrated into the social realities of the people, but in terms of composition, it must meet energy, mineral, vitamin and nutritional needs.

When the ingredients are available, the purchasing power is there too, and when the right foods are lacking, there will be no change. So, to ensure food security, the whole system must work together.

11. Conclusion

This study helped to clarify the expectations regarding food security. In addition, to ensure food security, it was necessary to assess the role and conditions of women in their daily lives. Thus, in food security, Africa requires women at every stage of food production. Therefore, the paper advises an improvement through education, awareness raising, access to land, financial services and access to technology, not only for food security but also for women's conditions. Most peculiarly, the report recommends that if women had the same access to resources as men, it would benefit the household's food self-sufficiency.

Author details

Eriola Marius Charlot Adenidji[1] and Orhan Özçatalbaş[2*]

1 Institute of Natural and Applied Sciences, Akdeniz University, Antalya, Turkey

2 Faculty of Agriculture, Department of Agricultural Economics, Institute of Natural and Applied Sciences, Akdeniz University, Antalya, Turkey

*Address all correspondence to: ozcatalbas@akdeniz.edu.tr

References

[1] FAO. Measuring Food Safety – Indicators to Achieve Sustainable Development Goals (SDGs). Bangkok: FAO; 2021

[2] FAO, IFAD, UNICEF, WFP and WHO. The State of Food Security and Nutrition in the World. Rome: FAO; 2020

[3] Wood W, Eagly A. Gender. In: Fiske ST, Gilbert DT, Lindzey G, editors. Handbook of Social Psychology. Hoboken, NJ: John Wiley, Sons; 2010. p. 629667

[4] UN. Goal 5: Achieve gender equality and empower all women and girls, [undated]. [Online]. Available from: https://www.un.org/sustainabledevelopment/gender-equality/

[5] Eurostat. The life of women and men in Europe, A Statistical portrait. 2018. [Online]. Available from: https://www.ine.es/prodyser/myhue18/images/pdf/WomenMenEurope-DigitalPublication-2018_en.pdf?lang=en

[6] Eurostat. How do women and men use their time-statistics. 2019. [Online]. Available from: https://ec.europa.eu/eurostat/statisticsexplained/index.php?title=How_do_women_and_men_use_their_time_-_statistics&oldid=463738

[7] Eurostat. The life of women and men in Europe. 2021. [Online]. Available from: https://ec.europa.eu/eurostat/cache/infographs/womenmen/img/pdf/WomenMenEurope-DigitalPublication-2021_en.pdf?lang=en

[8] Cheryl D. If Women Hold up Half the Sky, How Much of the World's Food do They Produce? Rome: FAO; 2011

[9] Farnworth C, Fones Sundell M, Nzioki A, Shivutse V, Davis M. Transforming Gender Relations in Agriculture in Sub-Saharan Africa. South Africa: Swedish International Agricultural Network Initiative (SIANI); 2013

[10] FAO. Declaration on World Food Security. Rome: FAO; 1996

[11] FAO. World Food Security: A Reappraisal of the Concepts and Approaches. Rome: FAO; 1983

[12] WFP. Gender Policy (2003-2007): Enhanced Commitments to Women to Ensure Food Security. Rome: FAO; 2002

[13] Staatz J, D'Agostino V, Sundberg S. Measuring food security in Africa: Conceptual, empirical, and policy issues. American Journal of Agricultural Economics. 1990;**72**:1242

[14] Maxwell S, Frankenberger T. Household Food Security: Concepts, Indicators, Measurements. A Technical Review. New York and Rome: United Nations Children's Fund and International Fund for Agricultural Development; 1992

[15] Doustmohammadian A, Omidvar N, Shakibazadeh E. School-based interventions for promoting food and nutrition literacy (FNLIT) in elementary school children: A systematic review protocol. System Review. 2020;**9**:87

[16] Aktaş N, Özdoğan Y. Food and nutrition literacy (Gıda ve Beslenme Okuryazarlığı). Harran Tarım ve Gıda Bilimleri Dergisi. 2016;**20**(2):146-153

[17] Özçatalbaş O, Gürgen Y. Tarımsal Yayım ve Haberleşme. Adana: Baki Kitap ve Yayınevi; 1998

[18] Morgan M, Arrowood J, Farris A, Griffin J. Assessing food security through

cooking and food literacy among students enrolled in a basic food science lab at Appalachian State University. Journal of American College Health. 2021 Feb 12:1-6. DOI: 10.1080/07448481.2021.1880414. PMID: 33577420

[19] AGRA. Africa Agriculture Status Report. Feeding Africa's Cities: Opportunities, Challenges, and Policies for Linking African Farmers with Growing Urban Food Markets. Nairobi, Kenya: Alliance for a Green Revolution in Africa; 2020

[20] FAO. Strengthening National Food Control Systems: Guidelines to Assess Capacity Building Needs. Rome: FAO; 2006

[21] Gerstel N, Sarkisian N. Sociological perspectives on families and work: The import of gender, class and race. In: Pitt-Catsouphes M, Kossek EE, Sweet S, editors. The Work and Family Handbook: Multi-Disciplinary Perspectives and Approaches. Mahwah, NJ: LEA; 2006. pp. 237-267

[22] UN. Gender Mainstreaming: An Overview. 2002

[23] Jhpiego. Gender Concepts and Definitions. 2020

[24] Milazzo A, de Walle van, Dominique P. Women Left Behind? Poverty and Headship in Africa. World Bank Policy Research Working Paper No. 7331. 2015. p. 2-3. Available from: https://ssrn.com/abstract=2622317

[25] SOFA Team, and Cheryl D. The Role of Women in Agriculture. FAO Agricultural Economics. Working Paper No 11.02. 2011

[26] Oluwatomipe AP, Adebanke OI, Babatunde AA, Salau Odunayo P. Exploring the hindrances to women entrepreneurship, development and prosperity in Nigeria. Journal of Entrepreneurship: Research & Practice. 2015;**2015**:148163

[27] Ferrant G, Pesando LM, Nowacka K. Unpaid Care Work: The Missing Link in the Analysis of Gender Gaps in Labour Outcomes. Paris: OECD Development Center; 2014

[28] Akilova M, Marti YM. What is the effect of women's financial empowerment on intimate partner violence in Jordan? Global Society of Welfare. 2014;**1**:65-74

[29] NEPAD. African agriculture, transformation and outlook. NEPAD, 2013

[30] FAO, CEA et CUA. Afrique—Aperçu régional de l'état de la sécurité alimentaire et de la nutrition, Statistiques et tendances. Accra, FAO, 2021

[31] FAO. The State of Food and Agriculture: Lessons from the Past 50 Years. Rome: FAO; 2000

[32] Shimeles A, Verdier-Chouchane A, Boly A. Building a Resilient and Sustainable Agriculture in Sub-Saharan Africa. Springer International Publishing. 2018

[33] FAO. Decision Tools for Family Poultry Development. FAO Animal Production and Health Guidelines. Rome: FAO; 2014

[34] Kararach G et al. ACDF, Africa Capacity Indicators: Capacity Development for Agricultural Transformation and Food Security. South Africa: ACDF;

[35] AfDB. Feed Africa: Strategy for Agricultural Transformation in Africa 2016-2025. 2016

[36] C.T.A. The role of smallholder farmers in seed production systems. Report and recommendations of a study visit to Zimbabwe. 1999

[37] Ogunlela YI, Mukhtar AA. Gender issues in agriculture and rural

development in Nigeria. The Role of Women. 2009;**4**:19-30

[38] Bjornlund V, Bjornlund H, Van Rooyen AF. Why agricultural production in sub-Saharan Africa remains low compared to the rest of the world – a historical perspective. International Journal of Water Resources Development. 2020;**36**:S20-S53

[39] Giller KE. Small farms and development in sub-Saharan Africa. Farming for food, for income or for lack of better options? Food Security. 2021;**13**:1431-1454

[40] Douglas G. Smallholder Agriculture in Africa: An Overview and Implications for Policy. London: IIED;

[41] Frelat R, Lopez-Ridaura S, Giller KE et al. Drivers of household food availability in sub-Saharan Africa based on big data from small farms. Proceedings of the National Academy of Sciences. 2015;**113**(2):458-463. DOI: 10.1073/pnas.1518384112

[42] Abdullahi MR. Women in Agriculture: The Role of African Women in Agriculture. Zaria, Nigeria: National Agricultural Extension and Research Liaison Service;

[43] De Bruyn J, Wong J, Bagnol B, Pengelly BC, Alders R. Family poultry and food and nutrition security. CAB Reviews Perspectives in Agriculture Veterinary Science Nutrition and Natural Resources. 2015;**10**:1-9

[44] Padhi MK. Importance of indigenous breeds of chicken for rural economy and their improvements for higher production performance. Scientifica (Cairo). 2016;**2016**:2604685. DOI: 10.1155/2016/2604685. Epub 2016 Apr 7. PMID: 27144053; PMCID: PMC4838803

[45] Webb P, Kennedy E. Impacts of Agriculture on Nutrition: Nature of the Evidence and Research Gaps. Food and Nutrition Bulletin. 2014;**35**(1):126-132. DOI: 10.1177/156482651403500113

[46] Masset E, Haddad L, Cornelius A, Isaza-Castro J. Effectiveness of agricultural interventions that aim to improve nutritional status of children: systematic review. BMJ. 2012;**344**(d8222):1-7

[47] Semba RD, de Pee S, Sun K, Sari M, Akhter N, Bloem MW. Effect of parental formal education on risk of child stunting in Indonesia and Bangladesh: A cross-sectional study. The Lancet. 2008;**371**(9609):322-328

[48] United Nations Economic and Social Council. Thematic Issues before the Commission on the Status of Women: Report of the Secretary General. New York: United Nations Economic and Social Council; 2002

[49] Bagnol B. Gender issues in small-scale family poultry production: Experiences with Newcastle disease and highly pathogenic avian influenza control. World's Poultry Science Journal. 2009;**65**(2):231-240

[50] Kitalyi AJ. Rural Chicken Production Systems in Rural Africa: Household Food Security and Gender Issues. Rome: FAO; 1998

[51] FAO. The State of Food and Agriculture 2010-2011: Women in Agriculture: Closing the Gender Gap for Development. Rome: FAO; 2011

[52] FAO. La lutte des femmes rurales pour faire évolué favorablement les inégalités entre les sexes. 2017

[53] Njobe B, Kaaria S. Women and Agriculture, the Untapped Opportunity in the Wave of Transformation. 2015

[54] INSAE, (Institut National de la Statistique et de l'Analyse Economique). Monographie de la filière « coton » au Benin. 2020

[55] GFSI, (Food security index). Food security index. Global Food Security Index (GFSI2020). The Economist. 2020

[56] WDI (World Development Indicators). The World Bank Data Catalogue. 2018

[57] Turkish Statistical Institute. TurkStat National Education Statistics, Population and Demography. Ankara: Turkish Statistical Institute (Türkiye İstatistik Kurumu); 2015

[58] Özçatalbaş O. The agricultural extension in Turkiye. Innovative Processes in Extension: Problems and Prospects. The National University of Life and Environmental Sciences of Ukraine. 2011;**631**:115

[59] TOB. Education and extension: Rural women and family services, Republic of Turkiye Ministry of Agriculture and Forestry (Eğitim ve Yayım: Kırsalda Kadın ve Aile Hizmetleri, Türkiye Cumhuriyeti Tarım ve Orman Bakanlığı. 2019

[60] Özçatalbaş O, Ozkan B. The role of women in agriculture and rural development in Turkey. Asian Journal of Women's Studies. 2003;**9**(4):114-124

[61] Özçatalbaş O. Kirsal Alanda Kadin ve Kalkinmadaki Rolu ("The Role of Women in Rural Development"). Journal of Agriculture and Engineering. 1999;**60**:41-47

[62] UNCTAD. Unctad, Least Developed Countries Report. 2011

[63] Diiro GM, Seymour G, Kassie M, Muricho G, Muriithi BW. Women's empowerment in agriculture and agricultural productivity: Evidence from rural maize farmer households in western Kenya. PLOS ONE. 2018;**13**(5):e0197995

[64] Lecoutere E, Spielman DJ, Van Campenhout B. Women's Empowerment, Agricultural Extension, and Digitalization: Disentangling Information and Role-model Effects in Rural Uganda. Washington, DC: International Food Policy Research Institute (IFPRI); 2020

[65] Bain C, Ransom E, Halimatusa'diyah I. Weak winners' of Women's empowerment: The gendered effects of dairy livestock assets on time poverty in Uganda. Journal of Rural Studies. 2018;**61**:100-109

[66] Vanhaute E. From famine to food crisis: What history can teach us about local and global subsistence crises. Journal of Peasant Studies. 2011;**38**(1):47-65

[67] Delarue J, Mesplé-Somps S, Naudet JD, Robilliard AS. Le paradoxe de Sikasso: coton et pauvreté au Mali. Paris: DIAL; 2009

[68] Balakrishnan R, Fairbairn-Dunlop P. Rural Women and Food Security in Asia and the Pacific: Prospects and Paradoxes. RAP Publication; 2005/30

[69] Climate Change, Agriculture and Food Security (CCAFS) and FAO. Training Guide – Gender and Climate Change Research in Agriculture and Food Security for Rural Development. 2013

[70] Eze NM, Maduabum FO, Onyeke NG, Anyaegunam NJ, Ayogu CA, Ezeanwu BA, et al. Awareness of food nutritive value and eating practices among Nigerian bank workers: Implications for nutritional counselling and education. Medicine. 2017;**96**(10):e6283

[71] FAO. The State of Food Insecurity in the World. Rome: FAO; 2012

[72] Touzard JM, Temple L. Sécurisation alimentaire et innovations dans l'agriculture et l'agroalimentaire: vers un nouvel agenda de recherche? Une revue de la littérature. Cahiers Agricultures, EDP Sciences. 2012;**21**(5):293

Chapter 15

Non-Timber Forest Products as an Alternative to Reduce Income Uncertainty in Rural Households

Luz María Castro, Diana Encalada and Luis Rodrigo Saa

Abstract

Rural households face uncertain income due to several risks associated with markets, climate and productive uncertainties. In South Ecuador, subsistence seasonal agriculture constitutes the main livelihood strategy for local farmers. Non-timber forest products, such as *Caesalpinia spinosa* locally known as tara, constitute an alternative to diversify income. Tara is collected from natural areas, by mostly women, during male migration periods, which coincide with the dry season. To identify farmers' income composition, a field survey was conducted among 125 farmers, who also happen to collect tara. Prevalent agricultural options for the region included maize, beans, cattle ranching, pigs and poultry. To calculate risk-efficient combinations, we applied Markowitz's portfolio theory, which combines options based on their income and risk performance. The results revealed that tara is only part of low-income portfolios, despite the low correlation between the markets. The exclusion in tara from high-income portfolios might be a consequence of its lower returns compared with other options such as maize and cattle ranching. Collectors need to improve efficiency during harvest and post-harvest processes to reduce loss, which is above 50%. If appropriately managed, tara could contribute to raising household income, alleviating agricultural risks and boosting gender equality.

Keywords: risk, income, diversification, NTFPs, sustainability

1. Introduction

Farming activities are often exposed to several sources of risk that are faced by households without enough information to support their management decisions. Production, marketing, financial, institutional and human risks are experienced by most farms, either independently or interrelated. Various socio-economic conditions affect people's livelihoods, such as the availability of employment opportunities, access to markets, agricultural development, the degree of linkages with urban areas and labour migration [1].

Several studies focused on the livelihood strategies in developing countries have highlighted the relevance of diversification to reduce risks [2]. Rural households diversify their livelihoods and combine various strategies to obtain food, goods and income. With increasing exposure to national and international markets, new opportunities are emerging. People at the forest fringe combine the exploitation of natural resources with farming, off-farm employment and labour migration [1]. The contribution of forest for income diversification is underrated though. Forests provide a wide range of goods and services to local dwellers that create opportunities to address many sustainable develop goals (SDGs). Sustainable forest management might promote economic growth and productive employment in rural communities, especially the poorer ones [3]. Moreover, the broad branch of non-timber forest products (NTFPs) can create greater inclusion of women in the field of sustainable forest management, boosting women empowerment [4].

Crop diversification and the inclusion of alternative options such as NTFPs can enhance the performance of low-income farms [5–7]. Over the past few decades, NTFPs have been playing a significant role in the improvement of livelihoods of communities around the world through cash income, food security, health care, nutrition, and other social and cultural ecosystem services [5, 8]. Besides the potential for income diversification, NTFPs have an important insurance role for farmers. In many developing countries, NTFPs are considered as a safety net that fills the gaps during emergencies and shortfalls in agricultural production [5]. In case of crop damage, the households harvest NTFPs for supplementary income [6].

Since the early 1990s, the role of NTFPs for sustainable forest use and poverty alleviation has received increased attention [1, 2]. Nevertheless, the income-generating capacity of NTFP extraction in natural forests is restricted to factors such as product availability, density and irregular distribution of valuable species. NTFP harvesting is mostly a part-time, seasonal and subsistence-oriented activity, complementary to farming [1].

According to the IPCC [9], climate change is expected to significantly impact the provision of NTFPs, especially in mountain regions. The increase in global average temperatures and change in precipitation patterns will impact the provision of NTFPs. Scientists foresee that warmer temperatures can drive to shifts in plant species distribution and richness, and some of them alert a rise in local extinction risks due to the competitive replacement of slow-growing plant species. Moreover, climate change is expected to increase exposure to other risks such as more frequent and severe forest fires, storms, landslides and floods [10]. Faced with this situation, it is necessary to implement long-term adaptation practices that guarantee the availability of NTFP and provide local communities with suitable livelihoods [11].

NTFPs have the potential to improve the livelihood of rural dwellers in tropical countries. In Ecuador, the market for NTFPs is rarely considered a profitable option because the accountability of harvesting and trade is deficient. *Caesalpinia spinosa* (Molina) Kuntze, locally known as tara, is a native species in the Andes [12]. Tara has been traditionally appreciated due to its multiple uses as firewood, construction material and fog catcher [13]. More recently, it has become a valuable NTFP alternative attributable to its high commercial value on growing international markets. Its pods and seeds are used for medicine, food and industry sectors based on their antimicrobial and antioxidant capacity [14–16]. The pods are used in the leather industry and for the manufacture of dyes because of their high content of high-quality tannins. In addition, the seeds are rich in a hydrocolloid called tara gum highly appreciated for the manufacture of food thickeners, cosmetics, varnishes, paints, etc. [17, 18].

In southern Ecuador, tara is distributed in wild populations along with dry tropical mountain forests, which is one of the most threatened forests in the world [18, 19]. This area is also affected by poverty, around 29% of the population is regarded as poor according to the National Institute of Statistics [20]. In this region, tara is mainly collected by local women [21]. Females are more likely to be engaged in NTFP collection [8]. Thus, policymaking and land-use planning must consider that NTFPs are part of the overall livelihood strategy of the people involved [6].

In order to understand the livelihood strategies of poor households in dry forest areas and their interactions with NTFP, as a source of supplementary income, this research was conducted in southern Ecuador. Throughout this study, we aimed to determine the current composition of farmers' income. We also analysed the share of the tara on overall farmers' income and the harvesting strategies. Finally, we assessed the risk related to the main farmer's activities in the region and how NTFPs can be incorporated as part of diversification strategies to cope with farming risks.

2. Risk and risk management in agriculture

Landowners make decisions every day concerning farming operations that affect their income. Several factors may affect the performance of the activities, some of them are deterministic while others are rather stochastic [22]. Many of the factors that affect farmers' decisions cannot be predicted such as weather conditions, market price volatility, labour availability, machinery, equipment failure and government policy change. The most common sources of risk in farming can be divided into five areas (**Figure 1**) [23].

Farming has become increasingly riskier over the years, due to market liberalization and globalization. This situation affects smallholder farmers severely, they need to improve skills for both production and business management. Changes in prices are beyond the control of any individual farmer. Price movements follow supply and demand trends, which change unexpectedly affecting the market price [24].

Financial risk occurs when money is borrowed to finance farm businesses [23]. Risks are associated with future interest rates, lender's willingness to provide funds and the ability of the farmer accomplish loan repayment. Smallholder farmers who borrow money at high-interest rates are prone to fail debt repayments, especially in

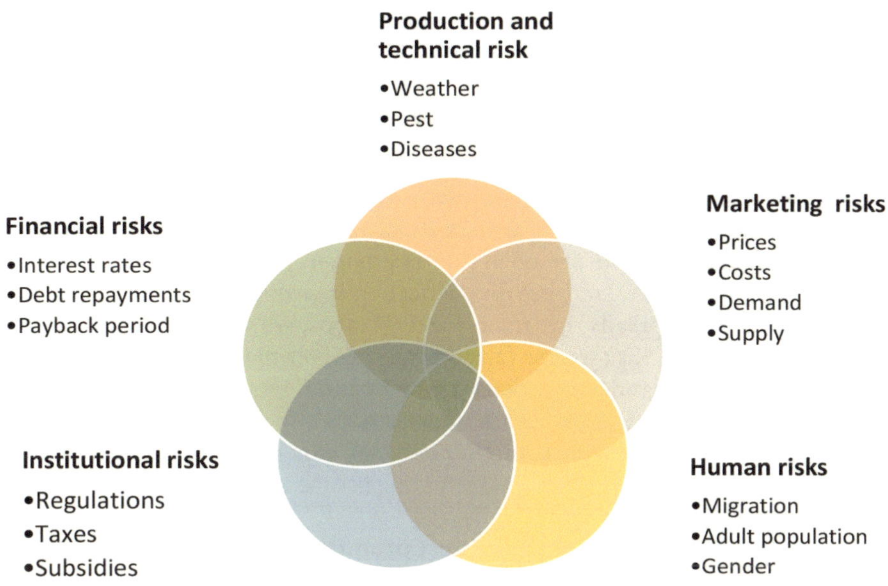

Figure 1.
Sources of risk in farming.

situations of lower than expected prices, combined with low yields, which can even lead to the sale of the farm. Another important source of risk comes from government policy affecting farming, such as price support and subsidies. Subsidies and regulations are examples of decisions taken by the government that can have a major impact on the farm business [24]. Finally, it is important to mention human-related risks because, in many regions, labour migration from rural areas is a common factor. Migration can cause labour shortages, farmers often face uncertainty concerning labour availability to meet demand during a farming season.

Risk management involves anticipating potential problems and planning to reduce their detrimental effects [23]. There are several strategies that farmers apply in order to cope with risks related to land use. Farmers can reduce risk by applying new technologies and practices designed to address specific risks, common to their area of production. Risk-reducing inputs (e.g. fertilizers and integrated pest management) might reduce the risk of low yields and crop damage. Similarly, irrigation reduces the risk of low rainfall. New seed varieties are being developed with certain characteristics to be resistant to drought, disease and pests. Nevertheless, not all inputs might reduce risk, even if fertilizer is used, the crop still depends on water availability, which may or may not be favourable. When soil moisture levels are low, using a fertilizer can still result in low yields. Understanding how farmers make land management decisions is critical to designing strategies to reduce risk exposure. Profitability of a particular land use obviously encourages farmers to allocate land to it; nevertheless, motivations behind decisions are often more complex than simple profit maximization [25].

2.1 Land-use diversification

Farmers' decisions about how best to use resources are driven by the goal of improving their well-being. Well-being is defined across many dimensions, including income, security of livelihood and health. Decisions about land use are influenced by the potential benefit of each activity, which, in turn, depends on the available technology, market and environmental conditions [26].

A key factor to consider is farmers' preferences towards risks. People generally do not become involved in risky situations unless there is a chance of making high profits. Many farmers around the world integrate crops and livestock to reduce risk and improve efficiency and sustainability. The risk involved with the farming options can be an essential factor in assessing preferences, because risk-averse farmers tend to choose the option with the lowest uncertainty, despite the fact that the potential reward may be lower as well [22]. Farmers are often regarded as risk-aversive, which means they give up on profits provided that a certain income is guaranteed.

For risk-averse farmers, land-use diversification improves the overall performance of the farm because it spreads risk among several crops, which should have a low market correlation [27]. Risk-averse farmers may achieve high levels of risk reduction by mixing two or more land-use options whose financial yields fluctuate independently from one another (low or negative correlations). In other words, in periods when returns from one asset drop, another one may generate unexpectedly high returns, thus moderating the effects of economic booms and busts [25].

2.2 The safety-net use of non-timber forest products

Evidence suggests the potential of NTFP to contribute to conservation of forest and to improve livelihood of rural landowners. Nevertheless, the exploitation of NTFP has often been neglected both in policy and research, because they are often regarded as secondary forest products. Recent research suggests that NTFPs are an important source of cash income for communities living in remote areas and safety net for diversifying income along the year [6]. Rural households, which have limited credit and insurance options, apply diversification strategies in order to reduce aggregate risk. The safety-net use of NTFP extraction may take the diversification strategy equivalent to a portfolio analysis, because the households use NTFP extraction as a risk-free asset [28]. Even though NTFP extraction might have a low annual value, it can provide insurance in the case of unexpected losses.

NTFP extraction appears to be efficient for poor rural households. Many NTFPs do not have strong positive correlation among themselves or with agricultural output [7] so they can be efficient risk-management instruments. Two characteristics of NTFP are important to note. First, there are low capital and skills requirements to NTFP extraction as well as open or semi-open access to the resource, so poor households can easily extract the resource. The poorest people are those who are the most engaged in NTFP extraction. Second, NTFPs habitually have low return to labour, so they have poor potential to alleviate poverty [28, 29].

3. Methods

The study area is located in mountain dry forest areas in southern Ecuador, where tara populations naturally occur, and the population is currently participating in the market of tara (see **Figure 2**), covering an approximate extension of 2396 km^2. The area is characterized by a temperate and dry climate, with two well-defined seasons: dry and rainy, the latter from December to April. The average altitude is 1385 metres above sea level. The annual rainfall has a mean value of 953 mm, and the annual temperature mean value is 18°C.

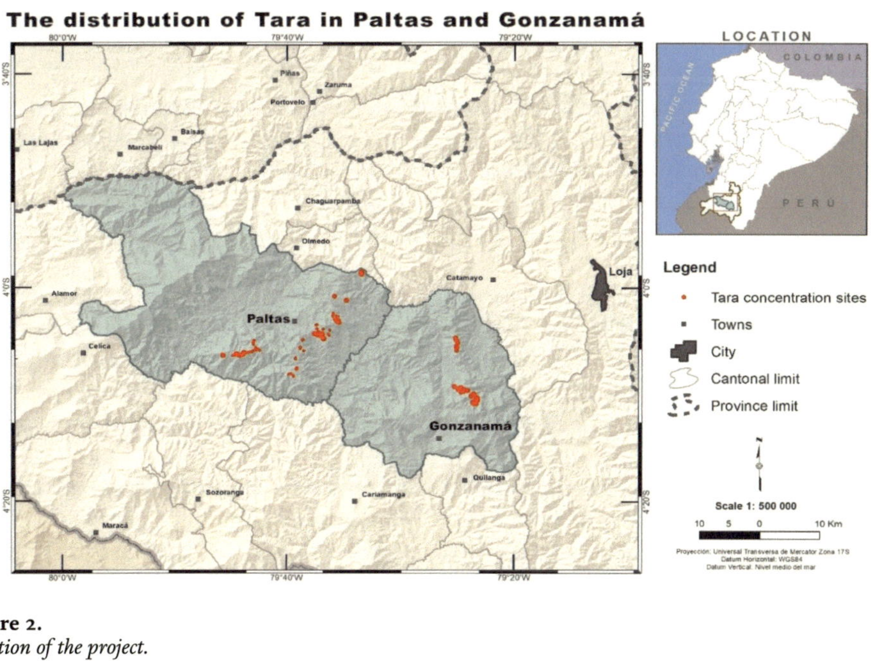

Figure 2.
Location of the project.

The estimated population in 2020 was about 34,424 people, with approximately 8200 households. More than 70% of the population live in the countryside, scattered across numerous small villages and communities. This region is characterized by subsistence agriculture and livestock production, with a poverty rate of 29%. It is estimated that 30% of households live below the country's poverty line [20].

3.1 Data collection

This study is based on primary data collection through a field survey to 125 rural households between January and July 2019. The questionnaire was focused on the socio-economic characteristics of the households, the assets and the main sources of income. Other questions include the participation of the different members of the household in the different income-generating activities, such as agriculture, livestock and the collection and sale of NTFPs.

Data were analysed both qualitatively and quantitatively to provide a deeper understanding of household production, management and use of NTFPS. The quantitative data were analysed using the Statistical Package for Social Science (SPSS) software and Microsoft Excel to obtain descriptive statistics such as percentages of responses, frequencies, means and standard deviations.

To calculate family income, volumes of production were multiplied by the current price at farm gate. Net household income was calculated as the difference between household income and production cost, main costs included labour and inputs (e.g. fertilizers, manure, seeds and pesticides).

3.2 Risks assessment and diversification

In order to assess risk, information on prices of the most valuable agricultural options for the study area was collected from Faostat for the period 1999–2009 [30]. As information of price for tara for the Ecuadorian market is scarce, we used data obtained from Peruvian institutions as a proxy, since the market for tara is better developed there.

Price values were simulated in an excel sheet based on their mean value and standard deviation using Eq. (1). This simulation generated normal random variables for price for each year of time horizon.

$$Y = \text{NORMINV}\left(\text{RAND}(), \overline{P}, \text{STDP}\right) \tag{1}$$

where Y is normally distributed random variable, NORMINV is excel-based assumed normal distribution, RAND () is the probability, P is average of price, and STDP stands for its standard deviation. We later performed a Monte Carlo Simulation to model the probability distribution of returns. Based on this information, we analysed the level of dependency of the markets through correlation and covariance analysis.

Diversification was performed through the application of the portfolio analysis [22]. Our work model farmers' options for balancing economic return and risk. This approach attempts, by means of the allocation of land to various land-use practices, to maximize the expected economic return, for a given level of accepted risk, which is represented by the standard deviation (SD) of economic return, through careful selection of the proportions of agricultural options. Those portfolios that provide the largest economic return for a given SD are termed efficient portfolios. All others are considered inefficient.

The expected economic return of a portfolio with two or more assets, Rp, is obtained by adding the expected economic returns, ri, weighted by their proportions, fi, of the single land-use options.

$$R_p = \sum_i f_i \cdot r_i \tag{2}$$

The SD of economic returns for the portfolio, σp, is quantified as follows:

$$\sigma_p = \sqrt{\sum_i \sum_j f_i \cdot f_j \cdot \mathrm{cov}_{i,j}} \tag{3}$$

with:

$$\sum_i f_i = 1 \quad f_{i,j} \geq 0 \quad \mathrm{var}_i := \mathrm{cov}_{i,i} \quad \mathrm{cov}_{i,j} = \rho_{i,j} \cdot s_i \cdot s_j$$

where i and j are the indices for the specific land-use options; fi is the proportion of a specific agricultural land-use practice in the portfolio; si is the SD of returns for land-use practice i; ρi,j is the coefficient of correlation between the returns for options i and j; vari is the variance and covi,j is the covariance between the economic returns for options i and j. Using this method, the effects of diversification can be identified for different combinations of land-use options, provided that the variability of their financial return is not perfectly positively correlated ($\rho \neq 1$).

The selection of the optimal portfolio can be made based on the reward-to-variability ratio (Eq. (4)) [31], where the portfolio return, Rp, minus the return of a riskless benchmark investment, Rri, is divided by the portfolio standard deviation σp.

$$\max R_p = \frac{\mathrm{Rp} - R_{ri}}{\sigma_p} \tag{4}$$

The riskless benchmark yield, Rri, is assumed as the interest yield that famers could obtain when investing in a safe financial asset. We assume that farmers can sell their land and invest that money in the capital market at an interest equal to 5%. Value of land has been set at a conservative value of US$2000 per ha, for which a riskless yield equal to US$100 can be obtained yearly.

4. Results and discussion

4.1 Livelihood strategies

Subsistence farming is the main economic activity in the region. Rain-fed subsistence farming relies on traditional crops such as corn and beans. Other representative crops are peas, coffee and peanuts on a smaller scale. Only 25% of the land is used for productive activities, due to the low soil quality and lack of labour, irrigation and financial resources. Regarding livestock production, landowners are engaged with cattle, pig and poultry production (**Table 1**). We observed high variation among the returns of households along with all the activities.

Concerning the two main farming options, a meaningful number of households depend on family labour (**Table 2**), nevertheless, hired labour is also required, especially during seasonal migration periods from June to December. Improved seeds are seldom used so far in the region, as most households admit to using their seeds. Technification of farming is also low, fertilizers either organic or conventional are occasionally used at the farms, the use of pesticides is more habitual though. The number of households that are reported to have loans and debt is also limited, most of them work with local banks and financial cooperatives.

4.2 Income from tara and other NTFPs

The data revealed that about 21% of the people collect fuel wood (e.g. *Eucalyptus globulus* and *Acacia macracantha*). An even smaller proportion collect NTFP other than tara, mainly *Marsdenia condurango* Rchb.f., locally known as condurango. The total amount of condurango harvested is later sold at the local market. There is an increasing demand for the product, but the collection occurs over a brief period, mostly by women. Tara, on the other hand, has a more established local market (**Table 3**). It is collected once a year in natural forests, mainly public areas. The

		Annual income per household		
Product	**Number of households**	**Mean USD**	**Max USD**	**Min USD**
Maize	107	670	9300	16
Beans	102	456	3445	24
Cattle	27	2181	15577	312
Poultry	92	282	1585	54
Pigs	37	550	2944	134

Table 1.
Annual returns of the main products in the study area.

	Number of households	
	Maize	**Beans**
Family labor force	57	37
Hired labor	64	35
Improved seeds	23	5
Organic fertilizers	20	10
Conventional fertilizers	27	15
Pesticides	60	37
Machinery	71	27
Credits and debt	22	14

Table 2.
Forms of production.

	Number of households	Total amount harvested (ton)	Total amount sold (ton)	Production (ton)			Income (USD)		
				Mean	max	min	Mean	max	min
Condurango	10	0.58	0.58	0.058	0.091	0.011	105	270	7.5
Tara	118	58.5	26.3	0.2	2.0	0.040	100	900	11

Table 3.
Local NTFPs harvested in the study area.

collection period ranges from May to December. Nevertheless, there is a waste of around half of the harvest caused by inadequate storage which produces losses due to fungus, under moist warm conditions. Training programmes and simple equipment can boost production, speed up processing times and reduce losses, they are essential tools for local farmers to enhance harvesting and post-harvesting.

In recent years, research on the role of forest-related income in rural livelihoods has been gaining momentum. Case studies around the world study the interactions between forests and livelihoods, and the contribution of NTFPs range from 6 to 45% [32]. In the case of tara, it currently represents less than 5% of household income. This activity generates an annual average of USD 100, which complements the income from agriculture. The potential of NTFP to improve farmers' livelihoods should not be exaggerated as poorer communities are the main actors in NTFP extraction [6].

4.3 Risk assessment and diversification

Farmers often choose productive activities that maximize their well-being, given the resources and opportunities available to them. To assess the risk involved in farming activities, five productive options were selected: maize, beans, cattle ranching, pig and poultry together with tara. **Figure 3** displays the prices of the selected products over the period 1999–2019.

Based on this information, we calculated the correlation matrix to observe the degree of dependency between the selected options. We observe a considerable dependency among the markets of beans and animal breeding as shown in the correlation matrix (**Table 4**). Tara on the other hand has a medium positive correlation

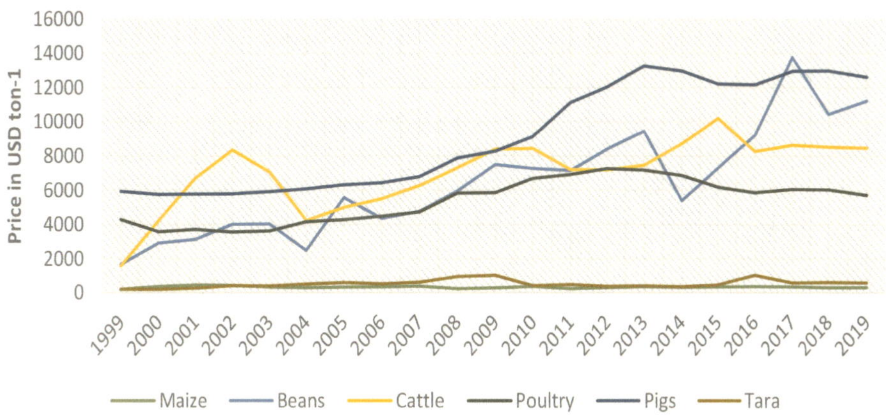

Figure 3.
Price volatility of farming options.

	Maize	Beans	Cattle	Poultry	Pigs	Tara
Maize	1	−0.043	0.292	−0.330	−0.155	−0.185
Beans	−0.043	1	0.668	0.687	0.844	0.409
Cattle	0.292	0.668	1	0.561	0.661	0.362
Poultry	−0.330	0.687	0.561	1	0.863	0.240
Pigs	−0.155	0.844	0.661	0.863	1	0.197
Tara	−0.185	0.409	0.362	0.240	0.197	1

Table 4.
Correlation matrix.

with all the agricultural options, except maize. NTFP collection is positively correlated with an agricultural shortfall and expected risk [33].

To achieve income diversification, combinations of five agricultural products and tara were tested by applying the portfolio theory. We calculated portfolios for given amounts of return subject to a minimum standard deviation, as a measure of risk. The resulting shares were highly dependent on the performance of each asset in terms of return and risks. Tara, for instance, was part of the portfolios with the lowest returns (**Figure 4**). It only was part of the portfolios with returns below 300 USD. As the amount of portfolio income rises, the share of tara drops to zero. This situation is a result of the low returns that tara harvesting delivers so far. Nevertheless, improved management of the species might enhance the overall financial performance of tara as an income generator option for rural households. So far, we have measured the risk of tara based on the volatility of the price, but little is known about how natural risks affect the supply of pods.

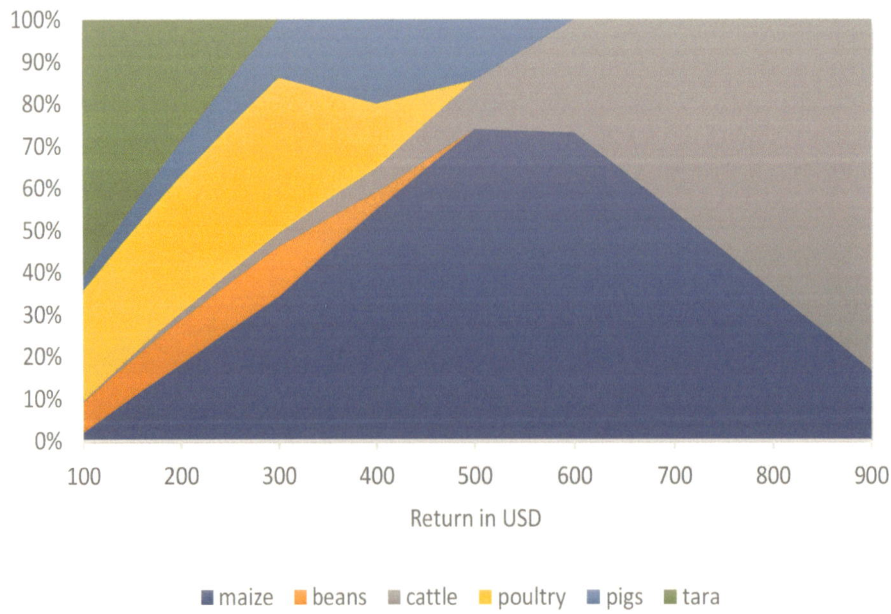

Figure 4.
Share of agricultural options in optimized land-use portfolios based on expected return.

		Portfolio Share %						
Return USD	**Sd USD**	**Maize**	**Beans**	**Cattle**	**Poultry**	**Pigs**	**Tara**	**Sharpe ratio**
100	32.24	0.02	0.07	0.00	0.26	0.03	0.61	0.620
200	37.77	0.18	0.11	0.02	0.33	0.08	0.29	3.178
300	58.43	0.34	0.12	0.03	0.37	0.14	0.00	3.765
400	86.89	0.55	0.04	0.06	0.15	0.20	0.00	3.683
500	122.86	0.74	0.00	0.12	0.00	0.14	0.00	3.418
600	190.53	0.73	0.00	0.27	0.00	0.00	0.00	2.729
700	286.94	0.54	0.00	0.46	0.00	0.00	0.00	2.161
800	393.37	0.36	0.00	0.64	0.00	0.00	0.00	1.830
900	503.52	0.17	0.00	0.83	0.00	0.00	0.00	1.629

Table 5.
Land-use portfolios for southern Ecuador.

Land-use portfolios were more diverse under 500 USD, above that income, portfolios were dominated by maize and cattle ranching. This behaviour is explained due to two main reasons, the financial performance of options and safety-net effects. First, cattle ranching delivers higher profits than any other option, despite the risk involved with the activity. Cattle also serve as a saving strategy used by farmers under distress. Maize, on the other hand, has a low risk compared with the other options and serves food security. Farmers who operate under subsistence conditions tend to be the most risk-averse. The provision of food for their dependants is an overriding priority for many of them. Activities with a monetary reward are frequently forgone in favour of meeting the objective of producing their own food.

We determined the optimal land-use portfolio by applying the Sharpe ratio, which is constructed based on the performance of the reward-to-variability ratio (**Table 5**). According to this method, the portfolio with the best performance was the one that achieved 300 USD. Above this bar all the portfolios deliver a riskier outcome, they are thus, inefficient.

5. Discussion and conclusion

Rural households relying upon subsistence agriculture face multiple uncertainties that so far are poorly understood. Moreover, they usually have poorer risk management strategies due to lack of training, limited access to credits and long distance to local and regional markets. This study, based on rain-fed subsistence agriculture, explores income diversification based on farm options and NTFPs, as a strategy to cope with risks.

Even though a body of literature has documented the potential of NTFPs' on poverty reduction [1, 5] livelihoods improvement and environmental sustainability [34], evidence suggests that the impact that NTFPs might have on household income should not be overestimated [6]. This is particularly important because landowners engaged in the collection of NTFPs generally live in poor conditions where even the most basic healthcare and educational services are lacking [35].

Even though the collection of tara is a commercial activity with potential for the economy of the study area and a source of income for women, we observed that its extraction has low poverty alleviation effects, due to its low return. Nonetheless, it can become a viable tool to compensate for the agricultural risk, if properly managed during harvest and post-harvest campaigns to reduce losses. The assimilation of tara as a viable business among farmers can boost the achievement of SDGs related to economic growth, decent work and gender equality [3, 4, 21].

Tara is being harvested from wild populations, a common pattern by poorer rural households in developing countries. As generally, harvest leaves the forest structure intact, it can also be promoted as an alternative for forest conservation. Sustainable commercial exploitation of NTFP could serve as a stimulus to sound forest management [13].

NTFPs are among the forest products that better serve as livelihood resources in the face of climate stresses [36], since NTFPs can provide income opportunities [37]. The use of NTFPs has also been recognized as important for the climate resilience of small production systems since natural forests are more resistant to climate change than monocultures [38, 39]. Migration is one adaptation strategy practised by farmers due to climate change as it provides off-farm income [40]. Nonetheless, we observed in the study area that it affects household performance because it reduces labour availability during the cropping season.

NTFP trade also face risks and challenges, which can lead to only short-term returns, rather than sustainable businesses [41]. They recommend to paying attention to the governance of resources, organizations and gender. All of these factors are extremely relevant for most developing countries, and certainly for our case study because collectors of tara are mostly women. The authors recommend considering on how men and women participate in the collection, who benefits and controls NTFP harvest and trade and how the benefits are shared within the household. Lack of knowledge can also affect the performance of NTFP, we found a waste of about half of the harvest due to fungus. Limited knowledge about storing technologies, processing opportunities, market information and how to domesticate NTFPs constrains its overall outcome as a generator of income.

Moreover, the contribution of NTFPs to improved livelihoods can be assured through a process of gradual domestication in man-made vegetation types such as forest gardens and plantations. There is often a gradual transition from the collection of wild products in natural forests to enrichment planting in secondary forests and managed home gardens. Tara is tolerated in the farms, where it provides shade and serves as a reference in the division of farms [13]. Future studies should assess the potential of tara as a commercial plantation to reduce the impact on wild areas and reduce the time spent by farmers for collection, as natural areas are located remotely.

Acknowledgements

The authors express their gratitude to Nature and Culture International for the financial support during this research and to Stefany Loayza, who collaborated during the field survey.

Author details

Luz María Castro[1*], Diana Encalada[1] and Luis Rodrigo Saa[2]

1 Department of Economics, Universidad Técnica Particular de Loja, Ecuador

2 Department of Biological and Agricultural Sciences, Universidad Técnica Particular de Loja, Ecuador

*Address all correspondence to: lmcastro4@utpl.edu.ec

References

[1] Ros-Tonen MAF, Wiersum FK. The Importance of Non-Timber Forest Products for Forest Based Rural Livelihoods: An Evolving Research Agenda. The Netherlands: Amsterdam Research Institute for Global Issues and Development Studies (AGIDS), University of Amsterdam; 2003

[2] Zoomers EB. Land and Sustainable Livelihood in Latin America. Amsterdam: KIT/Vervuert Verlag; 2001

[3] Campos Arce JJ. Forests, inclusive and sustainable economic growth and employment. In: Background Study Prepared for the Fourteenth Session of the United Nations Forum on Forests. UNFF14 Issue Brief 2. New York, USA: United Nations Forum on Forests Secretariat; 2019. Available from: https://www.un.org/esa/forests/wp-content/uploads/2019/04/UNFF14-BkgdStudy-SDG8-March2019.pdf [Accessed: January 28, 2022]

[4] Arora-Jonsson S. Forty years of gender research and environmental policy: Where do we stand? Women's Studies International Forum. 2014;**47**: 295-308

[5] Suleiman MS, Wasonga VO, Mbau JS, et al. Non-timber forest products and their contribution to household's income around Falgore game Reserve in Kano, Nigeria. Ecological Process. 2017;**6**:23. DOI: 10.1186/s13717-017-0090-8

[6] Delacote P. The Safety-Net Use of Non Timber Forest Products. Documents de Travail. France: INRA Laboratoire d'Economie Forestière (LEF); 2008. p. 30

[7] Delacote P. Forest Products as Safety Net, Deforestation and the Tragedy of the Commons. Documents de Travail. France: INRA Laboratoire d'Economie Forestière (LEF); 2010. p. 26

[8] Rahman H, Bishwajit R, Shahidul I. Contribution of non-timber forest products to the livelihoods of the forest-dependent communities around the Khadimnagar National Park in northeastern Bangladesh. Regional Sustainability. 2021;**2**(3):280-295

[9] IPCC. Climate Change 2014: Impacts, Adaptation, and Vulnerability. Part A: Global and Sectoral Aspects. Contribution of Working Group II to the Fifth Assessment Report of the Intergovernmental Panel on Climate Change. Cambridge, United Kingdom and New York, NY, USA: Cambridge University Press; 2014

[10] Kohler T, Giger M, Hurni H, Ott C, Wiesmann U, Von Dach SW, et al. Mountains and climate change: A global concern. Mountain Research and Development. 2010;**30**:53-55

[11] Gurung LJ, Miller K, Venn S, Bryan B. Climate change adaptation for managing non-timber forest products in the Nepalese Himalaya. Science of the Total Environment. 2021;**796**:148853

[12] Santos MB, dos Santos CHC, de Carvalho MG, de Carvalho CWP, Garcia-Rojas EE. Physicochemical, thermal and rheological properties of synthesized carboxymethyl tara gum (*Caesalpinia spinosa*). International Journal of Biological Macromolecules. 2019;**134**:595-603

[13] Villena JV, Seminario Cunya JF, Valderrama CMA. Variabilidad morfológica de la "tara" Caesalpinia spinosa (Fabaceae), en Cajamarca: descriptores de fruto y semilla. Arnaldoa. 2019;**26**(2):555-574

[14] López SA, Oré SR, Miranda VC, Trabucco J, Orihuela TD, Linares GJ, et al. Capacidad antioxidante de poblaciones sivestres de «tara» (Caesalpinia spinosa) de las localidades de Picoy y Santa Fe (Provincia de Tarma, departamento de Junín). Scientia Agropecuaria. 2011;**2**:25-29

[15] Chambi F, Chirinos R, Pedreschi R, Betalleluz-Pallardel I, Debaste F, Campos D. Antioxidant potential of hydrolyzed polyphenolic extracts from tara (*Caesalpinia spinosa*) pods. Industrial Crops and Products. 2013;**47**:168-175

[16] Guevara GJM, Guevara GJC, Guevara DJM, Béjar V, Huamán A, Valencia E, et al. Evaluación del cocimiento de diferentes biovariedades de Caesalpinia spinosa (tara) frente a cepas de Staphylococcus aureus sensibles y resistentes a oxacilina. Anales de la Facultad de Medicina. 2014;**75**:177-180

[17] Dc la Cruz Lapa P. Aprovechamiento integral y racional dc la tara Caesalpinia spinosa - Caesalpinia tinctoria. Revista del Instituto de Investigación FIGMMG. 2004;7:61 73

[18] Cordero I. Respuesta ecofisiológica de Caesalpinia spinosa (Mol.) Kuntze a condicionantes abióticos, bióticos y de manejo como referente para la restauración y conservación del bosque de nieblas de Atiquipa (Perú) [Tesis doctoral]. España: Facultad de Ciencias biológicas, Universidad Complutense de Madrid; 2015. p. 342

[19] Quesada M, Sanchez-Azofeifa GA, Alvarez-Añorve M, Stoner KE, et al. Succession and management of tropical dry forests in the Americas: Review and new perspectives. Forest Ecology and Management. 2009;**258**:1014-1024

[20] INEC (2019). Instituto Nacional de Estadísticas y Censos. Obtenido de: https://www.ecuadorencifras.gob.ec/institucional/home/ [Accessed: January 10, 2022]

[21] Castro LM, Encalada D. El rol de las mujeres en la bioeconomía: El caso del vainillo. Revista Tecnológica ESPOL. 2021;**33**(3):126-138. DOI: 10.37815/rte.v33n3.875

[22] Castro LM, Lechtaler F. The contribution of bio-economic assessments to better informed land-use decision making: An overview. Ecological Engineering. 2022;**174**:106449

[23] Kahan D. Managing Risk in Farming. Rome: FAO; 2013

[24] Lucas M, Pabuayon I. Risk perceptions, attitudes, and influential factors of rainfed lowland rice farmers in Ilocos Norte, Philippines. Asian Journal of Agriculture and Development. 2011;**8**(2):61-77

[25] Baumgärtner S, Quaas M. Managing increasing environmental risks through agrobiodiversity and agrienvironmental policies. Agricultural Economics. 2010;**41**:483-496

[26] FAO. The State of Food and Agriculture. Investing in Agriculture for a Better Future. Rome: Food and Agriculture Organization of the United Nations; 2012 Available from: http://www.fao.org/docrep/017/i3028e/i3028e.pdf

[27] Ogurtsov VA, Van Asseldonk MPAM, Huirne RBM. Assessing and modelling catastrophic risk perceptions and attitudes in agriculture: A review. NJAS-Wageningen Journal of Life Sciences. 2008;**56**:39-58

[28] Angelsen A, Wunder S. Exploring the forest-poverty link. CIFOR Occasional Paper. 2002;**40**:1-20

[29] Neumann R, Hirsch E. Commercialisation of Non-timber Forest Products: A Review. Technical Report. Bogor, Indonesia/ Rome: CIFOR/FAO; 2000

[30] FAO. Statistics Division. Rome. Available from: http://faostat.fao.org/default.aspx: Food and Agriculture Organization of the United Nations; 2022 Accessed: January 20, 2022

[31] Sharpe WF. Mutual fund performance. The Journal of Business. 1966;**39**:119-138

[32] Dash M, Behera B, Rahut DB. Determinants of household collection of non-timber forest products (NTFPs) and alternative livelihood activities in Similipal Tiger Reserve, India. Forest Policy and Economics. 2016;**73**:215-228

[33] Pattanayak S, Sills E. Do tropical forests provide natural insurance? The microeconommics of non-timber forest product collection in the Brazilian Amazon. Land Economics. 2001;**77**(4): 595-612

[34] Ahenkan A, Boon E. Non-timber forest products (NTFPs): Clearing the confusion in semantics. Journal of Human Ecology. 2011;**33**(1):1-9

[35] Ros-Tonen M. The role of non-timber forest products in sustainable tropical forest management. Holz als Roh- und Werkstoff. 2000;**58**:196-201. DOI: 10.1007/s001070050413

[36] Balama C, Augustino S, Eriksen S, Makonda FBS. The role of priority non-timber forest products in enhancing local adaptive capacity to climate change stresses in Kilombero district, Tanzania. Climate and Development. 2016;**9**:231-243

[37] Yadav PK, Saha S, Mishra AK, Kapoor M, Kaneria M, Kaneria M, et al. Yartsagunbu: Transforming people's livelihoods in the Western Himalaya. Oryx. 2019;**53**:247-255

[38] Ros-Tonen MAF, Freerk Wiersum K. The Importance of Non-Timber Forest Products for Forest-Based Rural Livelihoods: An Evolving Research Agenda. Bonn, Germany: International Conference on Rural Livelihoods, Forests and Biodiversity; 2003

[39] Smith P, Bustamante M, Ahammad H, Clark H, Dong H, Elsiddig EA, et al. Agriculture, forestryand other land use (AFOLU). In: Edenhofer O, Pichs-Madruga R, Sokona Y, Farahani E, Kadner S, Seyboth K, et al., editors. Climate Change 2014: Mitigation of Climate Change. Contribution of Working Group III to the Fifth Assessment Report of the Intergovernmental Panel on Climate Change. Cambridge: Cambridge University Press; 2014

[40] Basu J. Adaptation to climate change & non-timber forest products a study of forest dependent communities in drought prone areas of West Bengal, India. International Journal of Development Research. 2019;**09**(09): 29786-29795

[41] Awono A, Ingram V, Schure J, Levang P. Guide for Small and Medium Enterprises in the Sustainable Non-Timber Forest Product Trade in Central Africa. Bogor, Indonesia: CIFOR; 2013